Effective Scientific Communication

Effective Scientific Communication

The Other Half of Science

Cristina Hanganu-Bresch
Kelleen Flaherty

OXFORD
UNIVERSITY PRESS

Oxford University Press is a department of the University of Oxford. It furthers
the University's objective of excellence in research, scholarship, and education
by publishing worldwide. Oxford is a registered trade mark of Oxford University
Press in the UK and certain other countries.

Published in the United States of America by Oxford University Press
198 Madison Avenue, New York, NY 10016, United States of America.

© Oxford University Press 2020

Library of Congress Cataloging-in-Publication Data
Names: Hanganu-Bresch, Cristina, author. | Kelleen Flaherty, author
Title: Effective scientific communication : the other half of science /
Cristina Hanganu-Bresch and Kelleen Flaherty.
Description: New York : Oxford University Press, 2020. |
Includes bibliographical references and index.
Identifiers: LCCN 2019048258 (print) | LCCN 2019048259 (ebook) |
ISBN 9780190646813 (hardback) | ISBN 9780190646837 (epub) |
ISBN 9780190646844
Subjects: LCSH: Communication in science. | Communication.
Classification: LCC Q223 .H3484 2020 (print) | LCC Q223 (ebook) |
DDC 808.06/65—dc23
LC record available at https://lccn.loc.gov/2019048258
LC ebook record available at https://lccn.loc.gov/2019048259

1 3 5 7 9 8 6 4 2

Printed by Integrated Books International, United States of America

Contributing Artists

Max Beck is a cartoonist who resides in Houston, Texas, and can be reached at maxrobertbeck@gmail.com.

Kris Mukai is a cartoonist and illustrator living in Los Angeles, California. More of her work can be found at http://www.hikrismukai. com and she can be reached at Kr.mukai@ gmail.com.

Maritsa Patrinos is a Brooklyn-based illustrator, writer, and comic artist who has created materials for BuzzFeed, MTV, Marvel, and *The New Yorker*. She can be reached at maritsapatrinos@gmail.com.

Erica Perez is an Emmy Award–winning 2D animator and animation director from Brooklyn, New York. She can be reached at EricaPerezAnimation@gmail.com.

Max Beck is a cartoonist who resides in Houston, Texas, and can be reached at maxrobertbeck@gmail.com.

Kris Mukai is a cartoonist and illustrator living in Los Angeles, California. More of her work can be found at http://www.krismukai.com and she can be reached at Karmukai@gmail.com.

Marissa Paternos is an established illustration writer and comic artist who has created materials for Fox, ted, MTV, Marvel, and the ... ta... She can be reached at marin.tpatinov@gmail.com.

Erica Perez is an Emmy Award-winning 2D animator and animation director from Brooklyn, New York. She can be reached at EricaPerezAnimation@gmail.com.

Peer Reviewers

We would gratefully like to acknowledge and thank our friends and peers who · kindly performed reviews of our chapters: Mignon Adams, Phyllis Blumberg, James Flaherty, Amy Jessop, David Perlman, Andrew Petto, Matthew Schultz, and Ken Short.

From Cristina
For **Mihaela Cosma**, *the first professor who taught me to love language and everything we can do with it.*

From Kelleen
For my father, **James Flaherty**, *who taught me how to be a writer, and my auxiliary back-up college mentor father,* **Robert Knowlton**, *who taught me how to be a scientist. Thank you both for providing me with the passion that has fueled my life and my career.*

Contents

Contents

Note From the Authors

Scientific writing is hard, complex, maddening, frustrating. It is also fulfilling, innovative, productive, immersive, and a critical component of science and progress. Mastering writing in your STEM (Science, Technology, Engineering, and Math) field is central to your academic and professional success. Learning to communicate science well will contribute to your own personal satisfaction and sense of achievement.

A skilled writer in the sciences must have a solid understanding of audience, purpose, critical thinking, research skills, visual communication, genre, ethics, and how to communicate statistics, in addition to being able to master grammatical, citation, and stylistic conventions. In this book, born of our experiences teaching scientific writing to undergraduate and graduate students at a science-oriented institution, we wanted to cover as many of the topics that we felt to be important for the education of grounded, competent scientific communicators. We introduce these skills with the hope to foster curiosity and a spirit of exploration. Exercises provided at the end of each chapter are often deliberately open ended to encourage discussion and critical thinking.

While (sadly) there is no precise formula or recipe for writing the perfect report or paper or poster, there are steps you can take to manage your writing tasks and demystify the scientific writing process. We encourage you to use this text as a springboard for a deeper, more focused exploration of your own STEM field, and it is our fondest wish that you have a pleasant, successful journey.

Introduction

The Story of Scientific Writing

© 2020 Maritsa Patrinos

"Selfish scientists won't share new findings," ran one headline in *The Onion*. The story was about a group of rebellious scientists who made a groundbreaking, life-saving discovery but decided to hold on to it unless they were paid a ludicrous reward. Imagine that for a second: science happening but without anyone finding out about it. If a new species is observed in the forest but no one talks or writes about it, does it even matter—much like the proverbial fallen tree? Science would not be served in any way if scientific results were kept secret. Experimentation, observation, brainstorming ideas, and solving puzzles are only side of the scientific coin. The other side is communicating it. What happens in the lab needs to be understood, described, explained, and shared; otherwise, the most spectacular, groundbreaking experiment will remain inconsequential. Writing is, of course, the most common way we tend to communicate science: we write papers, books, lab reports, letters, posters, or articles. We talk about science too—although we usually need some record of that talk as well (in writing or as video or audio recordings). (It would be hard—although probably not impossible—to convey one's latest and greatest discovery through, say, interpretive dance.)

That being said, *Science* (the journal) organizes an annual "Dance Your PhD" contest (https://www.sciencemag.org/projects/dance-your-phd). No word yet on whether thesis committees accept submissions in the form of dance performances.

"Science exists because scientists are writers and speakers," geologist and science communication specialist Scott Montgomery memorably wrote. It doesn't matter if you're a budding mathematician, an experimental biologist, a doctor-in-training, or an environmental engineer: You will need to write in a manner that follows the conventions of your field and that will make your work known to the scientific community. You need to *write*—and write well—for professional (including academic) success.

What Is Scientific Writing?

Operational Definition

Scientific writing is the craft of effective communication in scientific disciplines. Note that in this provisional definition, we're keeping things pretty general, on purpose—we don't define "effective" or "communication" or "scientific disciplines." Here is a useful exercise: Let's think about what scientific writing is *not*. So far, you've probably been exposed mainly to these types of writing:

- *Creative writing*: fiction, poetry—from Shakespeare to Harry Potter.
- *Academic writing*: essays, reports, reviews, and other stuffy prose on scholarly topics.
- *Journalistic writing*: news reports, investigative coverage, interviews, and editorials.
- *Technical writing*: the ever-helpful "User's Guide" for your PlayStation or "help files" for your software.

How are they different?

Scientific writing differs from other forms of discourse in that it aims to be a perfectly transparent code that renders nature intelligible to us. (Note "aim" as an aspirational verb rather than reality. More on that later.) This is why every word in scientific writing *must* be precise and meaningful, and why the writing must faithfully and intelligibly render the scientists' ideas about reality. Because the sheer volume of research produced nowadays is overwhelming, scientists value brevity: Conciseness is both a signal of respect for your reader's valuable time as well as a sign that as an author you have thought about and chosen your words carefully for maximum impact. Concise prose is, of course, only as good insofar as it is clear and coherent. If your explanations do not proceed from the general to the particular, if you focus on details but lose the big picture, if you use redundant or meaningless phrases, if you fail to render your explanations in precise prose or neglect to include important information, or if you lose track of the point you were trying to make, you may end up producing a distorted image of reality, in which case you are not fulfilling your contract with your readers, who expect and deserve nothing less than excellence and objectivity. They also expect to either be able to replicate your findings or build and improve on them—yet another reason why your writing must be clear and precise. So let's refine our operational definition to say that *scientific writing is a style of writing that aims to disseminate scientific information in an objective, succinct, and effective manner.*

> *Scientific writing is a style of writing that aims to disseminate scientific information in an objective, succinct, and effective manner.*

Competing Definitions: Science Writing, Technical Writing, Grant Writing, Medical Writing

We often see science writing and scientific writing being used interchangeably, together with other terms that gravitate in the same sphere yet are not quite the same. Since such usage can create confusion, we feel compelled to straighten out some definitions from the beginning.

Scientific Writing Versus Science Writing

Science writing is *not* scientific writing (no matter how often you hear the terms used interchangeably). They may share some common generalities ("science"), but the audience, purpose, and style are quite different for each. *Science* writing aims to make scientific discoveries intelligible to a lay (or general) audience, while *scientific* writing is designed to cater to an audience of peers—specialists who usually don't need extensive background sections or explanations of basic concepts, who expect a certain paper organization and structure, and who are comfortable with lengthy formulas, polysyllabic

Latinate and Greek words, and complex charts. Science writing, by contrast, will usually favor open journalistic forms (unlike Introduction, Methods, Results, and Discussion [IMRAD] articles so common in the sciences) and a more informal or creative style. Science writers often borrow from journalistic conventions and from creative nonfiction to explain the natural world in a way that is both informative and entertaining. The target audience, primarily, is going to dictate style and format.

Scientific Writing Versus Technical Writing

Technical writing is sometimes used as an umbrella term to encompass everything remotely "technical"—which would make other specialized branches of writing, such as scientific or grant writing—subspecialties of "technical writing." There are many similarities between technical and scientific writing: formality, use of closed genres, and their informative purpose. However, technical writing often tackles topics that are not necessarily scientific, but technical, organizational, or procedural, and it usually has a utilitarian goal—to help a specific audience perform specific tasks or solve specific problems. Technical writing is also common in organizational and business culture (in fact, it has a lot in common with business writing and occasionally with legal writing). By contrast, scientific writing aims to inform and explain, is deeply rooted in the laboratory and research culture, and typically deals with experiment and observation.

Scientific Writing Versus Grant Writing

Writing scientific grant proposals is a subfield of scientific writing—and of technical writing. The bulk of scientific research depends on grants (from either governmental or private institutions), but not all types of grants are necessarily for scientific purposes. Grant proposals, are written in many other fields and for different purposes, (e.g., business management or service delivery proposals). Such project proposals, while sharing structural similarities with the type of proposals requested by federal or state agencies, differ from them in subject, scope, goal, authorship, and audience: They propose methods for solving a specific, often local and time-sensitive, organizational, administrative, technical, or economic challenge, and are often authored by organizations for approval by managers of other organizations.

Scientific Writing Versus Biomedical Writing Versus Engineering Writing Versus . . .

Once upon a time, scientists in all disciplines informed each other of the results of their observations and experiments in long letters, often read aloud before larger audiences. Today we recognize that science is too broad a field to fit in one epistolary format. The language of physics is completely different from the language of medicine, which has little in common with the language of environmental science. Each scientific field has its own conventions, conversations, and restrictions. Writing in the life sciences has constraints that writing for engineering does not, simply because life sciences deal, quite often, with, well, *living* organisms, which do not behave uniformly. The writing follows suit and also has to hew to a more complicated lexicon. Before biologists can even *start* their work—and certainly before they start writing about results—they have to be steeped in training in ethics, understand how the Institutional Animal Care and Use Committee (IACUC) works if they do research involving animals or understand how the Institutional Review Board works if they use human subjects in their research. Electrical engineers are unlikely to be subjected to IACUC scrutiny. Physicists are used to seeing complex, arcane equations in their papers. Engineers think in microseconds—geologists, in millennia.

There are some *basic* common elements of writing and researching in the sciences—a solid foundation that is common to *all* sciences and that will serve you well, no matter what field you choose. This book is designed to cover those basics, with forays into common scientific genres. However, all scientific fields of study are going to have specific variations on the genre of the scientific paper, variations that we cannot possibly cover in their entirety in this book. It will be up to your instructor, your professors in your major, and you to focus on the specialized writing conventions of your field. In other words, you need to become card-carrying members of your specific *discourse community*.

Discourse community, a concept popularized by linguist John Swales, is a group of people who link up to pursue common goals (other than socialization and solidarity), using specific communication tools to achieve those goals. For example, scientific discourse communities are united in their desire to pursue and communicate knowledge in their respective disciplines. They develop their own linguistic conventions (terms, notations, abbreviations, etc.), their own genres (such as the scientific article), and their own channels of communication (such as scientific journals, professional associations, etc.). To become a member of a scientific discourse community, you must acquire the required expertise and be vetted by community-specific mechanisms (such as peer review).

A good place to start learning your field's conventions would be to consult the relevant style manual(s) put out by a variety of professional and scientific societies, such as:

AMA (American Medical Association)—for medical fields

APA (American Psychological Association)—for psychology and social sciences

ACS (American Chemical Society)—for chemistry and related fields

CSE (Council of Science Editors)—for biology and related fields

IEEE (Institute of Electrical and Electronics Engineers)—for engineering

AMS (American Mathematical Society)—for mathematics

AIP (American Institute of Physics)—for physics

ANSI (American National Standards Institute)—for technical writing

This is *far* from an exhaustive list, and you'll have to consult the publication guidelines for the journals in your field very carefully to figure out exactly what style they follow in case you decide to pursue the path of publication. It will serve you well to zoom in, as early as you can, on the guide that is most often used by professionals in your field and become familiar with its conventions. As you progress in your field, you'll be reading more and more materials: papers, books, reports. *The more you read, the better your writing is likely to become.* One of the most effective ways to pick up the writing conventions of your field is to see them all the time.

Scientific Writing: Past, Present, and Future

The complexities of modern science usually require teamwork, and nowadays single-authored articles seem to be the exception rather than the rule. Most importantly, scientific articles are evaluated by a community of peers who vet the research to be published and ensure that it conforms to rigorous, universally agreed-upon scientific standards. This process is called *peer review* and is a sacrosanct component of scientific (and, generally, academic) publication process.

> They say "well, science doesn't know *everything*."
> Well science *knows* it doesn't know everything. Otherwise, it would stop.
>   —Dara Ó Briain, in a comedy routine

Scientific Journals

The first scientific journal, the *Philosophical Transactions of the Royal Society*, was published in 1665. The goal of the *Transactions* was to publish the correspondence of the fellows of the Royal Society of London, on the assumption that timely communication was the single most important factor in the advancement of science. It only published *original research articles* using a system of peer review (not blinded at the time, but you have to start somewhere). The "philosophical" in the title was used much more broadly then than it is today and essentially meant "dealing with the entirety of human knowledge."

The *Transactions*, much like other contemporary journals, was a general scientific journal, covering all aspects of science, from astronomy to biology to chemistry to medicine and physics. As each of these disciplines evolved and many others started to take shape, there was an increasing need for specialized journals, which started to appear by the end of the 18th century in England, France, Germany, and the United States, among other countries. Stricter guidelines for blinded peer review were also developed by the mid-18th century, ensuring a more objective quality control of the papers published. Further specialization and professionalization gave rise to more and more specialized and interdisciplinary journals (usually sponsored by a professional society or association or an academic institute) as we entered the 20th century. Curiously, we have not yet figured out precisely how many active scholarly journals are in existence today, as new journals originate every year, all over the world, and others are phased out or experience a hiatus. A 2019 report by the International Association of Scientific, Technical and Medical Publishers estimated that about 3 million peer-reviewed papers were published annually (in 33,100 journals in English), up from 1.8 million in 2012.

The Development of the Scientific Article

Today, the scientific article is the preeminent genre of scientific communication. To stay current in your fields, you will have to read articles constantly and eventually maybe even write some. The scientific article came a long way from its humble, somewhat narrative and open-form beginnings in the *Transactions*, where authors followed no discernible format or style and would sometimes interject personal circumstances in their descriptions. Each feature of the scientific article that we take for granted today has taken decades and sometimes centuries to evolve. For example, Louis Pasteur is credited with the introduction of a Methods section in his papers in the 1860s, which effectively gave birth to the famous IMRAD format that is now universally accepted as one of the international standards of reporting in the sciences.

Throughout its 350-year history, the scientific article gradually acquired other formal and substantive features as well; among the most important are:

- increasing reliance on visuals (e.g., tables, figures, charts)
- increasing emphasis on rigorous citation and references to the current literature (sometimes called "citational density")
- modular arrangement of parts (with headings and subheadings, for easy "scanning" of the article)
- a plain style, increasingly formal and impersonal
- inclusion of abstracts (not until the second half of the 20th century)
- increasing emphasis on argumentation and theoretical explanation rather than simply observation and experimentation.

The Fight for Plain Style

Even if you've never read a scientific journal article, you've surely had the chance to study a scientific textbook, so you probably have an idea that scientific writing favors precision and concision and generally avoids flourishes or emotional appeals. This is not the style usually favored in the typical English classes you've taken so far, and, in fact, as instructors, we have had to curb on occasion some of the more poetic or dramatic tendencies of our students. In rhetorical terms, this is simply a matter of knowing your audience and delivering what they need and expect. We mercilessly ask students to edit out introductions that appeal to pathos: "Imagine that you have to wake up every day with excruciating back pain, unable to perform basic activities of daily life" is not the way to start a review on robotic-assisted physical therapy interventions for lower back pain. We explain that some audiences *may* want to read about the ethereal appearance of the pale blue butterfly resembling a speckled, lonely hydrangea bud in its dewy bed of leaves but that such descriptions (appropriate elsewhere—science writing, perhaps?) will not serve well the specialized audience (entomologists), who needs a straightforward description of color patterns, appendages, periods of prepupae and pupae, peak days of eclosion and pupation, and all of the other exact entomologic detail they need to push back the frontiers of their science.

The Future of Scientific Communication

The most significant recent impact on the evolution of the scientific article is the Internet. Most, if not all, articles published today are also published electronically, and sometimes *exclusively* on the web, as in the case of open-access journals (many of which are also refereed but forgo the editorial and managerial supervision of a publisher to make the content freely available to the public). This development engendered a new set of rules for scientific publishing: from rules governing citations of web materials, to the type of data and appendices that can be included (e.g., supplemental data are commonly included on the web but almost never in print due to lack of space, increased expense, or color limitations), to rules governing features unique to the Internet, such as the inclusion of hyperlinks for citations and other textual and graphic elements such as video abstracts or self-playing slides. As open-access journals gain more visibility and respectability, some of the major publishers have started offering delayed access (meaning that one can access some articles for free after an "embargo period" has elapsed) and hybrid access (meaning that one can purchase individual papers rather than subscribe to the entire journal).

Web publishing has also cut drastically the time between the acceptance of a paper and its actual publication, as new articles can be made available almost instantaneously on the web. Tools such as email, blogs, and social media (Twitter in particular) allow for rapid reactions and critiques of those articles. This unprecedented publication and response rate has had some unforeseen consequences—for example, it has been tentatively linked with the rising rate of journal article retraction. The web is almost certainly the future of scientific publication, and time will tell whether the new open-access journals will gain the same status and prominence as some of the more established publications. The "more established publications" almost all have online presences now—often their materials are available in e-format and frequently "ahead of print." Unfortunately, there is also a growing number of open-access "web journals" that are not legitimate, preying upon authors who want to be published . . . published *anywhere*. Much like "fake news" or "viral" content that turns out to be a hoax, predatory scientific publications will publish anything for a fee, leaving it to the reader to discern whether the *American Journal of Scientific Research* is of the same nature as, say, *Science* (they are not. Can you tell which one is a reputable publication and which one is not and why?).

Ethics in Scientific Communication

Partially as a result of the forces at work in our current scientific publishing landscape, we devote an entire chapter of this book to ethics in scientific communication. Ethical concerns regarding scientific research in general also have implications in the way science is written and disseminated;

societal standards and expectations have (fortunately) changed significantly on these matters since the days of the *Transactions*. For example, in the 17th century one could—and did—experiment on transfusing sheep blood to humans with little concern for human consent and well-being or for animal welfare. Today you need to pass rigorous, federally supervised ethics panels at the institutional level to conduct any sort of invasive procedures (the human-animal blood transfusion would not pass muster, we're afraid). Plagiarism and fraud in scientific research and communication have also become more prevalent. As a future scientist, you need to be equipped to make ethical decisions as well as recognize unethical situations in both research and writing.

Dispelling Myths About Scientific Writing

Before we get started, let's dispel some of myths about scientific writing that we've heard over and over.

Scientific Writing Is Only Done by Scientists

Wrong! As students, you engage in scientific writing every time you write a report or a paper reporting research. In fact, it is more accurate to say that researchers *must* engage in scientific writing—whether or not they are in training. In addition, professional writers, such as technical and biomedical writers, are often called upon to produce manuscripts in collaboration with scientists. Here is a little-known fact: Scientific writing *can* be a lucrative profession, whether it is done academically or in the corporate world.

Scientific Writing Is Only Done in Academic Settings

Not really! Pharmaceutical companies, biomedical companies, private research and development labs, clinical research organizations (CROs), professional engineering firms, marketing firms, and even legal firms employ professionals with scientific writing backgrounds to produce reports, articles, protocols, grants, marketing reports, historical supporting documents, and anything else that might be required to support any decision involving science. You might be the only scientist at a market research company that creates marketing reports for the pharmaceutical industry. Or you might be the only writer at a CRO, which also has to prepare reports for its clients. Academia is, of course, one of the biggest generators of scientific articles, but it is by no means the only one.

Scientific Writing Is Only for People Who Want to Become Professional Writers

By now you know that communication is half of science. If you become a scientist, you're not going to be a good one unless you can communicate what you've done. *Scientists must write* to demonstrate productivity, show results, get funding, and get—or keep—their jobs.

Scientific Writing Is a Mere Tool, and It Is Devoid of Cognitive Power

Scientific writing has rigors that are unlike most types of writing you've been asked to do so far, and you'll learn those rules only with practice. Occasionally, you'll have to do some unlearning of what you've been told about writing in the past. The proliferation of guidebooks for writing in the sciences, most of which target scientists who have finished their training, suggests that there is a need for more writing instruction in this field—and the earlier, the better. Again, we're not the first to have this idea. Back in 1967, a worried F. Peter Woodford warned readers of *Science*, one of the preeminent scientific journals in the world, that educational reform was needed to strengthen the writing skills of scientists-in-training. In a bit of a reversal from traditional wisdom, he believed that good writing promotes clear thinking, which is why he "advocate[d] a course on scientific writing as an essential feature in every scientist's training," a course that would focus, among others, on "logic, precision, and clarity." He was one of the first scientific writing instructors to emphasize that writing has heuristic and explanatory powers that can improve thinking—and we couldn't agree more. Having a clear, organized idea of the writing processes and genres specific to scientific writing will very likely help you organize your thoughts about your subject matter, force you to be systematic in your approach, and teach you to be succinct, coherent, and precise. When you can write about a topic with ease, that means that you have mastered it. *Learning how to write will help you learn how to think.* This is why we believe that "scientific writing" should be, in fact, be considered a scientific skill.

The Format of This Book

This book is divided in two parts: Part One is devoted to fundamental topics that pertain to contemporary scientific writing and writing in general (we call it The Tao of Scientific Communication), and Part Two is devoted to common scientific genres. Thus, we start Part One by talking about good writing habits (Chapter 1)—where we, essentially, remind you of writing habits and routines that are universally true, with a few adaptations specifically for the task of scientific writing. Chapter 2 focuses on targeting your audience, probably the most important skill a writer (and communicator) needs to master at any level and in any capacity. We foreground the role of ethics in Chapter 3, as we have found it of critical importance especially in the context of scientific communication, especially in our current age in which science is under intense scrutiny. While critical thinking skills are emphasized in many college-level courses, we have decided to devote an entire chapter to them because critically understanding and interpreting scientific information is foundational to producing cogent scientific prose (Chapter 4). With Chapter 5, we depart somewhat from the standard fare in scientific communication texts and provide a primer regarding *writing* and *communicating* what has become the backbone of most sciences—statistics. No textbook is complete without talking about visual communication (Chapter 6), which presents particular challenges and opportunities in the sciences. Finally, Chapter 7 discusses references and documentation to shed some light and clarification on a topic that often stumps not just students but seasoned writers. We have found all of these topics essential in laying the foundations of a competent scientific writer and critical reader/consumer of scientific literature.

In Part Two we address staple scientific genres such as the research article (Chapter 8), review paper (Chapter 9), abstracts and summaries (Chapter 10), proposals (Chapter 11), and presentations (Chapter 12). We recognize that, if you choose to pursue science-related careers, you will likely have to engage in other genres as well, some of which we explore in Chapters 13 (we call them "minor genres" because they do not form the bulk of scientific writing); finally, in Chapter 14 we tackle scientific writing for multimedia and the web. Rounding up the book, a stand-alone "Writing Toolbox" section answers perennial questions about such topics as pronoun usage, dangling modifiers, and nominalizations—but also offers practical tips on crafting effective sentences, paragraphs, and papers. We intend this as a reference section to be used often and on an as-needed basis.

We have based this book on our collective professional experience in the classroom and in the field, and we hope it is useful to you as well. In the words of the illustrious physicist Richard Feynman: "You're unlikely to discover something new without a lot of practice on old stuff, but further, you should get a heck of a lot of fun out of working out funny relations and interesting things." We hope you get to practice your skills based on the modest knowledge we offer in this tome and then move on to discover bigger, better things.

Exercises

1. Identify the style manual most commonly used in your field of study. In groups, discuss:
   a. What sort of advice does it dispense—is it merely about grammar, formatting, and citations, or are other areas covered? Are there any areas *not covered* that maybe should be?
   b. What part(s) of the manual is/are most commonly used? notation? referencing? abbreviations? format?
   c. Finally, can you find one guideline or item in the manual you have chosen that is new or surprising to you?
2. Force yourself outside of your comfort zone: Pick a single broad topic (e.g. "chlorophyll," "crystals," "electromagnetic spectrum," "evolution") and do a keyword search—across at least three different disciplines. Pull up a biology paper on the electromagnetic spectrum, and an engineering paper, and a geology paper. How do these papers approach a similar (broad) topic? How different is the writing? Can you tell in what kind of journal the writing occurs? In groups, select a paragraph from each of the papers you have found and present the three paragraphs to the group. Can you guess the general provenance of the paragraphs?
3. How do you access scientific resources at your institution's library? What does your library have? How do you know? How do you find out?
4. Identify a top-tier journal in your field (again, your instructor or a professor in your field may help you with that). Find out how long it has been in circulation and locate an article that was published close to the beginning of the publication but no more than four to five decades ago. Compare it to an article containing similar content in the current issue. How do the two articles differ? (Pay particular attention to the organization of the articles and their style.)
5. In groups of three or four, choose one of the following writing prompts (alternatively, pick your own or use a topic suggested by your instructor),

then write a brief paragraph (100 words or so) about the topic, to the best of your abilities, in a particular style: creative, technical, journalistic, scientific. Compare notes at the end. What kind of stylistic choices did you make to render your topic suitable for the purpose and audience required by each of these styles? Was it an easy choice for each? Discuss similarities and differences.

Suggested topics:

- a warm summer's day
- a murder
- a walk in the park
- a pet (your choice)
- a drug

PART I
THE TAO OF COMMUNICATING SCIENCE

1

Cultivating Good Writing Habits

You've seen this coming from several miles away: the dreaded deadline. It was right there, on the syllabus, and this time you vowed you would not let it sneak up on you. You would do the research (fungal phenotypic heterogeneity). You would write that paper in advance. You would get feedback and revise conscientiously, and your graphs would stun Dr. Stecklery, your bio instructor, with their clarity and brilliance. You would smugly hand it in *before* deadline, just to make a point.

Of course, it is now the night before it is due and all you have is a few disjointed paragraphs. Wait—as you look more closely one isn't even a paragraph—it's a shopping list. You were pretty sure your notes were right there in your green notebook but only ½ page? Surely you took some more, but where? After turning the contents of your backpack upside down produces no results, you sigh, whip out your phone, and tweet about it:

> @biogeek
> And they told me you can't write an assignment the night before.
> #challengeaccepted

In chaotic caffeine-fueled bursts punctuated by social media rants, you haltingly produce what might generously be called a draft, before you pass out, exhausted, head on the keyboard. You dream of spilled petri dishes and rogue sentient fungi taking over the lab. Waking up should be a relief but you have a headache and still have to proof, add a graph and references, format it, and print it out. You have a long day ahead of you.

Does this sound even partially familiar? What do *your* writing practices look like? How do you go about writing a research paper, review, or lab report? Be honest.

In our years in the classroom, we found that many of our students entertained the idea that good writers are born, not made, and that the corollary of this idea was that scientists, by definition, are not good writers. This is, of course, a myth born out of a romantic notion of the writer locked up in a dark chamber, prone to fits of inspiration occasioned by a capricious muse. Such writers don't exist. We teach instead the value of writing practice as a training skill akin to training for a sport or to learning a craft. Inspiration or talent do not make you a better writer; writing habits do.

One of the most important (and prolific) science fiction writers of the late 20th century, Octavia Butler, wrote in an essay titled "Furor scribendi" ("Mania for Writing"): "First forget inspiration. Habit is more dependable. Habit will sustain you whether you're inspired or not. Habit will help you finish and polish your stories. Inspiration won't. Habit is persistence in practice. Forget talent. If you have it, fine. Use it. If you don't have it, it doesn't matter. As habit is more dependable than inspiration, continued learning is more dependable than talent. Never let pride or laziness prevent you from learning, improving your work, changing its direction when necessary."

—*L. Ron Hubbard Presents Writers of the Future*, Vol. IX.
Los Angeles: Bridge Publications, 1993.

Develop dependable habits. As students and future scientists and professionals, you *have to* write in your daily jobs. While you may not be writing science fiction, you will have one thing in common with all writers: a writing *habit*, which will give you the tools and confidence to tackle any writing task.

Is It Really True That Scientists Must Write?

"What's the big deal?" you may mutter under your breath. "I want to be a scientist, not a writer." To better understand the role of writing in the life of a scientist, we asked Liz, a postdoctoral fellow in a biomedical engineering program to explain what, when, why, and how often she had to write to advance her studies and her career. Here is her answer:

> I would say that writing is so intertwined with my profession I sometimes hardly even notice it as a separate entity—I never thought of writing as something *separate* from what I am doing. It's just part of the science. To start with, I keep a log of my experiments in a lab book, including instruments, subjects, procedures, hypotheses—and I even scribble research questions and notes in the margins as I go along. My lab books are the repository of all my raw data . . . except for data I keep in my computer. I take notes on articles I read—that's writing, right?—because I know I will need them later when I discuss my results, or write the literature review portion of a grant, for example. I had to write papers for my classes, of course, and a research proposal for both my dissertation and postdoc application. My life as a postdoc basically consists of writing grants and groveling for money. The "writing half" of my scientific career is practically a full time job in of itself. Recently, I was asked to write a review article that involved quite a bit of synthesis of *other* research, which was different from writing your basic IMRAD [Introduction, Methods, Results, and Discussion] paper. I write papers for school, abstracts, meeting posters, lab protocols, results, presentation slides . . . even a weekly "report" for the Monday lab meeting. I have my tablet with me all the time. . . . I'm writing in the lab, in the library, at Starbucks, on the train . . . everywhere. All the time. I'm used to it. We're all used to it. Further, you need to get published to get into a postdoctoral program, and then you need to publish to get a teaching job, and then you need to publish to get tenure, to get grants, to get a name for yourself. Frankly, I don't see it as separate from science.

While maybe not all of you are preparing for the exciting life of a postdoctoral student in biomedical engineering, the fundamental point you should be getting out of all of this is: *writing is not separate from science.* On the contrary, it is symbiotic with sciences in *all* of its aspects, and it is something you are expected to do, and do well if you want to advance in your career. Scientific articles are how discoveries are heralded, so that other like-minded researchers can replicate and build upon your findings. Scientists know that if *you didn't publish it first, you didn't discover it*—and being "scooped" is, in effect, one of the fears researchers have to live with. A good example is the famous Watson and Crick 1953 paper in *Nature* describing the double helix structure of the DNA, one in a series of revolutionary articles. This discovery eventually won them the Nobel Prize in 1962. Watson and Crick are now household names you should know well, but you've probably heard less, if anything, about Rosalind Franklin, a remarkable researcher (who went to Cambridge University and studied physical chemistry when she was 18) who made the initial critical scientific observation that enabled the elucidation of the double helical nature of the molecule. She was not part of the first publications, despite the contributions she made and, to add insult to injury, did not share in the Nobel Prize (the prize was rarely awarded posthumously; she died tragically young, of ovarian cancer, at age 37—four years before the Nobel Prize was awarded). Moral of the story: Recognition and awards are only usually granted for work that has been *published.*

So now that you know writing is not peripheral but central to science, what can you do to acquire good writing habits? First, you need to stop thinking of writing as something you're just "not good at," or as something only the likes of Shakespeare get to do. Instead, you need to reframe writing as any other skill you need to achieve your goals. A good analogy for this is athletic training. Nobody gets to finish a race or win a game without methodical, consistent training, and, in general, the more training you put in, the better the results. It is the same thing with learning a musical instrument: you have to practice your scales before tackling more complex music, and you have to practice daily and consistently. Writing is no different: it is practice. The more diligent you are about your writing routines, the better your results. Consider this textbook your playbook, if it helps: a set of instructions to create the conditions for victory. This chapter outlines the mental workout that will allow you to tackle the writing process with confidence, joy, awareness, and (very importantly) a sense of purpose. Because these routines should ideally become second nature to you, we call them writing habits—which, if practiced diligently, will help you reframe or refine your approach to writing.

10 Habits For Writing

Habit 1: Think of writing as a way to solve problems.

Habit 2: Find and use writing models.

Habit 3: Incorporate writing in every stage of research.

Habit 4: Establish rhetorical strategies and goals for yourself.

Habit 5: Develop strategies to conquer your writer's block.

Habit 6: Think of writing as a recursive process.

Habit 7: Find a writing process that works for you.

Habit 8: Cultivate good revising and editing habits.

Habit 9: Learn to write collaboratively.

Habit 10: Reflect on your writing practices.

Habit 1: Think of Writing as a Way to Solve Problems

At the core of every writing task is a problem waiting to be solved. As scientists, you are familiar with this way of looking at the world. The world around us offers us myriad puzzles: Why does the apple fall from the tree at a constant speed? Why are there spots on the sun? How do we cure the common cold? How do we fit a 100-story building on that narrow piece of real estate? How did the platypus evolve? Can we predict when the next tsunami will occur? Science offers us a method to tackle these questions without getting entirely lost. Scientists are engaged in a puzzle-solving game, an activity that is inherently rewarding for us as humans—to say nothing of the practical benefits. (The puzzle metaphor credit goes to Thomas Kuhn, who studied how science works and wrote a ground-breaking book about it [*The Structure of Scientific Revolutions*, 1962].)

In their role as writers, scientists or engineers need to solve a different puzzle: how to best communicate their findings to a community of scientists, to the general public, or to potential employers or financial supporters; the problems they face are how to inform, persuade, and achieve their goals. As scientists or professionals, you may pursue publication, grants, public understanding of a scientific phenomenon, appreciation, promotion, etc. Solutions might include crafting the argument to elucidate how your research responds to a particularly pressing societal need, sticking to a predetermined format, or finding a clever analogy to explain a technical concept. Writing, as it turns out, is also about solving problems.

Even as students, when you write you are trying to crack the puzzle of "How can I get an A? What does the professor want?" Consciously or not, you are already approaching writing with hopes of finding the formula to the perfect A. No magic formula exists, but you can try to reimagine your writing task as a problem or series of problems to be solved. What do you need to write—a research paper? lab report? grant? Do you have a specific format to follow or specific questions to answer? Tackle them one by one. For example, your "problem" may be to find the most succinct, clear, and compelling way to get an audience of peers (scientists) interested in your topic in the introduction—whether you're talking about the evolution of the spoonbill or the applications of cristobalite in bioengineering. What is the best way to present your research results in the Results section—a line graph or a table? Which will best help the audience understand your results? What do you highlight among the data you obtained that will make the reader understand the significance of your experiment? What other studies can you compare yours with in the literature review section to best show the reader the necessity, logic, or innovative nature of your study?

There is no clear-cut formula, and there are multiple possible and plausible answers to writing conundrums. That is perfectly okay. In your own writing practice, try tackling writing tasks as you would math or science problems: even though you may be struggling with the answers at first, the longer you practice, the better the outcome, and the better you will become in the process. (See Exercise 1 for practice.)

Habit 2: Find and Use Writing Models

Imitation is frowned upon in writing studies, and for a good reason—it may stunt your personal voice and growth as a writer and limit your definition of quality. But that only happens when you get stuck imitating the mannerisms of a given author and suppress your own personal voice and creativity. We trust you not to do that; instead, use good writing models as something to aspire to, as a way to test what is possible. In the words of T. S. Eliot, "Mediocre writers borrow; great writers steal." Writers (of fiction, nonfiction, or scientific articles) freely acknowledge their muses, idols, and models who shaped their craft or offered an ideal one can aspire to. You, too, can find the one that suits your inclinations and abilities. In the sciences, the added benefit of good models is to understand acceptable formatting (such as IMRAD), acceptable title formats, acceptable references in your discipline, ways to label visuals, techniques for abstracting, the scope of the introduction or of the recommendations section. And, of course, you should try to find models whose prose is admirable for clarity, precision, and conciseness.

"If you want to be a writer, you must do two things above all others: read a lot and write a lot."

—Stephen King

How, exactly, do you find good writing models? We're afraid there is no straight answer, though asking your mentors is a good start. Good writing exists in all disciplines; it takes some work and determination to find your models for yourselves. For this, you would have to start reading in your discipline, as much as possible. When you encounter an article that impresses you because the ideas are both bold and succinctly and clearly stated, bookmark it or save it in your research folder. Refer to it often. How do the writers organize their material? How do they address incongruous results? How do they present and interpret their data? What kind of recommendations do they propose in their conclusions? How much literature do they cite and where? What can you "steal" from those authors?

Generations of young scientists were inspired by great figures such as Carl Sagan to go into science and/or into the popularization of science—for example, Neil deGrasse Tyson or Brian Cox. They were attracted by Sagan's clear but entertaining style, by the passion for his subject, by the power of his ideas. Who is your science hero? Who inspired you? Go find something that person wrote—ideally, both for a scientific and for a general audience; hopefully, that piece of writing can get you started on building your own library of style.

How do you make use of models? Since we have already brought it up, how about one of the most famous papers in science, the Watson and Crick 1953 *Nature* article on the structure of the DNA? Here's how it starts:

We wish to suggest a structure for the salt of deoxyribose nucleic acid (D.N.A.). This structure has novel features which are of considerable biological interest.

From this simple and direct introduction you may learn that the authors do not shy away from using the first person ("We") and are extremely forthcoming about the goal of their paper, as well as about the importance of their research.

You keep reading and see that the authors address other proposed models and reject them succinctly, thus establishing their ethos as knowledgeable, careful, and well-informed researchers (by ethos we mean, in short, credibility and authority). Then, Watson and Crick put forth their famous double helix structure that is now the stuff of textbooks:

We wish to put forward a radically different structure for the salt of deoxyribose nucleic acid. This structure has two helical chains coiled round the same axis (see diagram). We have made the usual chemical assumptions, namely, that each chain consists of phosphate diester groups joining β-D-deoxyribofuranose residues with 3′, 5′ linkages. The two chains (but not their bases) are related by a dyad perpendicular to the fibre axis. Both chains follow right-handed helices, but owing to the dyad the sequences of the atoms in the two chains run in opposite directions. Each chain loosely resembles Furberg's model No. 1; that is, the bases are on the inside of the helix and the phosphates on the outside. The configuration of the sugar and the atoms near it is close to Furberg's "standard configuration," the sugar being roughly perpendicular to the attached base. There is a residue on each chain every 3.4 A. [angstrom] in the z-direction. We have assumed an angle of 36° between adjacent residues in the same chain, so that the structure repeats after 10 residues on each chain, that is, after 34 A. The distance of a phosphorus atom from the fibre axis is 10 A. As the phosphates are on the outside, cations have easy access to them.

—Watson, J. D., Crick, F. H. C. Molecular structure of nucleic acids.
Nature. 1953;171(4356):737–738.

This description (and the accompanying diagram) revolutionized biological sciences. Yes, it turned out to be correct; it was also described succinctly and clearly, with a powerful topic sentence announcing the goal of the paragraph and unfolding from general (the basic "two helical chains" model) to specific (distances and angles and other important details). Sentences were pithy, clear, and varied in length for rhythm; every word served a purpose.

Ask yourselves: Who could your model be in scientific writing? How would they inspire you—and why?

Habit 3: Incorporate Writing in Every Stage of Research

Excellent writing does not occur independent of content—although the reverse might be true (i.e., excellent ideas can take the shape of appalling verbiage). Thus—need we say it?—you should know your subject matter well in order to write well. Mind you, that does not mean that at the onset of writing you will know exactly what to write, from introduction to conclusion. In fact, one of the great joys of writing is that the very process can clarify some ideas for you. However, it means that you should have a general knowledge of your subject matter, a fundamental understanding of the principles that inform your field, and a pretty thorough background of the topic you are exploring through writing. Also, inasmuch as possible, you should use writing at every step of the research process—or, to put it in composition terms, you need to do a lot of prewriting before you actually write your paper. Writing involves discovery (or invention, as the ancients called this stage of the process), and discovery (you guessed it!) involves writing.

Prewriting may include note-taking to summarize other papers, outlining, freewriting, concept mapping, diagramming, or brainstorming. You should already be familiar with these concepts from previous writing classes. Freewriting, for example, is a great way to manage your writer's block (more about that shortly). As well, researchers use literature reviews often in order to help clarify their ideas or methods, or the role of their own research in their field; many compile such reviews before and all through the writing process in fact. Outlining, concept mapping, and even doodling (as long as it's focused on your topic!) may help you if you are a visual learner. We cannot overemphasize the importance of reading in your field and of taking notes. You should find a prewriting method that you like best (it could be diagramming, index cards, or perhaps text or audio files that you can access on any digital device you use) and stick with it. Always carry with you something you can write on or record your thoughts on the project; this may be anything from a simple notebook to your mobile phone.

Writing does not occur in one sitting where you "pour" the undistilled, perfectly formed thoughts you had about your research onto the paper. It starts way before you create that first draft, and it continues well after it through multiple layers of revision and polishing. In the process of crafting what you want to say into written prose, you are actually clarifying your ideas as well. It may be rewarding, even fun, to perform experiments in the lab or make field observations, but without writing what the results of the experiment mean or interpreting your observations, no one else would understand their significance or benefit from them. Thus, writing *is* part of scientific research in the most fundamental sense.

Habit 4: Establish Rhetorical Strategies and Goals for Yourself

If you approach writing as a problem-solving process, your solutions should take the form of establishing your rhetorical strategies and goals for your project. By rhetorical we mean pertaining to discourse (including text, visuals, or speech) and targeting a particular audience in order to achieve your communication goals, which you can find on a continuum between informative to persuasive. Learning to target your audience is so important that we are devoting an entire chapter to it (Chapter 2). Audience and other constraints (such as genre) should dictate your rhetorical goals. Here are some examples of rhetorical goals that you may have for various sections of any given paper:

- Provide an overview of the topic
- Introduce the goal of the study
- Describe the importance of the study
- Emphasize the novelty of the approach
- Point out gaps in previous research
- Describe limitations of current study
- Summarize results
- Indicate how the study builds on previous research
- Provide recommendations for future research
- Emphasize implications for practice/research/teaching, etc.

To get there, you should ask yourself some basic questions:

How do I identify my rhetorical goals?

1. Why am I writing? (What is the purpose of your communication?) For example, are you describing an experiment? reviewing the literature? asking for grant money?
2. Who is my audience? (Be specific!)
   a. Why are they reading my piece?
   b. What are their expectations of my piece?
   c. How do I best address these expectations?
3. What are the technical specifications of my paper?
   a. How long does it have to be?
   b. What format does it have to observe? (Does it need an abstract and/or title page? What about typeface and font size, spacing, margins, page numbers, headers/footers, headings, references? What style manual do I use? Do I need figures/tables? What other conventions do I need to consider?)
4. What are the main components of my piece? Do I need simply an intro, main body, and conclusion? Do I need to follow an IMRAD structure + abstract? Do I need to break my paper down into main sections discussing several related topics or use some other format? In other words, what is my plan?
5. Can I sum up the gist of my paper in one to two sentences? (this is called "nutshelling" or the "elevator pitch": in 1 minute or less, can you define the project?)

Question 5 will force you to go back and think globally about whether your communication has achieved its purpose as set in question 1, while questions 2 through 4 allow you to ponder elements of the project and important considerations such as audience. This very general template can be adapted to each writing task. For example, if you need to write a cover letter for a prospective job, you would have to pay attention to the conventions of the genre related to both form (e.g., a strict, formal formatting including address, date, salutations, etc.) and content (e.g., addressing questions such as why you are interested in the job and why you would be a good fit for it). Depending on the nature of the application, you might also need to address other elements specifically requested in the job description. We discuss more specific goals and strategies for more scientific genres in other chapters of the book. (See also Exercise 4 at the end of this chapter.)

Habit 5: Develop Strategies to Conquer Your Writer's Block

Writer's block can afflict writers everywhere, so don't panic. You're not alone! If you find yourself unable to focus on the writing task; staring at the blank page/screen for hours waiting for the muse to pry open the vein of creativity; browsing social media instead of writing; or feeling guilty for your lack of progress, frustrated, or stuck, it's time for a different approach. Ask yourself instead:

- *Am I bored?*
  Perhaps you chose the wrong topic, or the topic no longer excites you. Perhaps your experiment failed or produced lukewarm results. Or maybe your hypothesis is disproven, or the papers you reviewed failed to reveal anything new or interesting about your topic, yet you are still stuck with writing that review paper? These are all possible hurdles on your way to scientific stardom, so if your boredom is related to them, take a step back and reevaluate what first attracted you to the project and why—and focus on the aspect that interested you the most. Remember that a failed experiment is as instructive as a successful one. And, most of all, seek advice from your mentor, professor, or colleague who is familiar with your research or at least with your field. Discuss these issues and ask for guidance. The mere process of talking about this with someone else may bring the fresh, new perspective you need.

- *Am I tired?*
Okay, that's an easy fix. Take a break, try some relaxation techniques—sleep, exercise, play with your dog, or perhaps go out for a walk—and then return to your paper with renewed energy.
- *Do I know enough about the subject?*
Sometimes the prose doesn't flow liberally from underneath your fingertips because you simply didn't do enough research. You know what to do in that case, right? You will be astonished at how much faster and easier the writing comes when you know what you're talking about.
- *Do I know too much about the subject?*
The reverse is also possible: you are paralyzed with the amount of information you need to convey and you simply don't know where to start making sense of it. In that case, we suggest you pick the one to two most important concepts about your topic (the ones without which comprehension would be impossible) and start with those. Imagine you have to explain the topic to your Aunt Bernice, for example, or, if that's going too far, to your former lab partner—where do you start? The rest of the paper should eventually fall in line.
- *Am I afraid my writing won't be good enough?*
Depending on how exacting your definition of "good" is, it may not be. But perfectionism is one of the most pernicious sources of writer's block. Just repeat after us: THERE ARE NO FIRST DRAFT MASTERPIECES. Or, if you prefer, learn from the wisdom of prolific, accomplished, award-winning writer Anne Lamott, and, like her, write "shitty first drafts" as a necessary step to getting to write the piece you want to write. Accept that your first draft is just that—a draft that will go through multiple revisions. You will have plenty of time to polish your draft. For now, you just need to write.
- *Am I overwhelmed?*
Ask yourself why. If the problem is that you can't possibly see yourself writing this whole paper in one sitting, we have good news for you: you don't have to, and, in fact, you shouldn't. You should, instead, write for minutes, not hours at a time. Use what we call the "kitchen timer" method: set the timer to 5 to 10 minutes and write without interruption until the bell rings. We mean that: no distractions should be allowed (TV, Facebook, trip to the fridge) during that time. Surely you can maintain focus for 5 minutes at a time! When your 5 or 10 minutes are up, look back at what you've accomplished and you'll likely be amazed: you should

have anywhere between 100 and 500 words already. Take a break before you commit to another "intensive writing" period. Writing in short spurts like this, totaling about 30 minutes of your day, should not constitute a big commitment and should allow you to make progress—even impressive progress—if you keep it up for weeks at a time.

The Pomodoro technique is cleverly named after a kitchen timer in the shape of a tomato (which in Italian is "pomodoro"). It articulates some basic productivity principles, such as working in 25-minute chunks, divided as follows: 5 minutes to review goals and previous work, 15 minutes to do the work, and 5 minutes to review your progress and set new goals. It applies not only to writing but to any task you need to do that involves learning, creating, or producing something for work/study. You can find more about it on the Pomodoro technique website or in the eponymous book.

To sum up: Use a heuristic process like the previous series of questions to identify what keeps you from starting that paper, and take appropriate steps to manage your writer's block.

Habit 6: Think of Writing as a Recursive Process

Writing is not an assembly line in which some words go in, some tweaking goes on, and some paper emerges at the other end from all those words put together. If it were that easy, robots would have replaced writers a long time ago. Writing is, on the contrary, a creative and recursive process, with lots of feedback loops. Any point in the loop can be interrupted to do some drafting, go back to research (or prewriting), get some feedback, revise, reflect, go back to drafting, and so on. And that is okay. If the scenario at the beginning of this chapter sounds familiar, some of you might have had an episode in which, 6 hours or less before a writing assignment was due, you leapt up and by sunrise wrote a 5-page paper in one fell swoop. Some of you may have even gotten high grades on some such projects. Anything is possible. That outcome may mean a number of things: that you are very smart, or the professor was distracted, or the other papers were quite atrocious and the class was graded on a curve; it does not, however, mean that you are a great writer, and it definitely doesn't guarantee you good results in the future. Good writing is forged in the furnaces of feedback, revising, and rewriting. To borrow a phrase from writer and Pulitzer Prize winner James Michener, "Good writers are, in fact, excellent rewriters."

This recursive process is not unique to writing. Science functions much the same way: Hypotheses are tested, retested, discarded, and modified, and tested again. Results are peer reviewed, revised, published, and tested against the larger scientific community and society at large. New theories and experiments build on those results, and so it goes.

Writing, similarly, moves nonlinearly back and forth between prewriting (planning, researching, taking notes, analyzing) and drafting, between revising and peer review, between incorporating feedback and revising again, and between editing and proofreading/polishing. Approaching writing is not substantially different from how you approach science, technology, or math. You will not build the perfect engine or bridge based on your first draft; you will not get perfect results the first time you perform a test; and you will not write a great paper the first time you try your hand at it. Allow enough time, be patient, and be willing to accept constructive criticism from peer review, write multiple drafts, and revise as often as necessary until you are satisfied.

Habit 7: Find a Writing Process That Works for You

Throughout your academic career, you have been exposed to and have practiced the three main stages of the writing process: prewriting, composing, and revising. As we have indicated, this process is recursive—and bound to get messy rather than linear and neat. Most papers are not written in the same order that they are read—for example, it is not unusual that the Methods section is written first and the Introduction, with the abstract, last. We urge you to find your own process that works for you: In other words, if you need to write the abstract of the paper first, before it's completed, because it gives you a sense of purpose and it keeps you organized, by all means do so. If you cannot stop outlining and diagramming throughout the last stages of revision, then so be it. In general, however, you do need to make sure of three things: (a) that you write multiple drafts; (b) that you get feedback on your drafts; (c) that you revise those drafts substantially in terms of structure, content, vocabulary, and grammar.

There are many writing guides out there that also prescribe when and where and for how long you should write: mornings, in an airy quiet room, etc. Realistically speaking, the ideal conditions and "perfect" time to write rarely, if ever, come by. Besides, many of us have idiosyncratic ways of approaching writing—some work best in the early morning and some only late at night; some prefer a dark screen to a light screen, or composing in coffee shops rather than in the library, computer lab, or kitchen. What works with your own creative rhythms? Do you have a designated writing space and time that is most conducive to productivity? Perhaps even a prewriting ritual (e.g., hot cup of coffee and logging out of social media accounts)? There is one constant that most writing guides agree on: You need to make writing a *habit* (this applies to anything you want to get better at, like running or learning a new language). This means you have to write every day, even if it is for 5 minutes at a time. Above all, writing (especially writing *well*) needs *time*.

Habit 8: Cultivate Good Revising and Editing Habits

To perform a good revision, one needs to learn to accept criticism with equanimity and approach revising as both a global and a local process. Theories and advice regarding levels of edit abound in technical writing literature, but for the purposes of this chapter, we stick to a true and tried approach, according to which there are three main levels of edit:

1. **Substantive editing:** As the name implies, this type of editing looks at the paper globally and looks for major issues with the text. Questions you may ask at this level are:
   - Do all parts fit together in a coherent whole?
   - Is the argument or order of information logical?
   - Does the paper include all the necessary information for the target audience?
   - Does the paper include any extraneous information for the target audience?
   - Are the visual aids useful, appropriate, and correctly labeled for the target audience?
2. **Copyediting** (or language edit): This level of edit assumes that the paper is more or less coherent, cohesive, and complete and moves on to address any of the stylistic issues we address in The Writer's Toolbox—for example:
   - Sentence complexity
   - Use of active or passive verbs
   - Conciseness
   - Clarity
   - Transitions

- Use of jargon or technical terms appropriate for the intended audience
- Parallelism

3. **Proofreading and polishing:** When you get to this level, you can assume that you have a coherent, complete, clearly written, and grammatically correct text that might still need some work. At this level you are looking for any issues related to:
   - Capitalization
   - Punctuation
   - Spelling
   - Formatting (including headings, subheadings, etc.)
   - Lists
   - Tables and figures (double-check the order and the numbering)
   - Title page
   - Headers and footers
   - References (make sure they adhere to the required documentation style).

An invaluable tool in your arsenal should be the *peer review*. As a rule, do not submit anything, ever, without it being peer reviewed. Cultivate a relationship with reviewers whose comments effectively help your paper reach the next level. When you act as a peer reviewer, focus your attention on the substantive-level questions, and only address copyediting and proofreading issues after you address global issues.

A Note About Peer Review

Learn to be a good peer reviewer—it is a useful skill and helps you set expectations of others as well. As you review, keep in mind:

- The assignment description (genre, scope, length, etc.)
- The overarching goal of the assignment (What does the instructor want you to prove with this assignment? Don't be afraid to ask directly if it was not clear from the instructions.)
- The research question (What is this paper actually investigating and why?)
- The knowledge level of potential second-party readers (i.e., other than your instructor.)

Richard Straub, in "Responding—Really Responding to Other Students' Writing" (in *The Subject Is Writing,* 1999), provided several tips on how to be a supportive, helpful peer reviewer. The following list is partially based on his advice:

- Read deeply—several times and/or aloud if you have to.
- Mark the paper as you go with "?" wherever your understanding was unclear.
- Ask yourself: what's the bottom line? What is the main message—or answer to the research question? Make sure the paper offers you a clear answer.
- Avoid dwelling on minor stylistic issues (commas, capitalization, typos, etc.) or do not make them the focus of your review.
- Write your comments in complete sentences, rather than cryptic words or sentence fragments. Take your time: more is better.
- Be friendly: don't sound like a teacher or a judge but rather like your regular self who is trying to support a fellow classmate; conversely, don't be an undiscerning cheerleader who thinks everything about the paper ranges from flawless to great.
- Keep a balance between critique and praise. Whether you critique or praise, explain why (i.e., what is good about the opening? the structure? the connection between x and y? the support? other?)
- Offer advice, but do not rewrite the paper for the author. You do not want to rob her of the opportunity to improve.
- Ask questions—real questions—related to the substance of the paper. Exercise your curiosity in order to genuinely engage with the author's material. Your considered attention is probably the best way for the author to feel taken seriously and to seriously consider your suggestions.
- Identify gaps that could potentially confuse a second-party reader.

Habit 9: Learn to Write Collaboratively

Science is intrinsically a collaborative game, and scientific writing is, usually, authored by teams rather than by individual authors. Learning to work and write collaboratively is essential in the sciences. In fact, the ability to work well in teams ranks second on the list of desirable skills expected of *any* college graduates on the job.

The report *Are They Really Ready To Work? Employers' Perspectives on the Basic Knowledge and Applied Skills of New Entrants to the 21st Century U.S. Workforce* shows that employers seek applicants who can document

For new entrants with a four-year college diploma, applied skills are the top five "very important" skills in combined ranking with basic knowledge and skills.

Rank	Skill	
1	Oral communications*	95.4%
2	Teamwork/Collaboration*	94.4
3	Professionalism/Work Ethic*	93.8
4	Written Communications*	93.1
5	Critical Thinking/Problem Solving*	92.1
6	Writing in English	89.7
7	English Language	88.0
8	Reading Comprehension	87.0
9	Ethics/Social Responsibility*	85.6
10	Leadership*	81.8
11	Information Technology Application*	81.0
12	Creativity/Innovation*	81.0
13	Lifelong Learning/Self Direction*	78.3
14	Diversity*	71.8
15	Mathematics	64.2
16	Science	33.4
17	Foreign Languages	21.0
18	Government/Economics	19.8
19	History/Geography	14.1
20	Humanities/Arts	13.2

Basic and applied skills rank ordered by percent rating as "very important".

Number of respondents varied for each question, ranging from 382 to 409.

*Indicates an appiled skill

Top applied skills expected from college graduates according to the *Are They Ready to Work?* (2011 report).

experience and success in professionalism/work ethic (i.e., taking responsibility for the completion of tasks, time management, working productively with others), teamwork/collaboration (i.e., working with diverse groups, working collaboratively from a distance), communication skills (i.e., oral, written, and technology mediated), and critical thinking/problem solving (i.e., analytical and applied thinking). Key to success in all these areas is the assumption that one is technology savvy and has the life-long learning skills to continue to adapt to the changing face of electronically mediated communication. These skills are absolutely essential for any professional in the science field. You need to be skilled in global, instant, and constant communication—and work well as a functional member of long-distance teams.

Luckily, we live in an era of unprecedented expansion of online collaborative tools: Google Docs, Dropbox, wikis and blogs, Zoom, Skype, GoToMeeting, Slack, and even Twitter can be great ways to share information and collaborate, often in real time. Other useful collaborative tools like Zotero, Mendeley, or CiteULike can allow you to manage your citations (for shared group projects this is an enormous boon). We are far from having a standardized platform for group collaboration, but don't let that deter you. Establish what works for you and take advantage of the available technologies. They are no longer a "nice" addition to your learning toolbox: They are a necessity, and an indispensable tool for group work.

> You may be aware of forums for scientists (such as sciforums.com or the nakedscientist.com), but you should also familiarize yourself with the best and most frequently updated forums or listservs in your discipline. They can be a great way to see what is current in terms of research, gauge the opinion of a scientific community on a topic, ask questions of your own, find out about research opportunities, and even find collaborators.

Managing any collaborative group is difficult, and this is especially true of online groups, but it can be done with a clear system of responsibility and accountability. Try assigning roles in each group (e.g., Leader, Reporter, Devil's Advocate, Editor, etc.); make sure that responsibilities are fairly divided; agree on a calendar of deliverables, a means of communication, and a sharing platform for your output; and establish a system of notifications, reminders, and penalties to keep stragglers or procrastinators on task. Some have also suggested drafting up team contracts, which can be signed by all parties, in which all these rules are spelled out and agreed upon beforehand (your instructor may provide you with a template to get you started).

© 2020 Maritsa Patrinos

Habit 10: Reflect on Your Writing Practices

You finished your writing project and submitted it, and now you've learned the outcome. Whether it's an A or a C–, whether it's been accepted or rejected for publication, whether you've received the grant or not, you have a golden opportunity to learn from the experience. What do you think worked, and what didn't? If you are lucky, you will get feedback from your teacher or peer reviewers, and if you wrote a grant, you will surely get feedback from the granting agency (sometimes with specific instructions on how to revise and resubmit). It is a good time for you to reflect on your writing practices in order to improve them. Ideally, you would keep a journal or a blog, which you would update regularly with the lessons you learned and strategies that will help you become a better writer. Just sharing your thoughts on your writing process with someone else can also be useful (see Exercise 4 at the end of the chapter).

Take-home

Writing is essential in the sciences. Through writing you are often solving the problem of informing and persuading a particular audience of an important (science-related) issue—an issue you know better than others do. You should always try to find writing models that inspire you. The time to write your first draft should not be the first time you record an idea related to your writing project. As with any other project, set clear writing goals and strategies, and learn to master your writer's block, if you happen to be occasionally afflicted with it. Accept that writing is a recursive process and master the drafting and revising process. Finally, learn how to work as a team with your collaborators and reflect on your practices often.

Exercises

1. Practice writing as problem-solving. Think of a past or current writing task and "divide" it into discrete problems you need to solve correctly in order to be successful. For example, if you had to write an essay on the uses of metaphor in Poe's "The Raven," a cover letter for a potential job, a lab report, or a history presentation, what were the tasks that you successfully needed to accomplish in order to succeed? Think rhetorically: who was the audience for each of these texts? the context? the purpose? What means did you have to employ to accomplish the task? How did you divide your work? Did you have any limitations? If your efforts were less than successful, can you reframe your obstacles in terms of problems that you couldn't solve at that particular moment?

2. Think of a recent hazardous situation that required the intervention of scientists and engineers (e.g., the Fukushima nuclear meltdown, the oil spill in the Gulf of Mexico, loss of crops or epidemics caused by *E. coli*, the devastating fires in Australia, the hurricanes and flooding in the Caribbean, an earthquake, etc.). Imagine you are the scientific consultant for a media outlet that needs to convey information to the public about that disaster. Identify two to three core issues in such a situation that need to be communicated to the public and come up with a communication solution for conveying those issues to the public. This solution should include some key wording, a genre/delivery mode (e.g., a press release, an interview, an update for an online news site, an article in a local or national newspaper, a social media posting, etc.) and a strategy of discourse (e.g., what to emphasize in the given situation). Work in groups. Share your solutions with the entire class, explaining why and how you came up them.

3. Find and study models. Find the written work(s) of a great scientific figure whom you admire or who inspires you. What is about the person that inspires you? What is it about that person's writing that you can "borrow" or "steal" in your own writing? Identify the specific characteristics that you would wish to emulate in your own writing.

4. Writing project management. Create a "Writing Project" template sheet that you can adapt to any of your major writing assignments, describing the genre, audience, basic requirements, calendar, and to-do list. For example:

Writing Project

Writing Task:	[*insert type of paper/genre*]
For:	[*course/instructor*]
Topic:	[*topic/tentative title*]

Requirements:

Length:	[*pages/words*]
Formatting:	[*typeface and font size; spacing; page numbers; justification; title page; abstract page; etc.*]
Visuals:	[*tables/figures*]
References:	[*no. and type*]
Style Manual:	[*APA, ACS, AMA, etc.*]
Delivery:	[*electronic/hard copy; if electronic, where, and what format; if hard copy, whether full color, etc.*]

Deadlines:

Final Draft:	[*date*]
Draft 1:	[*date*]
Draft 2:	[*date*]

To do:

1.
2.
3. . . .

Reviewers:	[*best qualified/available person to get feedback from*]
Notes:	[*any other info you have at this point?*]

At the very least, this sheet can provide you a start when you don't know where to start and you need to write the paper. You can adapt this sheet depending on your style, specific course requirements, specific assignment instructions, etc. It could be a useful tool to overcome writer's block, and it can also help keep you organized and on task.

Manage your productivity. Try the Pomodoro technique or the kitchen timer technique. Give yourself 5 or 10 minutes, clear your desk, turn off phone/social media alarms, pick a science topic you know well, and write about it for that duration, without stopping or acknowledging interruptions. How much could you actually write? Use it as a benchmark for your basic abilities to churn out prose. You can pair up with another colleague and then revise each other's paragraphs using the three levels of revision hierarchy (start with substantive, move on to copyediting, and finally to proofreading and polishing). How long did revision take? And could you write a better paragraph or so based on the feedback?

5. Reflect. Throughout the semester, keep a weekly journal or blog in which you reflect on the scientific writing you have accomplished each week. What have you learned about yourself as a writer? What do you need to improve? How will you address your weaknesses? What made you successful? Alternatively, write a reflection after each writing assignment you have to submit, commenting on your writing techniques and how you attempted to meet the audience's needs through your prose. What worked for you in completing this assignment and what would you do different next time?

2
Target Audience

Every day people, people who enjoy commuting into work, are taunting death, hurtling toward the metropolis in a metal tube that could, hypothetically, derail any minute, plunging its hapless passengers into flaming chaos. Fortunately, there are emergency instructions posted prominently at the end of each car. This signage is a marvel to behold! Oh, the poetic beauty of that verbiage!

"*Egress.*" Why can't *I* write like that?

We'll tell you why. If you have a flaming rail car filled with humanity in chaos, the word you want is "*out,*" not "*egress.*"

Yes:

> *Our prison strong, this huge convex of fire,*
> *Outrageous to devour, immures us round*
> *Ninefold; and gates of burning adamant,*
> *Barred over us, prohibit all **egress**.*

> —John Milton, *Paradise Lost*, Book II, 1667

No:

In the event of a critical emergency, your **egress** choice should be to stay calm and proceed in an orderly fashion through the parlor door, and then continue through the vestibule door. If no other means of exit is available, remove the window by pulling the emergency handle and zip strip; be aware of the 6' to 8' drop to the roadbed below.

—Emergency sign on train, 2009

Who rides on a train? a kid with a sixth-grade education? a young, high school–educated single mother of two? a Korean immigrant? your parents/ grandparents/neighbors, maybe? an executive visiting from Europe? This would be your *target audience* on a train. You may be assured that not everyone on a train is going to know what "egress" means, and, for that matter, what is the "parlor door" and the "vestibule door" or a "zip strip"? If a rail car is on fire, its inhabitants will be disinclined to "stay calm," and this unease is likely to be compounded by the information that if they do indeed make it off the car, they might plummet six to eight feet to the roadbed. "Emergency Exit" is what's called for here, not that staying calm and exiting in an orderly fashion and looking before you leap onto the roadbed isn't a good idea.

What Is an Audience?

A *target audience* is the person or persons with whom you intend to communicate, the audience that is going to read what you write. If there is *one single cardinal law of writing* that you need to learn—*any* writing, not just scientific writing—it is that you must *know your target audience*. Accurate identification of target audience is the single most important skill for effective writing. Period.

"That can't be right," mutter the naysayers. "When you're writing, you mostly need to know how to put words together and know where the commas go, and make sure your participles aren't dangling precipitously, and you don't mix up 'there's' and 'theirs' and 'its' and 'it's,' and spell things like 'eschatology' correctly. You have to know that introductions go first and conclusions go last—important things like that." Well, that too, but you can be the Comma King of your lab and it won't matter one tiny bit if that comma is in a treatise on the physics of sound waves but is supposed to be in a fourth-grade elementary school unit on music. In other words, knowing how to use commas only matters if you got the audience right in the first place.

We have all met people—professors, even!—who think that grammar and style, along with (possibly) organization and critical thinking, are the most important things you need to master to write well. Of course, if you're going to be doing a lot of writing, you probably ought to learn grammar and vocabulary and the other assorted components of the writer's armamentarium. *But master your target first.* It doesn't matter how well you write if your intended audience isn't going to or can't read it. If you must write to advance in your career and you can't write to your target audience, you will not get very far. Editors can fix transient lapses in grammar, but they can't fix an off-target piece.

Writing to target can be very hard if you're writing to an *unfamiliar target.* All of us subconsciously assume that the target will understand the things *we* understand. *Never assume this.* This is a tough thing to not assume, since it is subconscious. The difficult part is lifting your target audience consideration *out* of your subconscious and moving it *into* your consciousness. Your writing skills can prevent tragic fiery death on a train. It's true!

How Do I Find Out What My Target Audience Is?

A useful first step is to assume you don't know your target audience, even if you've been writing to it for years. You may have an idea about your writing project (it might even be a specific assignment from, say, a professor or a journal editor), but don't even make an outline until you sit and think about your target audience. You don't have to spend a lot of time on this, particularly if you're used to writing to the same audience ("all I write are grant proposals"), but give it some *conscious* thought. Practice Zen targetry. *Be* your audience before you *write* to them.

Zen Targetry: Becoming Your Target Audience

If you want to be your audience, you have to know who is in it. Do a little research. Are you be contributing to a textbook for college freshman or fourth-graders? Are you supposed to write a lay article for the science section of the newspaper or a draft an article for a scientific journal? Do you have to create a PowerPoint presentation to be delivered at a meeting or a blog post for the department's website? These are all good starts, but you could use some more information. Don't just say to yourself "Okay, fourth-graders." Close your eyes. Think back to fourth grade. Think about sitting at your desk, what your room looked like, what your teacher was like, what the other kids were like in your class, the way you spoke to each other, what you did during recess, what you had for lunch, where you put your books, how your day was structured, what kinds of things you studied. You're reaching into your desk . . . you're feeling around for your science book . . . you put it on your desk. Do you look at that book and think "Oh no, not this again," or do you look at that book and say, "Today Mr. Wolfe said we were going to do the exploding chemicals experiment! *Excellent!*" You want your fourth-grade target to *want* to read your book and not think it's "too dumb" or "too babyish" or "too boring." You want them to think it's *awesome*. You're a bench chemist. Why did you ever agree to write part of a science book for fourth-graders? Because you're passionate about getting kids excited about science, it will look great on your CV and, well, you can write about blowing things up for a change instead of all that deadly p-chem stuff.

If you're writing for peers, put yourself in their shoes. How do you even pick an article to read? Is the title catchy? Are you reading it online or in print? Are you scanning for the "good" parts? What is likely to draw your attention? How much time do you really have to read up on the literature in the field and what would you value in, say, an article, to help you get through it faster? Is there something of value that you can use in the article? You must respect your audience and be considerate of their time and needs. Is the bottom line clear? Does the abstract correctly mirror or summarize the contents? Are the figures accurate—and do they make a point?

Creating a Target Audience Profile

Practicing Zen targetry might not be the most practical advice, but it's the most important. If you need more *practical* advice for creating a target audience profile, take into account demographics: Is your audience comprised of a specific age, sex, or level of education? Is it mostly women (e.g., you're writing a Mayo Clinic patient web page on menopause)? younger kids (e.g., chapter in a chemistry science book, including an awesome experiment involving exploding chemicals)? Are you delivering a paper at an engineering conference where everyone in your audience is an electrical engineer with an advanced degree? What about race and cultural characteristics? If you're creating a poster (either in Spanish or to be translated into it) targeted to Hispanic families in a border town about the importance of vaccinations, it would be wise to take into account any potential cultural concepts that might need to be addressed. The Black Barbershop Health Outreach program that started in 2007 is a spectacularly effective initiative wherein information about health issues of particular importance to black men (originally, prostate cancer but now expanded to other areas such as cardiovascular disease) was first disseminated and discussed—*in barbershops*. This lay effort resulted in an increase in the number of black males going for prostate cancer screening. Now *there's* a specific target audience for you: male, black, adult, urban, primarily high school educated—and sitting in a barbershop. *And it worked.*

How would you do it?

What kind of grassroots health or public science (e.g., environmental) initiative would you create? How, to which target audience, and where?

Oliver Sacks, a neurologist and award-winning author of several popular lay books, wrote about neurologic topics to a variety of audiences. Here, he writes about dementia:

To physicians (*Neurology*, 2005): *"Other cases of long-lasting dementia syndromes following GC withdrawal have been recorded, with residual impairments occurring in predominantly hippocampal-mediated functions."*

To patients (*Neurology*, 2005): *"While Alzheimer dementia is the most common type of dementia, it is not the only type. In general, dementia is a decline in thinking that most often includes problems with memory."*

To a lay audience (*The Man Who Mistook His Wife For A Hat and Other Clinical Tales*, 1970): *"Perhaps there is a philosophical as well as a clinical lesson here: that in Korsakov's, or dementia, or other such catastrophes, however great the organic damage and Humean dissolution, there remains the undiminished possibility of reintegration by art, by communion, by touching the human spirit: and this can be preserved in what seems at first a hopeless state of neurological devastation."*

Where is this mysterious thing you're writing going to be seen or read? all over the country? in the Southwest? rural areas? urban areas? There is a reason you see advertising for specific products in specific places. You are far more likely to see ads for John Deere tractors out in rural America than, say, scantily clad 20-somethings modeling Calvin Klein underwear. Geographically, you could be creating materials for an area that's conservative, an area that's more liberal, or an area that's countercultural. A conservative, rural small town will have to have the ecologic virtues of a windmill farm explained to them in a carefully crafted way. Toned supermodels delivering a treatise on green energy is not going to be quite as effective as a friendly small-town businessman talking about how his electric bill went way down when the windmill farms came in. You're an engineer. How do you explain a windmill farm to a small, conservative town?

Learn the Knows, Wants, and Needs of Your Target

Now that you've practiced Zen targetry and you know your audience and your topic, you need to keep in mind your purpose (e.g., convincing your target audience that they need to approve a windmill farm near their town), and you need to consider what your audience *already knows*, what they *want to know*, and what they *need to know*. Your target audience is the populace of a small country town, not a seaside resort area or an urban area or a mountain community.

What Does Your Audience *Already* Know?

What do they *already know*? In this example, since you've done your Zen targetry due diligence and have optimally spoken to some of the townspeople, you've ascertained that they know what a windmill is, that it doesn't look like a Don Quixote windmill, that it makes "cleaner energy" than fossil fuels, and that there is some controversy over them. Of course, some people in town— say, politicians, the occasional scientist, or teacher—might know a bit more, and some may know less. Keep in mind that there might be things your audience already knows—but doesn't realize it.

> "Now look, your grace," said Sancho, "what you see over there aren't giants, but windmills, and what seems to be arms are just their sails, that go around in the wind and turn the millstone."
> "Obviously," replied Don Quijote, "you don't know much about adventures."
> — Miguel de Cervantes Saavedra, *Don Quijote de la Mancha,* 1605

What Does Your Audience *Want* to Know?

Since you know the plan is to convince the town to approve a windmill farm 5 miles outside of the town limits, and you're the engineer preparing the technical report and/or printed materials to be distributed to the townspeople along with a PowerPoint presentation by the windmill farm rep, you need to anticipate what they're going to *want to know*. How will it affect their town? Will their electric bills go down? Will there be a lot of people moving into their town? Is that a good thing (income) or a bad thing (keep our town small)? Will it damage any crops potentially growing near it? What is the town going to get out of it? Does the town get paid for it? Does the town get a tax break? Do the windmills make noise? Will they interfere with hunting season? Are they dangerous? How come people have complained about other windmill farms, like those ones built way out in the water? What if it isn't windy enough? They're metal—what about the tornadoes we get out here? What about thunderstorms? Who's going to maintain them? It is important for you to close your eyes and Zen yourself into the town hall meeting and anticipate what different kinds of citizens are going to demand answers: farmers, hunters, store owners, teachers, parents, ministers, older people, younger people, politicians. Each of these targets is going to *want to know different things*, and they're going to want to hear the answers to the *other* questions that the *other* targets are asking. *What will they want to know?* You want to provide this information up front, rather than being taken off guard later when your audience asks questions.

What Does Your Audience *Need* to Know?

If you can divine what they know and what they're going to want to know, you can craft your report so that it addresses what they *need to know*: Here's a picture of a windmill, here are the parts, here's how they're assembled, here's what they do, and here's how the whole windmill works. The energy company building the windmills will pay a $250,000 stipend to the town each year. The windmills are safe and clean and will be maintained by the energy company that builds them. They can be shut down during inclement weather, and perhaps an arrangement can be made so that they are stilled during daylight during duck-hunting season. Electricity bills are going to go down 25%. During construction, the town will benefit from increased spending by work crews, but it is unlikely that anyone would move to town for the sole purpose of maintaining windmills. The turbines do make a quiet humming noise, but at a distance of 5 miles, you won't be able to hear it. Sometimes other communities have complained that they can hear the windmills, but they live much closer to them—more like one-quarter of a mile. Some communities don't like the way they look and think that will harm tourism or property values.

Part of identifying your target is delivering information that is relevant to them. A highly technical report on turbine kilowatt-hours—your forte—won't give this audience what it needs.

---

10 Tips for Identifying Your Target Audience

1. Think carefully about your target audience *before* you start writing.
2. Imagine your audience reading (or hearing, etc.) what you write.
3. Consider the tone, voice, and level of sophistication of the writing.
4. Consider the specifics of your audience (e.g., age, gender, ethnicity, geography, etc.).
5. Never assume your target audience knows what you know.
6. Show your writing to someone in your target audience if possible for feedback.
7. Identify what your target audience *already* knows (or thinks they know).
8. Identify what your target audiences *wants* to know—even if you have to guess.
9. Identify what your target *needs* to know—don't waste their time.
10. Read everything you write out loud when you're done writing it, so that you can hear the voice and level of sophistication.

Writing to Multiple Target Audiences About the Same Topic

Your Tilapia Adventure

You're in luck! Your BS in biology got you hired by the World Wildlife Fund (WWF)! One look at the website shows that WWF scientists are a highly versatile bunch, comfortable with both high-tech scientific information or teaching elementary school kids. They could be writing something as simple as poster text to something as complicated as governmental guidelines for tilapia fishing. They have videos—videos need scripts. The WWF needs grants; grant writing is a critical skill for a scientist, and the grantors themselves represent different targets (the government, organizations, and individuals, each of which being motivated by different reasons: political, cultural, ethical, or environmental). Content experts are needed who can write or guide writers to write compelling copy to educate people (and get people to donate). And someone has to be supervising content on the WWF web page. Will you have to do all of this, necessarily, if you're brought on as a scientist for the WWF? No, but if you're a resident zoologist you'll be doing a lot of it, or at least contributing to it. You might be working with writers or editors, but projects will originate with you. You might have had a concentration in ornithology when you got your degree, but you'll be writing about tiger legislation, wetlands conservation, activism, pollution, and extinction. You'll be writing or overseeing grants to a variety of different organizations or individuals, and you'll be collaborating with other scientists, experts, communications professionals, technicians, and the lay public—and, of course, those fabulous tilapia farmers. There are so many targets, so many different kinds of media (e.g., web, print, and video, among others) that good target identification skills are a must. Is it going to be that tough for you if you are only writing reports for, say, the *Congressional Record*? No (although a quick search in the *Congressional Record* only yielded 17 hits for "tilapia," so maybe *all* scientists should learn to write about this fish).

There's nothing like a couple of paragraphs of inflammatory fish rhetoric to liven up the *Congressional Record*. ©2020 Maritsa Patrinos.

"What a fantastic CV," yelps HR at the WWF, not noticing your concentration in ornithology, "you're on Tilapia Patrol for the next six months!"

"Thank heavens," you say (sotto voce), "that I learned so much about tilapia in that scientific writing course." When you talk to Dr. Stuyvesant at the

Scripps Institute, however, the leading expert in the world on tilapia, she will *not* be impressed that your total knowledge of tilapia is that (a) it's farmed and (b) it's mentioned in the *Congressional Record* 17 times.

Exemestane Is Exciting to a Tilapia *Expert*

As the resident tilapia go-to person, you're going to communicate a lot about tilapia, to a variety of target audiences, and you're going to start by working with a tilapia expert. You positively are *not* going to call her or meet with her before you can talk to her about how exemestane is not an effective steroidal aromatase inhibitor in *Oreochromis niloticus*.

In other words, *your expert is your first target, and you're not even writing anything yet.* Your level of understanding about this fish and the sophistication with which you communicate about it is *substantial* when you talk to this target. Sure, you might get the occasional research scientist passionate about his work who will out with "Is this a totally cool fish or *what*?!" but this enthusiasm is unlikely to be communicated in this format in a scientific brief provided by the WWF. How do you prepare for your meeting with the tilapia expert? You do your research. You look at basic references on ichthyology and fish farming, and then more detailed review papers and government reports on tilapia and perhaps even *more* detailed scientific papers on this beleaguered fish. You are starting to amass your tilapia dossier.

Tilapia *Farmers* Really Don't Care About Exemestane

A tilapia farmer, on the other hand, unlike your tilapia expert at Scripps, is extremely unlikely to know about exemestane and, even more unbelievably, is unlikely to care about it. He is staring down into the murky depths of the water, fuming over the fact that the Secchi disk visibility limit in oligotrophic receiving waters above which tilapia production is not certifiable has just been reduced from 10 meters to 7 meters; he just read that in the regulatory update you wrote. Your expert probably doesn't care about Secchi disk visibility in oligotrophic receiving waters and probably is not remotely familiar with the new *International Standards for Responsible Tilapia Aquaculture.* Neither are the seafood wholesalers or restauranteurs who read your *Farmed Fish Newsletter*: they're worried that that Secchi disk thing, whatever it is, will somehow increase the cost of fresh tilapia. There is such a demand for tilapia! It is indeed a tasty, tasty fish! Fish farmers reap tons of profits from selling them.

So they are going to cut corners and hope the International Tilapia Task Force doesn't catch them. The WWF doesn't like this, so they will do what they do best: raise consciousness. They will turn to you for the lay-targeted blurb on the WWF website, pleading the heartrending cause of the beleaguered farm-raised tilapia. You won't mention (a) exemestane, (b) Secchi discs, or (c) fluctuating market price of fresh tilapia. A lay target is different. This is pathos.

How would you do it?

How would you explain what a Secchi disk is and how it's used to a restauranteur? a layperson? (Caveat: this may require research on your part about Secchi disks.)

Tommy Lee Jones Also Doesn't Care About Exemestane (So Far as We Know)

Fabulous news! The WWF has lined up Tommy Lee Jones to narrate a one-hour special for Discovery Channel's "The Plight of the Tilapia." Jones is a great guy, and really smart, but it is extremely unlikely that he's going to go around tilapia farms or interview exemestane-obsessed tilapia experts at the Scripps Institute and improvise on camera. He's going to need a script. Your boss comes into your office over the moon about this unbelievable stroke of luck and tells you that now you're going to be a scriptwriter. You lean back in your creaking chair and say, "Can't I just write another report about exemestane?" And then your boss laughs, gives you a companionable slap on the back, and tells you they need a draft by the end of the month. When you finally produce a draft, it reads: "It is the dawn of another day in Tilapiaville, Louisiana. The clouds cast leaden shadows over the rows and rows of tidy square concrete pools, in each of which languish upwards of 5,000 tilapia hatchlings." Is this the kind of text that would go in the wholesaler's newsletter? If you wrote it in the regulatory update for the fish farmers, they would cancel their subscriptions.

This noble, noble fish—this once-proud, gonochoristic fish. Crowded. Suffering, with significantly suppressed plasma estradiol-17 beta, and increased plasma 11-ketotestosterone. It is a pretty fish, and it likes to live in clean water. And it tastes great in a white wine sauce with capers.

☐ Good script   ☑ Bad script

Tommy Lee Jones will read this script and think you're writing a comedy.

The Tennessee DNR Doesn't Think *Kids* at the State Nature Centers Really Need to Know About Exemestane

While you're finishing up the last draft of the script for the Discovery Channel, you're reviewing page proofs you just got back for the WWF kid's magazine. This month, the front cover shows a grinning, gap-toothed tyke, holding aloft a fish half his size, with

"Let's Fish!"

splashed gaily across the front. Tilapia has a whole page! Where it comes from, what it looks like, what its eggs look like, and what the baby fish look like when they hatch—and as a special treat, WWF visits a fish farm! The kid's magazine is full of facts—the *fourth-grade* facts. Would you write "tilapia have to grow in clean water," or would you write "Secchi disk visibility is now mandated at a maximum of 7 meters"? The coloring book that is your responsibility for the Department of Natural Resources for the state of Tennessee will be easier to write, since its text will be limited to "Matching" and then "fish," "bird," "turtle," and "squirrel." Crayons in hand, your preschool target will draw a line from "fish" to (hopefully) the *picture* of the fish, blissfully unaware that it is a plasma 11-ketotestosterone-saturated tilapia that lived a miserable, short, crowded life in sediment-choked water that exceeded a 7-meter Secchi disk cut-off and died a tortuous death so that it might be arranged artfully on a plate drizzled with white wine sauce and capers, its entrée price tragically increased from $19.95 to $22.50, due to the vagaries of fluctuating market price.

If You're Helming the Tilapia Bureau, You Could Get Handed Anything

You may only end up writing monthly reports. You may be mostly out in the field and writing primarily scientific journal articles. You might be focusing primarily on grants. And then the next thing you know you're in a meeting with Tommy Lee Jones who smiles and shakes your hand and wants to talk to you about your fish. Can you write all of this? You know your stuff—you really, really know your stuff—but can you communicate what you know to all of these different targets?

So the next time you sit down to write something, ask yourself this: "How well do I know my fish? And what if it is stuck on a burning train?"

Our planet has seen its share of devastating disasters: tornadoes, hurricanes, earthquakes, tsunamis. The Centers for Disease Control and Prevention (CDC) in Atlanta takes disaster preparedness very seriously and does its level best to provide educational guidance. But who reads it? The CDC decided to get creative about it and produced survival rules for the zombie apocalypse. Word got out, and traffic increased. If you were a ninth-grader, would you prefer to read a standard web page, or would you prefer reading a comic about zombies? Visit the CDC's web page and search for "zombie preparedness." It's educational!

How Do I Know If I'm Writing to Target?

After you *be* your target, write something, and then, if you can, *read* it to them, or even better, ask them to read it. Watch your readers and their expressions. Ask them if they understand it. Remember, you're writing about *science*; the topic may be of little interest to them, or they may think they understand it when in fact they don't. If you watch your target listening to you, and they're nodding politely with glazed eyes, take a clue.

You don't want to see glazed-over eyes. You want a captive audience. Use the right word for the right context. When you're boxed in by the individual in the aisle seat and the train is on fire, your conversation is likely to have more elements of urgency to it than "excuse me, our seat is on fire. Might you consider getting up so that we can effect egress immediately through the vestibule door and jump down to safety on the roadbed?"

"Why, yes, I completely concur. I would, however, prefer to continue this dialogue once we are, in fact, on the roadbed, being as how my sneakers are on fire."

Borrow a target if you can. Borrow a colleague, a professor, a teenager, an engineer, a stay-at-home parent, your doctor, your pharmacist. *Read what you've written aloud to them* and watch them. Don't look to see only if they "get it"—look to see if they're interested or if they're bored. Getting such feedback is priceless.

Reading what you write aloud is also an excellent way to hear where you're not being clear, or where you're being awkward, or where you've left something out, or where your grammar has lapsed. If someone is listening to you read something, they, too, will pick up on things they might miss while reading. Keep in mind that the feedback you get is as good as the audience you've got for your test-drive. If you're reading aloud the Conclusions section of a paper you've just written that is specific to exemestane biology in gonochoristic fish to a colleague who is also an ichthyologist specializing in gonochoristic fish endocrinology and the paper is destined for the *Journal of Applied Ichthyology*, that's great. This same colleague, however, might not be your best assessor if the paper is destined for *Scientific American*. Time to bring in another zoologist or biologist from your department.

Despite your best efforts, you might not always have access to your particular audience, whether it is a fourth grader, a USDA scientist, or a senator. This is why it's important to get as many people as possible to read what you've written. If you can't get a fourth-grader but you *can* get your dad, and he happens not to be an ichthyologist, you have half of what you need right there: a lay audience. Tell your reader about the audience you're trying to reach. No USDA grantor? Talk to someone else who's had to submit a grant somewhere. They can school you in the finer points of pleading. No senator handy? All you have to do is look at a small subset of your 562 friends on Facebook to find the super-politically active ones. Pick one—or several—to test your story for various political leanings. Even if you do not find the perfect person to read what you wrote, make sure at least one other person has read it, knowing your intended target. Finally, *get a peer to read it:* Peer review is one of the most powerful tools for writing to target (see Chapter 1).

Tired of fish yet? What if you were with a federal agency trying to assess who was actually reading your Fish Fact Sheet? Check out Burger J, Waishwell L. Are we reaching the target audience? Evaluation of a fish fact sheet. *Sci Total Envir.* 2001;277(1–3):77–86.

How Do I Start Writing to an Unfamiliar Target?

If you are accustomed to writing to a specific target all the time, you pretty much already know what you're doing (always stop and think about your target when you start, though, no matter *how* long you've been doing what you do). But suppose you are now a bench scientist in research and development (R&D) in a pharmaceutical company. You've done only a modicum of writing—short meeting summaries and two conference posters as an undergraduate. Now you're required to write your first scientific journal article. You're supposed to work with the principal investigator (PI) of a clinical trial to write up the results in a paper for publication in a medical journal. How hard could that be? You've talked to the PI, you have the clinical study report from the pharmaceutical company, and you have everything you need—except experience, which will occur to you when you get all nice and organized, create a nice new word processing document, and settle down to stare despondently at your blank screen. The cursor blinks steadily, mocking you. You've read a *ton* of those kinds of papers; you know what that target wants and needs to see: *why is it so hard to start?*

Why *Is* It So Hard to Start?

We offered some tips on dealing with writer's block in Chapter 1, so it's time to put them to use. You should go back and read those tons of papers again, but not for content, the way you've read them in the past as research for writing projects. Read them as a *writer* this time. Look at how those authors were communicating with their targets; look at the vocabulary, the voice, the detail, the level of sophistication, the organization and structure. Go back and read the exact same *kind* of paper you're trying to write (e.g., an IMRAD paper), to the exact same target audience. Is your statin paper (statins are drugs that lower blood cholesterol) going to be targeted to a more general audience ("Management of dyslipidemia with miraculostatin") or to a more specific audience ("Clinical benefit of miraculostatin pretreatment in patients undergoing percutaneous coronary intervention: A collaborative patient-level meta-analysis of 13 randomized studies")?

Where Is My Paper Supposed to Go?

Figure this out first. Typically, your boss/professor/PI will tell you what your target journal is supposed to be: for example, "We want this paper submitted to *JAMA*." If that's the case, read clinical trial papers in *JAMA*. If you're given an actual journal target, that's great guidance, and less guesswork for you. If you're *not* given an actual journal target, talk to your collaborators (i.e., your boss and content experts) to determine the audience *before you start*. A paper on statins could go in *JAMA* or it could go in *Arteriosclerosis, Thrombosis, and Vascular Biology (ATVB)*. It's likely that a large percentage of individuals that read *ATVB* also read *JAMA*, but the reverse is not true: far fewer people who read *JAMA* are going to be reading *ATVB* on a regular basis. They're mostly all clinicians, sure, but what kinds? "Clinician" doesn't cut it for target audience.

Format Theft: It Can Be Your Little Secret

Next step: *Use this paper as a model for your own* (see Chapter 1 for using models). You may borrow its format, voice, level of sophistication, and organization. *Nothing* is more useful than an example. Create your own template populated by the kind of sections and headings that are most commonly used in the genre you are writing (this is also helpful for following instructions to authors with regard to typeface and font and referencing and other mechanical details). Do you need to write a web page for a drug manufacturer? monographs? regulatory updates? blogs? webcasts? Find a good model for each and study it if you need to write any of these genres. Pay attention to details—does an article in *Discover* or *Scientific American* have superscripted reference numbers in it and an extensive references section correctly formatted in APA or ACS or any other style guide? No. Does a technical document have an arty newspaper graphic in it? No, it shouldn't. Do slides in a webcast have massive tables of data in them? No, but those can go in a paper. What would photos of patients go in? What about radiographs? algorithms? diagrams? micrographs? Design your writing with your *target* in mind, not the way it would appeal to *you* or in the language that would make the most sense to *you*.

> "Don't look at me in that tone of voice."
>
> — Dorothy Parker (1893–1967)

Mind you, when we say "borrow," this does not mean that you should not interject your own approach or creativity to creating a final product, nor should you, necessarily, hew exactly to the model template your using. "Format borrowing" does, however, help guide you in initial construction of a project designed for a specific target—particularly if you're stuck trying to get started.

How About Genre?

The genre may also have considerable influence on *how* you communicate your topic. A live slide activity and a journal article can both be targeted to the exact same audience, and cover the exact same topic, but the two genres play by different rules: Their organization, selection and presentation of information, inclusion of visuals, overarching goals, and delivery of the material among other things are obviously very different. From a 200-page clinical study report you can create a 15-page journal article, a 120-slide training kit, or an 8-page detailing piece for sales reps. Genre *is*, in fact, intimately connected to its audience.

Don't Put "Hydrochlorothiazide" in a Newspaper Article: What Goes Where

Your genre dictates the medium, which in itself is going to be doing the communicating *to* your target audience and determines the level of complexity and the type of jargon that can be used, if at all. For example, a physician listening to a sales rep is not going to read a lot of text. She will want to see clear, attractive graphs she can scan in 60 seconds before she signs for samples, boots out the rep, and goes on about her day. Those slides are designed accordingly. A commuter on a morning (hopefully not on fire) train into the city reading a newspaper article provocatively titled "Hypertension: Will It Kill Us All, Including Our Pets?" will start to zone out immediately if diuretics are listed by name: "hydrochlorothiazide" is the wrong word to use in the newspaper, especially if the reader hasn't had his coffee yet. How about a table of raw data? or a study disposition? These are great things to include in a clinical trial paper, perfectly lethal to put in a slide kit, and never, ever included in a newspaper article. Keep your target in mind, and then *consider how you're delivering your message.*

I wrote a few children's books. Not on purpose.

— Steven Wright

Not Everything You Write Will Be on Paper. In Fact, Most of It Won't

What are the kinds of media for which you might be writing? Back in the day, the things scientists wrote got printed on paper. Dissemination of information has changed with the advent of computers, the Internet, animation, PowerPoint, interactive education, virtual reality, online databases, and a host of other ways to reach the masses. Keep in mind that you will learn how to write to these media primarily through experience. You can be introduced to the concepts in a class and do a project, but your expertise comes only with practice. It may well be that you're only working in one medium. You're a physics professor and write the monthly newsletter for the American Association of Physics Teachers. You've done it for 12 years, starting out as a freelancing physics undergrad. It's just writing to one very specific target, all the time, 16 pages a month, every month. It gets mailed to the members. But then it's available online. And after a few years, it's *only* available online. Soon, it includes useful animations that can be used in physics classes. And then video interviews with prominent physicists. And then downloadable slides and study materials. And then podcasts. PDA delivery. URLs. And then, of course, they set up a Facebook page and Twitter and Instagram feeds. You could be writing paragraphs of highly technical updates in the world of physics, or you could be writing a two-sentence Twitter update. You might have started out writing things that got printed on paper and then mailed, but over your tenure helming the newsletter, you've evolved—you've had to evolve—with changing technology. In fact, the newsletter genre evolved, gradually, to something very different than it was 20 years ago—although the audience stayed the same. ("Same" is arguable if you accept that the passage of time and the dynamics of progress are sufficient for turning a group into a different audience by significantly altering their expectations and tastes.)

So How Do I Write to Different Target Audiences in Different Kinds of Media?

Knowing your target audience is more than half the battle—perhaps the crucial part of the battle. Suppose you are doing research in hypertension and the materials you create are targeted to *doctors*. You could be writing:

- Scientific journal articles
- PowerPoint presentations for live meetings (for yourself or others to present)
- Scripts for videos
- Detailing materials
- Meeting posters

Let's say your job includes communicating with *lay audiences* about hypertension. You could be writing:

- Lay science magazine articles
- Consumer websites
- Patient education brochures
- Posters that go in doctor's offices
- Scripts for television

Or suppose you are working in the pharmaceutical or medical education industries and need to provide information useful to a *marketing audience* about hypertension. You could be writing:

- Grant proposals
- Marketing analyses
- Drug reviews
- Competitive intelligence analyses
- Gap analyses

Or maybe you are in an R&D department in a pharmaceutical company, sending materials to the *federal government* to get a new hypertension drug approved. You could be writing:

- New drug applications
- Clinical study reports
- Investigator brochures
- Periodic safety update reports
- Package inserts

You would not hand doctors a brochure called *Hypertension and You* to educate them about the latest drugs in hypertension, you would not submit a *Scientific American* article on antihypertensive drugs to the Food and Drug Administration in support of a new drug application, and you would not hand a patient a scientific editorial on clinical preference for the thiazide diuretics for the management of uncomplicated hypertension as compared with the somewhat less effective angiotensin-converting enzyme inhibitors.

> Remember the CDC "Zombie Preparedness" graphic novel? To whom was that targeted? And why?

Let's look at our clinical target audience. Same topic, different media. In a scientific journal article, you can use sophisticated doctor-speak: "ACE inhibitors restore the balance between the vasoconstrictive salt-retentive and hypertrophy-causing peptide angiotensin II (Ang II) and bradykinin, a vasodilatory and natriuretic peptide." A PowerPoint slide for a live conference presentation, however, might have "MOA of ACE Inhibitors" as a title, and then a series of bullet points—simple sentence fragments, not narrative—such as "Lowers TPVR" or "increases Na^+ excretion," and "increases K^+ retention." A doctor target understands this perfectly well, and the presenter elaborates on the materials on the slide ("talks to the slide"). Detailing materials, or the slick materials that sales reps take to show doctors, are going to be very light on text and heavy on simple graphics such as bar graphs showing improved blood pressure readings with use of their company's drug over a four-week trial period, and perhaps simple, colorful bullets extolling the marvels of improved comorbidities such as heart disease, renal disease, and diabetes. Does their drug make patients *feel* better too? There might be a subjective improvement scale on there too ("on a scale of 1 to 10, most patients ranked their subjective improvement at 8 or better"). What about a meeting poster? There you would find well-executed graphics showing how angiotensinogen is converted to angiotensin I by renin: same topic, same target audience, different wordsmithing—to match the delivery medium.

Take-home

Writing to the appropriate target audience is one of the most critical skills for good scientific writing (or any writing). Follow a methodical process to first identify your audience, choose how you're going to write to that audience effectively, and then adapt your voice and sophistication to the delivery medium you are going to use.

In other words, if you know your fish, you won't be responsible for the deaths of dozens of people when their rail car catches on fire.

> *Know your audience before you start writing.*
> *Know your genre and medium.*
> *Know your audience in your medium.*

Exercises

1. Who's your target?
   Example:

To doctors	In a newspaper article	On a lay patient website	In a text message
Psychopharmacologic substances such as the SSRIs and nonbenzodiazepine hypnotics may skew PET results and should be discontinued 2 weeks prior to imaging.	Some antidepressants and sleeping medications may interfere with brain imaging studies and therefore should be discontinued 2 weeks before the test.	Antidepressants like Prozac or Paxil, or sleeping pills like Ambien or Lunesta, might make your brain scan test results unreliable. You might have to stop taking your medications before you have a brain scan. Check with your doctor to see if you are taking any medications that might interfere with test results, and how long before your test you should stop taking them.	Dude R U takin Ambine doc sez quite 2 wks B4 brain thng >:((Note: *never* write this way, even in text messages. Proof your text messages!)

Your turn. What goes in the rest of the table on the topic of oil spills?

To NOAA* scientists	In a NOAA fact sheet for the lay public	In a legend for an oil spill diorama in museum	In a children's book on the environment
		The presence of discharged oil in the environment caused decreased habitat use in the area, altered migration patterns, altered food availability, and disrupted life cycles.	
* National Oceanic and Atmospheric Administration			

2. Who's your target? How about this one, about avoiding a fiery, chaotic death on a train:

In a report to the NTSB*	In a journal article abstract	In a "Train Safety" brochure for a lay audience	On an emergency sign on a train
			This Way Out ➲
* National Transportation Safety Board			

3. How would you write about the same topic (string theory), to the same audience, using different *genres* and *media*?

Physics journal	PowerPoint presentation for live meeting	Newsletter for physics teachers	Professional meeting poster
		"String theory is an evolving theoretical effort that seeks to describe the known fundamental forces of nature, including gravity, and their interactions with matter in a single, mathematically complete system."	

4. Think about it: How would you *not* write about the topic of psoriasis to these targets?

To physicians	In a patient education brochure	In **Discover** magazine	On a poster for kids in a pediatrician's office
			[ILLUSTRATION OF MICROSCOPIC SKIN SECTION] See this? This is *psoriasis* (so-RY-uh-sis). Note the marked hyperkeratosis with parakeratosis in this histopathologic micrograph.

5. See if you can do it:

Option 1 (working alone): Choose your own topic, in your own field. On that exact same topic, write:

(a) A paragraph targeted to a scientific journal audience (make it a general, rather than an applied or specialty journal within your field (e.g., biology, chemistry, physics, etc.), such as *JAMA* rather than the *Journal of Acquired Immune Deficiency Syndromes and Human Retrovirology*)

(b) A paragraph targeted to a lay audience interested in science (e.g., *Scientific American* or *Discover*)

(c) A paragraph targeted to a grantor

Option 2 (working in groups): As a group, pick a *single* topic (make it specific enough, e.g., not just "climate change" but "climate change and sea level change"). Write two 500-word documents, one piece targeted to specialists and one to the population of individuals in your group. What do you need to know before you start doing this? Hint: Group Zen is a great way to get to know your peers. When you are done, take turns and *read what you've written aloud* to your peers (if this is not possible, route what you've written to your peers for written evaluation). What works? What doesn't? Why?

3
Ethics in Scientific Communication

"You're doing great!" © 2020 Kris Mukai

Lucky you—you can't *believe* you managed to beat everyone out for the internship in Dr. Graham's physical therapy lab! Learning at the knees of a master! She's evaluating a bunch of physical therapy (PT) patients who have had total knee replacement (TKR) to see which kind of postoperative PT is the most effective: electrotherapy, hydrotherapy, massage, or sham therapy. She's bundled them into groups. You express surprise that patients would enter a study knowing that they could be receiving sham therapy. "Well," she says, "it's not exactly a formal *trial*. I'm just making observations." She just wants data. She didn't tell the patients that they were in a trial, much less ask their permission and get consent, or submit her study for review to the Institutional Review Board (IRB).

"Aren't these results *great*?" she says, hopping about, clapping her hands! "Trial" notwithstanding, she publishes three separate papers on the study results: one on electrotherapy, one on hydrotherapy, and one on massage, in standard scientific paper fashion (Introduction, Materials, Results, and Discussion). In her introduction, she cuts and pastes most of the introduction she wrote for a chapter titled "Postoperative Therapy for TKN" in "the" PT book on the market. She only mentions results that were statistically significant and, rather than addressing the serious worsening in the sham group, reports them as "no improvement."

Full disclosure: This scenario is *hypothetical*. Clinicians and professors *don't do this*. If anyone tried this, they wouldn't get away with it, and their careers would be destroyed.

Even you are vaguely uneasy, although you're not 100% sure why. After all, you're a sophomore. All you had to do was sign an "academic conduct" agreement when you started school.

"Um," you finally say, "I am unworthy to be your intern. Really, I haven't understood a whole lot of what I've been reading, and have kind of been thinking I want to switch my major to accounting anyway."

"Oh no!" she says, because she's really a very nice person, "I'm so sorry to hear that! I wish you the best of luck!"

You really wanted to study PT, so off you go, transferred to another university, lest the nice lady for whom you were interning finds out that you did not, in fact, change your major to accounting.

Why Do Some Things Make You Uneasy, Even If You Don't Know Why?

The answer is probably because you were likely to have been taught at some point in your life that you should not lie, steal, or cheat. You knew—somehow, vaguely—that the professor was doing all that, even if you couldn't exactly say why. She was:

- Performing a study using human participants without IRB review (which is against the law, incidentally)
- Using patients without any disclosure that they were in a study (even though she didn't call it that) or what might happen to them
- Not getting informed consent from her patients
- Subjecting some patients to certain dangers
- Publishing her results in separate (serial) papers rather than all at once

- Using data selectively
- Suppressing negative data or misrepresenting it
- Plagiarizing (plagiarizing something you wrote yourself is still plagiarism)

Again, legitimate scientists do not do any of this. Some of her transgressions may have been deliberate (if not ill-intentioned; I mean, she *does* care about her patients), and some might have been due to ignorance—although "ignorance" as a researcher cannot seriously be claimed as a defense.

She will not get away with this. In fact, she is unlikely even to make it through peer review. If the government gets wind of the IRB transgression, she can get the university fined. She could get sanctioned, suspended, or even fired. Her reputation will be in tatters. Is it worth it to take that chance for the sake of getting published? No, it is not.

---

10 Tips for Ethical Behavior in Scientific Communication

1. Don't cheat (bypass rules).
2. Don't lie (misrepresent data or results).
3. Don't steal (i.e., data or results from another researcher).
4. Don't fake things (make things up).
5. Don't leave out negative results (if you're doing something clinical, you can kill a patient by doing this).
6. Don't plagiarize (even from yourself).
7. Don't take things out of context.
8. Eschew ghost authorship (this is not the same thing as writing as a professional scientific writer).
9. Provide full disclosure of any research or writing conflicts.
10. Acknowledge collaborators; do not take full credit for anything you did not do by yourself.

Ethical Behavior Precedes Ethical Writing and Reporting

Suppose you stayed in Dr. Graham's lab and wrote the first draft of her "clinical study" paper. *You* aren't the one who conducted any misconduct. You are still somewhat unclean by association as a collaborator. And, from the standpoint of writing, if Dr. Graham—because she is genuinely such a nice person—lists you on that paper as a coauthor, you may face a little shunning later on down the path. You're just a student, with elements of innocence, so if the IRB or federal Office for Research Integrity (ORI) decide to make an example of Dr. Graham, you would in most likelihood escape with only first-degree burns. But those still hurt.

Good writers, whether or not they are scientists, need to understand the nature of what they are writing about and be able to ask questions about anything that raises eyebrows: absence of disclosure, suspicious statistics, text you've recognized from elsewhere, data that did not come from the experiment or study about which you are writing, things that make you uneasy, even if you don't know what is causing the ill ease.

You also need to be able to look at your *own* science—*before* you start research, whether on a small project or a dissertation—and say, "Wait a minute. There's something I'm missing here." Remember that ethical transgression can be deliberate or ingenuous. If you're a sophomore, and really excited about your project, any ethical transgressions are likely to be ingenuous. This is why we have advisors, professors, mentors, colleagues (yes, fellow students can be good advisors too), the IRB (for humans), the International Animal Care and Use Committee (IACUC, for animals), radiation safety officers, the Occupational Safety and Health Administration (OSHA), journal editors, peer review, and a host of other entities that can look over your shoulder and say, "You're right. Something *should* bug you about this. Can you guess what it is?" As you become more conversant with scientific ethics, *you* should be looking over shoulders and saying, "Now, can you tell me why this should bug you?"

If the topic about which you are writing contains any lapses at all (experimental or ethical), those lapses are going to make it into your writing. There is nothing you can do post hoc about not having an experiment or study involving human subjects which did not get IRB approval prior to starting your study. You're not going to get it post hoc, and then you can't publish at best—and get into serious trouble at worst.

Ethics in Conducting Research

If you are conducting your own research or are collaborating with others on a project, you must ensure that you are on the right track, ethically, from the get-go. This is why, once you write up your proposal, you talk to your professor or mentor about it. Your methodology and experimental design need to be sound, and your ethical considerations have to be thorough and sound as well.

Ethical Design of Studies and Trials

The research you conduct needs to be well founded, address gaps in know-ledge, contribute new information to the field, be based on solid scientific rationale (with preceding work to support it), and be ethically conducted. Studies and trials should not be designed to yield the results you want. Unfortunately, this happens a lot. If your outcome is efficacy and/or safety of a drug, "efficacy" should not be qualified so that it is guaranteed to be sta-tistically significant in the results. You need to have an appropriate number of subjects (not too few or too many), and they should not be "coached" or subtly nudged or conditioned toward certain responses. You should not use data selectively, nor should you leave out or manipulate data. Statistics *could* potentially be manipulated to "prove" anything. Even if you are doing re-search on a study you did not design yourself, you should be able to recognize flawed study design. This skill is especially important when reading and eval-uating other studies in order to write a review paper or the review section of an IMRAD article, a thesis, or a grant (for more on reviews, see Chapter 12).

> I can prove anything by statistics except for the truth.
>
> —George Canning (1770–1827)

Protecting Human Subjects and Subject Privacy

A "human subject" is any human being who is remotely involved in the study you are performing as a *subject* to an intervention (although the word "sub-ject" has fallen somewhat out of favor, preferred use being "participant" or "patient" if the subject is a patient). There are the inpatient clinical trials wherein experimental oncology drugs are being administered to terminal patients, but there are also studies where you give a bunch of college students a questionnaire about condom use, or a study where you don't actually in-teract with patients at all but look at their charts, or studies where you don't interact with people at all but are instead just handed a bunch of preassem-bled, anonymized data for analysis. Which of these have to get IRB approval in some form or another? The answer is *all of them,* although this can vary by institution. Obviously, asking students to fill out an anonymous online questionnaire about condom use is a far cry from injecting cytotoxic drugs into a dying cancer patient. The one thing, however, that all of these studies have in common is that they involve *humans*—directly, indirectly, distantly, or anonymously. Before embarking on any of these studies, you'll need to submit your proposal to the IRB and complete their proposal questions. If you're doing an anonymous online survey about condom use, you'll plug in a paragraph with the general proposal, explain why you're doing this (what gap in knowledge is being addressed), how you plan to distribute the survey, the exact verbiage you are planning to use for disclosure (e.g., identifying yourself and your institution, why you're doing the survey, all responses will be anonymous, survey can be accessed at such and such link, etc.), and the survey questions themselves. If you're injecting a cytotoxic drug into terminal cancer patients or giving athletes fake physical therapy for a knee injury, you'll be providing *volumes* of additional data, about the drug, patient selection, patient population (including vulnerable populations like pregnant women, children, or prisoners), interventional groups, known risks, delivery mechanism, disclosure, patient consent, and a host of other pieces of information. IRBs typically meet monthly, but that varies tremendously by institution. When proposals come in to the IRB, a single member of the IRB may review the proposal and decide if an expedited review is warranted or if it is a simple enough decision to render an exemption. If it is not "exemption material," it goes for full board review. Even studies that are deemed exempt can be kicked back for clarification, adjustments, additions, or recommendations. The IRB may feel that your disclosure is inadequate, informed consent language is weak, or your data collection methods will not ensure adequate anonymity. Again, IRB procedures vary tremendously across institutions. Some may require more or less information, and some studies may not even necessarily require IRB review at all.

One of the most famous cell lines used in cancer research is the HeLa lineage of cells. These were cancer cells harvested from a patient during surgery for cervical cancer. This particular line of cells was exceptionally hardy and has been immortal—since the early 1950s. This was surgery. Cells and tissues (biopsies or excision) are taken all the time during surgery. Is informed consent necessary for that? Read about Henrietta Lacks and her remarkable line of immortal cells in Rebecca Skloot's 2011 book *The Immortal Life of Henrietta Lacks*. Hint: Those cells were harvested in 1951, and it took until 2013 for the National Institutes of Health (NIH) to restrict access to the cell line without a grant.

Disclosure and anonymity are key. Each time egregious research (typically with regard to nondisclosure) is conducted, regulations get tighter, and then more and more (national and international) guidelines get developed. It is also very important to protect patient privacy. One of the most recent enactments with which you might be familiar debuted in 1996, the Health Insurance Portability and Accountability Act (HIPAA—with one P and two As). This is hard to miss, because frequently you're handed the HIPAA statement every single time you go to the doctor or dentist—and no matter how many times you've received it, they give it to you again (unless they have your signature on file in an electronic health record). HIPAA was crafted to ensure that personal health information is not freely shared without adequate anonymization. This includes sharing of health records but also applies to spoken discussion of private health information within earshot of noninvolved individuals.

When you have the blessings of the IRB, you can start your study or research or experiment, and when you write up your results, you indicate in your paper (or whatever you're writing) that IRB approval (or something similar, such as "university ethics committee," etc.) was secured prior to starting the study. Note that if you are collaborating with a fellow student in another university, or if the PT patient population you're studying is at another institution, it is very likely that you will need IRB dispensation (or some kind of official agreement) from your institution *and any other* institution involved in the study. Protection of human subjects is overseen by the federal Office for Human Research Protections, part of the US Department of Health and Human Services (HHS). Protection of human subjects is not just an institutional policy—*it is federal law.*

Want to read all 42 glorious pages of the part of the Code of Federal Regulations (45 CFR 46) under which protection of human subjects is covered? A dry read by any means, but the ultimate authority, which you can find on the HHS website. Have a look at what the feds are doing on behalf of the populace!

Protecting Laboratory Animals

"Well," you hesitate, because you are working in a laboratory mincing up mouse brains, brains that had been removed from mice which had been arguably killed ("killed" is the preferred terminology over "sacrificed"), most likely without informed consent. You can't exactly ask a mouse if it wants to take part in an investigational preclinical trial on a cytotoxic chemical, disclose that dosing groups are going to be used to determine the LD_{50} (lethal dose at which 50% of the animals will die), and then get informed consent. Even a low-IQ mouse would be likely to say "I'll pass, thanks." However, this does not mean that the mice (or any other laboratory animal) can be subjected to inhumane living or transportation conditions, cruel or poor treatment, or unnecessary pain and suffering. But how can you conduct a study where you're investigating pain thresholds? How long a mouse can go without eating? How long it can live without sleeping? How many of them die when they get a high enough dose of a toxic drug? These studies are conducted on a routine basis, but beyond study design, animals need to be treated with care.

An IACUC is required in all institutions that perform animal research. If audits or complaints (whistleblowing) turn up instances of maltreatment of animals, fines can be imposed, research halted, or extrainstitutional intervention may occur. IACUCs are mandates of the Animal Welfare Act and Public Health Service Policy. These are *federal* mandates, overseen by the Office of Laboratory Animal Welfare at NIH.

Does this apply to *all* animals? Even fruit flies? No. IACUCs generally review protocols involving animals capable of sensing pain (nociception)—typically only mammals (there might be rare exceptions). Of course, this is difficult to determine in invertebrates—what is stimulus-response and what is nociception? If there is any question about the potential for nociception in laboratory animals, it isn't a bad idea to check with the IACUC, if only to see if a proposal needs to be submitted. Different institutions will have different policies on this.

Animals cannot be kept in squalid, unclean, or overcrowded cages and need to be handled with care when they are transported (including if they are shipped). Where appropriate, enrichment programs are put in place to keep them stimulated (e.g., toys for them to play with). These are violations that are hard to hide, and ethical scientists should rightfully balk at them. Treat your animals ethically and with respect.

Registering Clinical Trials

A *clinical trial* is one in which drugs or devices are tested in humans (as compared with a *preclinical trial*, in which animals or cells or tissues are used). In 1997, Congress established the Food and Drug Administration Modernization Act (FDAMA). FDAMA mandated that HHS, through the NIH, establish a clinical trial registry (database) for all federally and privately funded trials. Registering a clinical trial (in a publicly accessible place, such as clinicaltrials.gov) is required before the first subject is recruited— this is not only a federal mandate, it was established as one of the tenets of the Declaration of Helsinki. Publicly available clinical trial information contributes to the evidence-based development of medical intervention and helps guard against publication bias and selective use of data. Research is built upon preceding research; if the preceding research is ethically flawed in some manner, then subsequent evidence-based research will also be flawed—if unknowingly. Further, with a searchable database of planned or ongoing trials (clinicaltrials.gov tells you the stage or phase at which the trials are presently active), it helps researchers, funders, and scientists see at a glance how many trials are out there doing the same or similar things. Do you really want to try to get a grant from NIH on cytotoxic drug uptake in terminal cancer patients when there are already three or four similar trials out there studying similar drugs? Would that be worth it to you? Would your data seriously be addressing gaps in knowledge that are not already being addressed? Would you get funded if there were already active research going on in the same area? And most importantly: Is it worth it to put additional patients at risk to do the same kind of research? But you can also look at what's out there and see where there *is* a gap in knowledge, something that *would* be worth pursuing, research that could be improved upon or expanded with different or better study design—and you can find out who else is doing this research, increasing the likelihood of collaboration. Collaboration is a fundamentally important component in trial design and execution—and, ultimately, important in writing about it too.

The Declaration of Helsinki was an international effort to characterize the ethics of medical research on human subjects. Drafted in 1964, it sought to provide an internationally agreed-upon set of principles for human experimentation. It was a necessary and more comprehensive expansion of the earlier and much shorter Nuremberg Code, a short list of ethical principles created to protect human subjects after the medical atrocities committed by Nazi physicians during World War II; the first tenet of this code was a requirement of voluntary consent.

No research group jumps into a clinical trial carelessly or without considering the possible implications of the study. Even if you are involved in a less-dire clinical trial, you go in eyes wide open. Know what's out there. Publicly available databases are often even used by individuals trying to get into clinical trials themselves as patients—they can get information on recruiting and perhaps find out if they are eligible or can be considered.

Accessing a trial on clinicaltrials.gov (which lists both national and international trials; CenterWatch [www.centerwatch.com] is another good international resource) will tell you if the trial is recruiting, active, completed, or terminated; identifies the condition and the intervention; and tells you if the trial has results. The database can be searched by multiple keywords (e.g., condition, drug, location, trial phase, participants, etc). Clicking on the link of the trial in which you are interested provides a wealth of details such as who is conducting the trial, purpose, study type, study design, and other information. If you're interested in insomnia trials because you're doing research in that area but only on drug interventions, and you're only interested in double-blinded trials, you'll get that information in the link for the specific trial. This freely available information contributes to ethically robust research: necessary, filling the gaps in knowledge, identifying what has already been done and assessing its efficacy, and, most importantly, ultimately benefiting a patient.

Making Trial Research Results Publicly Available

The whole point of research is to contribute to the growing body of scientific knowledge. When a trial or study or research project is completed, those results should be shared publicly. This includes making data available that you might not like. If you get statistically nonsignificant results in a trial you've conducted, it might not be the outcome you wanted, but the results are still important: Those results are *information*. You may have spent a significant amount of money on an experiment for which you were sure you were going to have ground-breaking results, but if you don't get those results, that is information other scientists—including yourself—can use in the future. You can't just publish the good stuff. You can't let the "bummer results" languish in a file folder buried in multiple subfolders in your computer. If, of course, you have any kind of grant (federal or private) you won't have a choice anyway: All results must be disclosed. Withholding negative results is unethical.

Scientific Misconduct

The most egregious kind of unethical behavior is the deliberate kind. Ignorant or ingenuous ethical infractions can sometimes possibly be rationalized in some ways into some kind of forgiveness, particularly for new scientists. But outright lying or stealing is supremely reprehensible. Apart from the fact that lying and stealing are bad in general, when you start lying about scientific research involving humans in any way, directly or indirectly, people can get hurt or killed, and the evolution of scientific knowledge may be derailed for decades, impairing progress in the field in which the ethical breach occurred.

One of the most famous historical examples of egregious medical misconduct was that of Andrew Wakefield, a British physician who conducted research on a small number of children (12, and without telling their parents or obtaining any kind of consent), concluding in his now thankfully retracted paper: "Ileal-lymphoid-nodular hyperplasia, non-specific colitis, and pervasive developmental disorder in children," published in *The Lancet* in 1998. This was the paper in which the unforgivable "link" between autism and the measles-mumps-rubella (MMR) vaccine was made. Although *The Lancet* retracted this paper in 2010 *a full 12 years later*, the results were almost immediately disproven by the medical community, and Wakefield was exposed as a fraud and stripped of his license, the idea that there is a link between autism and vaccines prevails in a startling percentage of the population. This has resulted in fewer children being vaccinated, with a concomitant increase in preventable disease and death. For a good summary of this event and its sequelae, see Rao TSS, Andrade C. The MMR vaccine and autism: Sensation, refutation, retraction, and fraud. *Indian J Psychiatry*. 2011;53(2):95–96. This review contains a wealth of additional references.

There are a number of ways you can perform scientific misconduct. *Fabricating data* simply means to make stuff up. Did you leave some part out of your experimental design, thereby leaving you with insufficient data or statistical power to analyze your data? Adding in a little extra data to bump up your statistical power, make your results look better or your study more comprehensive, is a perfectly vile thing to do. By extension, you can also eliminate a few pieces of troublesome data, like outliers (tip: don't do that). If you decide instead to *falsify your data*, you simply take the data or results you have and change them—ideally, subtly enough to not get caught or to arouse any suspicions—so that you get the kind of results you want for your study. A subtle change here or there can nudge your *p*-value down into the $P < .049$ safety zone. Or how about *fraud*? Just take the results from another study and present them as your own. If you're creative about it, you can find some great experimental results in a study published somewhere in an obscure journal (preferably one not indexed in Medline or other journal databases). Better yet, get it from a non-English journal. Or take some solid American results and publish them in an obscure foreign journal. I mean, who's going to know, right? Apart from, of course, the original author—who is likely to evince no small measure of disapprobation when she finds out about the transgression. You should have done your homework and found out first that she was fluent in five languages before you stole her work.

Fabrication, falsification, and fraud, once discovered, become front-page entertainment in the Fourth Estate and the major medical and scientific journals across the board. Reputations are trashed. People are sued. Institutions are sued. Innocence or oversight is claimed. Fingers are pointed. The scientific field in which the infraction occurs, by extension, gains a notoriety it does not want, particularly from the public; the public trust in science gets eroded. Worse, poor or unethical science may also influence thinking, often with devastating consequences (witness the continued "debate" about vaccines and autism). Worst of all, if it is human research, people can be damaged permanently or die. If you are working on a clinical trial and falsify, fabricate, or leave out data when you report trial results, and 271 unanticipated postmarketing deaths occur, one cannot just say "oops." You cause untold human misery and attract billions of dollars in fines from the government. Prison is also a possibility.

Combating fraud, falsification, fabrication, and other ethical infractions requires at least two people: the villain herself, and the individual who uncovers her. It is hard to be a whistleblower; people lose their jobs, get harassed, don't get rehired in the field, are shunned, and, if they manage to survive all that, will not want to blow the whistle a second time. Often, whistleblowers stand to benefit financially from reporting fraud—which compounds the distaste with which they are often perceived. Many companies (including academic institutions) have whistleblowing policies in place, and some of them permit anonymous reporting or have policies to protect the whistleblower. Anonymity won't necessarily help you if you are the lone holdout in the lab heavily in disagreement with the principal investigator (PI) over a perceived infraction. In any university setting, though, you will almost always have an individual you trust with whom you can speak. Universities have ethics officers who oversee the IRB and IACUC. Sometimes there are even offices for ethical conduct. Even Human Resources can be a place to register a concern.

If you are concerned about what you perceive to be an ethical infraction, bring it to the attention of someone you can trust. Depending upon the perceived infraction and the relationship you have with your supervisor, you may even be able to approach her yourself: "I'm really sorry, but I'm confused. . . . I thought the spectrophotometric results we got were . . ." Maybe it *was* an innocent oversight. If the answer doesn't satisfy you, and you are afraid to push it, it is time for some advice from someone else. Ethics levels the professional field. An ethical sophomore is worth more than all of the unethical MDs and PhDs in the world combined. If you aren't comfortable speaking with your professor, find another professor you trust. Universities are full of wonderful professors you can trust, as well as instructors, lab instructors, graduate students, counselors, deans, librarians, and administrators. If you are troubled, you will be taken seriously.

Ethics in Writing

Research, studies, experiments, and other scientific pursuits are only half of "science." The other half is communicating what you find—positive or negative, good or bad, significant or nonsignificant, equivocal, interesting or uninteresting, new or confirmatory (or not). Remember, something that doesn't work is still an important piece of information; it still contributes to scientific progress, to the body of scientific knowledge.

You need to know how to communicate your results effectively or your work is for naught. A critical part of communicating your results is knowing how to communicate those results *ethically*. If the experiment or study you've conducted prior to the communication stage had unethical elements to it, remember that your communication will be doomed from the get-go. If the IACUC shut you down because you had mice living in squalid, inhumane conditions, they (and your grantor, if there is one), will take a dim view if you publish a paper on your research anyway. If you conduct a clinical trial with inadequate disclosure or no informed consent, you can write a brilliant paper that will slam to a halt in peer review, if your IRB doesn't come after you first. If you shuffle around some numbers in your raw data to improve your results, those "new and improved data" will end up in your paper.

So you've conducted flawlessly ethical research; now what are the ethical things you need to understand when you are writing about your results?

---

10 Tips for Ethical Scientific Writing Communication

1. Ask: Who's the author?
2. Establish author(s) and the order of their names prior to writing the paper.
3. Consider four criteria for authorship:
   a. Significant involvement with design of the experiment or acquisition or interpretation of results
   b. Significant involvement with drafting or critical and intellectual revision
   c. Approval of the final draft of the paper
   d. Accountable for accuracy and integrity of the results
4. Do not plagiarize (including self-plagiarizing).
5. Do not violate copyright—or allow anyone else to do it.
6. Avoid publication bias (i.e., do not selectively publish only positive results).
7. Avoid duplicate, redundant, and "salami" publishing.
8. Hew to confidentiality, propriety, disclosure, and conflict of interest issues.
9. Avoid "spin."
10. *Do not communicate results you know to contain unethical elements.*

Authorship

Authorship is one of the most hotly debated topics in scientific writing. You've been involved with the research, and now your lab director wants you to write a very early draft of a short paper to be submitted to *Journal of Orthopaedic & Sports Physical Therapy*. This is going to be a short paper, with early research findings. She's encouraging you to take a stab at the first draft of it. Does this make you an author or the first author? Not necessarily, although some PIs will nudge their undergraduate or graduate students into starting their publication record by making them an author or first (rarely) author.

But who *is* an author? Really? *Ethically, who is an author?* The best definition of an "author" comes from the International Committee of Medical Journal Editors (ICMJE) and can reasonably be extended for consideration for all scientific publication:

- Substantial contributions to the conception or design of the work; or the acquisition, analysis, or interpretation of data for the work; AND
- Drafting the work or revising it critically for important intellectual content; AND
- Final approval of the version to be published; AND
- Agreement to be accountable for all aspects of the work in ensuring that questions related to the accuracy or integrity of any part of the work are appropriately investigated and resolved.

Of course, the study with which you are involved, the individuals working on it, and ultimately the PI are largely going to determine authorship. Say you're a sophomore interning in a PT facility and are observing and recording data (data determined by the PI) on the comparative effectiveness of different postoperative rehabilitation techniques for individuals who have had TKR. Have you made substantial contributions to the conception and design of the study? Probably not. Substantial contributions to the acquisition of data? Possibly, but it is unlikely that you've been collecting *all* of the data. Are you the one who has been analyzing and interpreting the data? Probably not, although you are probably being coached by the PI and might be performing the odd statistical analysis. Are you the one drafting the article? In this case, yes. Revising it critically for important intellectual content? Doubt it—you're not an expert in the field. Are you the one who is going to be providing final approval of the final draft submitted for publication? That would be a definite "no." So, according to the ICMJE criteria, are you an "author?" No. If your PI includes you as an author, is that unethical? No. You've been involved, and the PI can make the call as to whether you merit authorship or acknowledgment (not acknowledging you at *all* would be unethical). Perhaps you worked really hard, spent untold extra hours, provided new ideas or creative interpretation, and wrote and redrafted multiple times a phenomenal paper, and your PI decides that merits authorship. If you were uninterested in the study, put in only enough work to get credit for the internship, and wrote a completely unacceptable first draft that had to be discarded and rewritten, your PI may decide that you merit only the barest of acknowledgements (it would still probably be unethical to not acknowledge you). Claiming sole authorship if others met authorship criteria is unethical, but it happens. "Official" trouble *can* happen if researchers put themselves down as sole authors, in effect taking credit for the work of others; in that case, the real authors may lodge complaints, ethics or conduct committees can get involved, and lawsuits might even ensue. It is not uncommon at all in college and graduate school that *any* paper coming out of a PI's lab (or your major advisor or whoever runs the lab) has the PI on there as an author somewhere (often last).

Multiple authors on a medical paper or their editors might adhere strictly to the ICMJE criteria, but that clearly isn't always the case. Witness the paper "*Drosophila* Muller F elements maintain a distinct set of genomic properties over 40 million years of evolution" published in *G3-Genes, Genomes, and Genomics* in March 2015. It had 1,014 authors. But biologists have *nothing* on physicists, who so far hold the record with "Combined measurement of the Higgs boson mass in pp collisions at $\sqrt{s} = 7$ and 8 TeV with the ATLAS and CMS experiments," published in *Physical Review Letters* in May of 2015—with 5,154 authors. This article comprises 33 pages, of which 24 of them are the authors' names and affiliations.

"Authorship" is a moving target in the sciences. Many authors (even by their own admission in surveys) confess that they had little to do with some papers. Some authors are so prolific it seems like they must be publishing more than once a month. Bigger names on papers get more reads and citations, and grantors pay attention to scientists with a lot of publications. "Contributorship models" are coming more and more into vogue (particularly in biomedicine—for example, *JAMA* follows this format), where authorship is left up to the authors, but what they contribute (e.g., study design, drafting the article, statistics, etc.) in particular is disclosed (often to journal requirements and specifications). The Council of Science Editors (CSE) recognizes that authorship criteria vary from discipline to discipline, journal to journal, and grantor to grantor but determined a general consensus not too dissimilar from that of the ICMJE (http://www.councilscienceeditors.org/resource-library/

editorial-policies/white-paper-on-publication-ethics/2-2-authorship-and-authorship-responsibilities/):

- Identification of authors and other contributors is the responsibility of the people who did the work (the researchers) not the people who publish the work (editors, publishers). Researchers should determine which individuals have contributed sufficiently to the work to warrant identification as an author.
- Individuals who contributed to the work but whose contributions were not of sufficient magnitude to warrant authorship should be identified by name in an acknowledgments section.
- All individuals who qualify for authorship or acknowledgment should be identified. Conversely, every person identified as an author or acknowledged contributor should qualify for these roles.
- Individuals listed as authors should review and approve the manuscript before publication.
- Editors should require authors and those acknowledged to identify their contributions to the work and make this information available to readers.
- The ultimate reason for identification of authors and other contributors is to establish accountability for the reported work. (CSE website)

"Whooooooo wrote it?"
© 2020 Kris Mukai

Ghostwriting and Ghost Authorship

If you've heard about *ghostwriting*, especially in the scientific field, you probably heard it referred to negatively, as unfair or maybe as misleading. This is not necessarily the case. "Ghostwriting" unfortunately is also used to characterize legitimate assistance with writing. Scientific writing (and medical writing) is a separate, well-regulated profession. Some people are professional scientific or medical writers and work for publishers, communications companies, or companies within the scientific industry itself. They collaborate with the experts, researchers, and other personnel involved in the generation of scientific documents. They may do research, but it is *research into a topic*, it is not *original scientific research*. They are handed data or other information that they use to generate the ultimate communication of the science. Think of them as highly trained "technicians." Your PI might have an NIH grant, a lab with 12 different undergraduate, graduate, and postdoctoral workers, four lab techs, and two animal techs, and he might be getting his statistical analysis done by an outside company or colleagues in another university. And he might be working with a scientific writer (although use of scientific writers is far more common in industry than it is in academia). Is the journal article going to have 21 authors? No, it will not (unless it's in physics, then maybe). Recall the definition of an author: an individual who drafts or writes the final product only meets one of those four necessary criteria. Is it nice to add the writer on as an author? Yes. Does it happen? Rarely. Is that unethical? No, although—not surprisingly—many professional writers feel that it is, and passionate epistolary feuding about transparency in Letters to the Editor periodically flare in journals. *Acknowledging* the writer (or company providing editorial support) is important, and specifying that the writer "helped with the preparation of the manuscript" or in other way *identifying the writer as the writer* is important—*not* doing that is unethical. Again, in today's environment, journals or companies working with scientific writers may require the writer's CV or résumé, as well as disclosures, and perhaps even a nondisclosure agreement if the writer is writing about something proprietary (i.e., for industry). If you are ever, for some reason, hired to write a scientific paper, do *not* expect to have your name on the paper as an author. *Do* expect to be acknowledged. Are you a "ghostwriter?" No, not by CSE criteria. A true "ghost" is not disclosed nor acknowledged. *This* is wrong. Professional writers are not ghostwriters, and they need to be acknowledged.

Well then, why does ghostwriting have such a bad reputation? When trial subjects end up being injured, why is *ghostwriting* dragged into court, included as evidence, attacked as unscrupulous by lawyers, and included in staggeringly expensive lawsuits?

The answer is because in many of these cases it's usually not professional assistance with writing—it's inclusion of a ghost *author. Ghost authoring* is ethically reprehensible. What's the difference? In ghost *writing* (i.e., disclosed and acknowledged professional writing assistance), if it must be called that, you're a professional writer legitimately working with the content expert. You are writing under the guidance of the individual(s) who will ultimately be the author(s). Some of this guidance is extensive, and some of it less so, but the point is, you're working *with* the researcher(s). In ghost *authoring*, a scientific writer is charged with writing something, say, a journal article, with no author in mind. The completed paper is then shopped around to individuals (usually "big names") to see who'll consent to having his or her name placed on it as an author. This happens from time to time in industry (very rarely in academia), although ethical vigilance and transparency continues to improve. The paper is typically commissioned by a company selling a particular drug or device. The paper is friendly to said drug or device, or contains a definite positive "spin" if clinical trial results are being discussed, or emphasizes the importance of treating the disorder for which the drug or device has been manufactured to treat. The paper is written and finished, and *then* an author is secured. The writer has no contact with the "author." Authors who have been accused of this practice (along with the company who commissioned the paper) defend themselves by claiming that when they received the paper draft they reviewed it thoroughly, suggested edits, and were 100% comfortable with its content before approving its submission to a journal. Still, they didn't write it, nor did they have any input into writing it, nor did they provide data for it, nor did they even provide any guidance or oversight for it. They do not meet authorship criteria, not in any branch of science. They positively cannot answer "yes" to "Did you write this paper?" This is unethical.

When a blockbuster pain drug was pulled from the market after serious cardiovascular adverse effects (including deaths) started surfacing, its manufacturer (not surprisingly) found itself in court. Investigation turned up assorted instances of malfeasance, one of which being "ghost writing." This, of course, was prime (justifiable) fodder for our friends in the Fourth Estate, with The New York Times reporting that the company had drafted dozens of studies, then lined up prestigious physicians for inclusion as "authors," even mentioning one draft of an article that simply had "external author?" listed as an author (*The New York Times*, "Ghostwriters Used in Vioxx Studies, Article Says," April 15, 2008). The *Times* was referring to an impending article in *JAMA*. Outrage ensued, on the part of the scientific and medical community and the public. The Justice Department was also displeased, and the company was fined $950 million (for the whole fiasco, not just the ghostwriting). The concept of "disclosure" along with other tenets of publication ethics suddenly came in vogue, or at least to the fore in public perception.

Fifty years ago, this was not necessarily perceived to be a bad thing if no harm was done; "ghostwriting" wasn't an "issue" and was frequently not discovered. Not anymore. Full transparency is required. An author who did not, in fact, craft the actual paper, no matter how involved she might have been, will find herself on defense. Full disclosure of use of professional writers is helping to cut down on this—but nondisclosure of assistance with writing can get you into trouble. Bottom line: Make sure the authors on the paper merit authorship.

Plagiarism and Copyright Violations

Do you know what plagiarism is? You may think you do ("don't copy or cut and paste right out of a source"), but you really don't. Cutting and pasting or copying anything without acknowledgment is only a fraction of the definition of "plagiarism." Undergraduate and graduate students sign documents when they enter school attesting to the fact that they've read the conduct code and will hew to the tenets of academic conduct. And then they don't—in the case of plagiarism, often ingenuously. Most plagiarize unintentionally, but "I didn't know it was plagiarism" is not an excuse, particularly in college and beyond; it demonstrates an incomplete understanding of both ethical and professional behavior. Suppose you get your degree without having been caught for (or without even knowing that you have been) plagiarizing. And then you get a job. Plagiarizing on the job, no matter what kind of job—bench scientist, lab tech, clinical scientist—can get you fired and can even get your company sued. "I didn't know it was plagiarism," while marginally understandable in an undergraduate, is a staggeringly feeble defense as a professional. If some sleuthing is performed into your

© 2020 Erica Perez
TERRIFIED BOOK KIDS, MARITSA

academic career and egregious plagiarism is found, you might even find your degree revoked.

---

10 Tips for Avoiding Plagiarism

1. Learn what plagiarism is. The ORI defines it as "theft or misappropriation of intellectual property and the substantial unattributed textual copying of another's work."
2. Cite/reference everything.
3. Do not copy (or copy and paste) word for word, unless you are using a block of text (indented or otherwise delineated) and cite it.
4. If you need to use verbatim wording (e.g., if you're quoting someone), do it sparingly, and *put the phrase between quote marks*. Providing a citation *only* is insufficient.
5. Do not use text and *barely* alter it with marginal paraphrasing.
6. Do not use text and *kind of* alter it with marginal paraphrasing. If you need to paraphrase, also limit it, and paraphrase substantially (with a citation).
7. Do not plagiarize by copying (or copying with slight alteration) tables, graphs, charts, or illustrations.
8. Read multiple sources, craft an idea in your head, and synthesize a new way to communicate it. After you write it, go back to your source(s) and make sure you're making the same kind of point. And reference it—with *all* the sources you use.
9. Do not self-plagiarize. If you wrote something yourself in another document, do not cut and paste your own verbiage, assuming "it's yours." This is particularly true if what you're copying is published. If you retain copyright for your article (more and more common these days), this still does not justify plagiarizing your own text.

10. Keep yourself educated. Visit your school's plagiarism education website. Take a plagiarism self-test. Many universities use paper submission platforms such as TurnItIn, which have educational modules; they also detect plagiarism and insufficient paraphrasing. There are many options online you can use to check your own work for plagiarism.

Your professors *are* going to check, or spot-check, for plagiarism. Materials you submit for publication are going to be evaluated in peer review. They're not called "peers" for nothing: They know the literature well enough to be able to recognize that a passage or idea might come from elsewhere. If you plagiarize just once, you may as well wear a scarlet "P" on the front of your shirt. To complicate things, something that you have impressively resynthesized might in of itself have been plagiarized in the source document. Vigilance is key.

"But," you ask, "I have to report on the methods and materials of these studies in this review paper I'm writing. How do I legally paraphrase 'N = 422, 112 randomized to placebo, 310 randomized to treatment, 48% female, ages 18 to 65' and other straight strings of data like that?" A straight string of data, obviously, can't be tampered with. You can mix it up a bit, making it a "very close" paraphrase, or you can ask yourself if rereporting exact numbers in the same fashion makes sense for your rhetorical goals. If you are reviewing more similar studies, consider putting such data in a table. There are creative ways to cite without having to quote verbatim.

Suppose, though, that you're trying to communicate something supremely complicated and there's really only one good description of it in the literature. Read about the complicated concept, and look away from the source document. Think about that concept. Think about how you'd explain it to someone else—preferably someone not knowledgeable about the field. Make sure you understand it yourself. Can you read about it somewhere else, or can you read supplemental or more expanded information about the topic? Read that too. Then write. Synthesize something *new*.

Do you have to cite basic textbook information such as "DNA is comprised of the four nitrogenous bases adenine, thymine, guanine, and cytosine," or "osmosis is the process by which water diffuses from an area of high concentration of water molecules to an area of low concentration of water molecules?" No, "common knowledge" does not have to be referenced; otherwise everything we write would have over 700 references. Your target audience may also dictate what kind of references you include. If you're writing a treatise on the systematic analysis of a caecilian subclade, your target audience is going to know what a caecilian is. If you're writing a book for third-graders titled *The*

Big Book of Snakey Things That Aren't Really Snakes, even then you just explain what caecilians are and do not cite 1,241 references dating back to Linnaeus.

If you are concerned that something you are writing or have written has been plagiarized or may have elements of plagiarism or may be insufficiently referenced or has too much paraphrasing, take it to your mentor or anyone else from whom you can receive counsel: peers, lab instructors, librarians, the writing center. It is a sign of strength—*not* a flaw or a weakness—to ask for advice up front. If you're wondering, check, recheck, and check again with intellectual and self-confident impunity.

Words are not the only thing we don't want to steal. We also do not want to steal any images or graphics: photographs (or micrographs), charts, graphs, tables, spectrophotometric tracings, illustrations, and, for online writing, entire embedded documents such as image, audio, or video files. You cannot just lift a beautiful immunofluoresence micrograph off the Internet or out of a published paper and breezily place it in your paper. Maybe you could for a school assignment, but you would have to ask your professor, and then, of course, cite it. You cannot do the same if you intend to publish. If images of any kind are published, the copyright is retained by the publisher and/or the original creator of the image. You cannot reuse it without permission, and, often, if you're using it for a public presentation or a publication of some kind, you might even have to pay to use it or license it. Keep in mind that acquiring permission may take time and cost money. A mere copyright symbol (©) next to it is inadequate. Be mindful of collaborators who are unaware of copyright rules. When in doubt, consult with your instructor, mentor, or publisher.

There are often images you can get that are in the *public domain*; that is, anyone can use them free of charge. Many of the federal websites, for example, have public domain photographs. Even if a public domain image is used, its provenance needs to be cited. You may not have to ask for permission (although asking is always a good idea), but there may be a preferred way to acknowledge the source of the image. More and more, there are sites that make free images and clip art available (e.g., Pixabay, Wikimedia Commons, etc.). This does not mean, however, that they can be used without determination of provenance or (in some cases, in some cases not) acknowledgement or a citation.

With regard to images, there is no such thing as a "common knowledge" counterpart to plagiarism. Even "common knowledge" things like the structure of ATP or a photograph of Charles Darwin or a table of the periodic elements have to be rendered somewhere, somehow, by someone. Can you get an image of the structure of ATP off the Internet? Sure, that's easy. If it is plain and basic and not particularly arty, will its provenance be traceable? Unlikely. But don't do it, unless you find it free of charge and copyright.

If you don't pay attention, you may end up "borrowing" something that turns out to be have been plagiarized or borrowed originally itself. "But I didn't know" won't necessarily absolve you. Of course, if a collaborating lab hands you a micrograph that they have in fact stolen from a different lab, well then, there's not a whole lot you can do about it. That's why we have release forms (particularly for photographs that have people in them—it is tremendously unethical to publish photographs of individuals without their consent; this is one reason stock photos are so popular: they come with releases) and permissions forms. When plagiarized material is used, and then cited and recited, the plagiarism snowballs and becomes amplified over time. This also happens with misinformation and is pernicious and undermining to science in general.

Publication Bias

Again, no one wants to work on a research project for six years and end up with negative, statistically nonsignificant, irrelevant, equivocal, lukewarm, or noncompelling results. Again, though, never lose sight of the fact that negative results are *still results*. You cannot only publish positive results or you are guilty of publication bias. This, unfortunately, is not uncommon in science.

Ben Goldacre, a columnist for *The Guardian* who is a delightfully entertaining and ruthless Assailant of Pseudoscience, eloquently wrote in his 2008 book *Bad Science: Quacks, Hacks and Big Pharma Flacks* (which should be compulsory reading for everyone in the sciences):

> *Rightly or wrongly, finding out that something doesn't work probably isn't going to win you a Nobel Prize—there's no justice in the world—so you might feel unmotivated about the project, or prioritize other projects ahead of writing up and submitting your negative findings to an academic journal, and so the data just sits, rotting, in your bottom drawer. Months pass. You get a new grant. The guilt niggles occasionally, but Monday's your day in the clinic, so Tuesday's the beginning of the week really, and there's the departmental meeting on Wednesday, so Thursday's the only day you can get any proper work done, because Friday's your teaching day, and before you know it, a year has passed, your supervisor retires, the new guy doesn't even know the experiment ever happened, and the negative trial data is forgotten forever, unpublished. If you are smiling in recognition of this paragraph, then you are a very bad person.*
>
> — Ben Goldacre, 2008

A subtle and dangerous relative of publication bias is *citation bias*. If you've done your study and you have results, you need to resist the urge to pick and choose the references you cite, choosing only those that support or are otherwise in accord with your findings. No one wants to publish a paper that says "valerian root is an effective treatment for insomnia, even though there aren't any other study results that support our findings." Citation bias is self-serving and unethical, as it does not present research in full context. Further, citing a paper but altering or distorting its findings, however subtly, is also unethical, as is selectively citing from other papers. While it would not be unethical to discuss an hypothesis from another paper, leaving the part out that it's an "hypothesis" *is* unethical. You reader will come away thinking you're talking about fact or real findings or truth. And again, papers with fraudulent or manipulated or otherwise unethical data get cited and recited, and a game of "publication telephone" ensues. An idea that started out as an hypothesis in a paper suddenly starts turning up as fact. There are no data to support it—just "everyone knows" (e.g., "everyone knows that vaccines cause autism"). This amplified ripple effect is damaging to the integrity of scientific communication and ultimately can have a deleterious effect on experimental design and future research. Do a little research on the popular "spinach myth," and read about the constellation of events that has resulted in a large chunk of the populace believing that spinach has one of the highest iron contents of all the vegetables, which was why Popeye ate it. (It does not, and that's not why Popeye ate it.) A large chunk of the populace *still* believes that spinach is super-high in iron as compared with other vegetables, and (hint!) it came from some wonky upstream science, a claimed misplaced decimal point, and amplified citation error.

Another surprising and somewhat disappointing bias seen in scientific writing—at least from the standpoint of health studies or trials published in journals—is peer review "positive outcome bias." In a 2010 study, editors from the *Journal of Bone and Joint Surgery* and *Clinical Orthopaedics and Related Research* sent virtually identical papers out to reviewers (Emerson G, et al. Testing for the presence of positive-outcome bias in peer review: A randomized controlled trial. *Arch Intern Med* 2010;170:1934–1939). The papers were fictitious: a "randomized controlled trial" evaluating the use of two different presurgical prophylactic antibiotics. The only difference in them was outcome: one paper had a positive outcome (one drug was superior to the other), and the other had an outcome showing no difference between interventions. The authors also included the same deliberate errors in both papers. The positive outcome paper was recommended for publication by 97% of the reviewers, while only 80% of the reviewers recommended publishing the "no difference paper." Further, the deliberate errors in the papers were caught more than twice as often in the "no difference" paper as compared with the positive outcome paper. This is publication bias—and the author has nothing to do with it! Remember, you are ethically bound to publish negative or equivocal results. You may have crafted the most exquisite, completely transparent, virtuous paper of your entire young life and it can *still* succumb to bias even after you're done with it.

Spin

A closely related cousin to publication bias is *spin*, that art of presenting, framing, or distorting information in your document in a positive way, when your results themselves are not positive or "technically positive." This is not uncommon in clinical trial reporting. A researcher (particularly one working with a pharmaceutical company) is not going to be pleased to have spent years and years and millions and millions of their grantor's R&D dollars and end up with results comprised of lukewarm or statistically nonsignificant safety and efficacy data. But you need to publish the findings to be ethical, right? Without distortion, right? Well, creative wordsmithing, while not *technically* illegal, is shady, even without outright prevarication, but it's *a* way to publish a paper without coming right out and saying "this *totally* didn't work." One can put a spin on the fact that the *p*-value was .06: it means it was *close* to working. It *almost* worked (more on the *p*-value in the chapter on statistics). Also, the side effect profile was *great*. So, you can spin your not-so-great results to mean that you've got a totally safe drug with a *p*-value that's trending toward significance—this, *this* my friends, is exciting news. Also, even if 60% of the study subjects in the interventional group didn't get better, 40% of them *did*, at least a little, or at least they didn't get any worse. Mind you, trends are useful things. But hinting around that a "trend" is statistically significant when it technically is not is weaselly.

Papers that discuss or summarize data in such a way as to play up any outcome that can be interpreted in a remotely positive way and downplay any negative outcome(s) are exercising spin, and it is not impressive. If the results were statistically nonsignificant, those bothersome data can be reported in the Results section and just kind of left there without elaboration, right? Maybe only the positive results, or the *most* positive results, can be displayed graphically, but leave the nonsignificant results as a sentence or two (or even worse, available only in a journal supplement). Fifty percent of the study subjects felt better? That gets discussed at greater length than those results showing that there was no statistically significant difference in reduction of viral load between placebo and experimental intervention. "Feeling better" is great if you're talking about an antidepressant, but if someone has a viral blood infection, you want to give them a drug that will decrease viral load, even if it makes them *feel* worse.

A great way to amplify the skew you've introduced is also to concentrate your spin in areas of the papers that get read most frequently: the abstract and the discussion or summary at the end of the paper. In a 2010 paper in *JAMA* looking at spin in clinical trial papers, the authors found that in their study sample, spin was detected in 18% of article titles, 35.5% of abstracts, 29.2% of the Results sections, 43.1% of the Discussion sections, and 36% of the Conclusions sections, with 40% of their sample having spin in two or more sections. The abstracts of 23.6% of these papers focused *only* on treatment effectiveness (Boutron I, Dutton S, Ravaud P, Altman DG. Reporting and interpretation of randomized controlled trials with statistically nonsignificant results for primary outcomes. *JAMA* 2010 May 26;303(20):2058–2064). The kicker is that this was a sample of clinical trial papers that all had *statistically nonsignificant results*.

This does not mean you cannot discuss the positive aspects of a study with "negative results." That information *is* useful, and it *is* interesting, but it can still be discussed ethically, in context. There are also ways in statistics to do specific subanalyses (which have to be determined up front in your study design) that highlight some of the positive aspects of your data, and it's fine to do these, but not if it is the only information you are presenting front and center as the primary outcome.

Understanding spin may make you a better critical reader. A paper or other document read by another scientist, researcher, or nonscientist may be given a fast perusal; without careful assessment of the facts, the reader may come away with the "spin version." Scientific writing should not have an agenda—or rather, *education* should be the only agenda for scientific writing. Be aware of spin when you read to summarize and evaluate other papers, and avoid it in your own writing.

Conflict of Interest

Disclosure is an important part of writing—not just scientific writing. When you hear of a study showing that smoking might not be as bad for you as previously believed, and the study is sponsored by the tobacco industry, your suspicions are automatically aroused. The egg industry, anxious to defend its pure, natural source of protein, goes after the cholesterol reports. The cheese industry is disturbed at those thoughtless individuals who come out saying "Hey, wait; cheese has a lot of fat in it and it's really bad for you." The soft drink industries come out and say "Now wait a minute—limiting the size of sweetened soft drink sizes to 12 ounces is an egregious infringement on the American pursuit of happiness, and freedom of choice, and besides, it's the other things in American diets that make them obese." Lobbies ramp up their efforts. Reports discrediting the offensive claims are generated, either denying the evidence or diverting attention to something else. Sometimes those reports are generated by the respective wronged industries themselves, and sometimes peer-reviewed papers start creeping into the literature, or independent trials evaluating the true perils of eggs, cheese, tobacco, and soft drinks start appearing. Well, while these industry-generated reports can be viewed askance by critical thinkers, those peer-reviewed papers and independent trials and studies are scientific *research* and have been legitimately published. The truly critical thinker or skeptical scientist (skepticism is a critical virtue for all scientists to have) should still consult other sources before adjusting her opinion. In the end, these industries want and need to work with experts. Someone has to do their studies, their trials—a pharmaceutical company cannot run a clinical trial in-house. Legitimate scientists are solicited for guidance, advice, and studies. This, in fact, makes a great deal of sense—on the condition that any conflicts of interests are disclosed. If any of the authors receive compensation (grant, salary, or any other kinds of support) from an organization that may stand to benefit from the results of the study, this needs to be disclosed. If you read any paper or other scientific document that has a pro-product bent, or seems to have a clear agenda, go back and consider the author(s) and any potential conflict(s) of interest. If potential conflict of interest exists but is not disclosed up front, it may still be brought to light later by skeptical scientists or journalists, with damaging effects on reputations and the industries themselves.

Confidentiality

Apart from the confidentiality that must be provided for any human subject in research, there is also confidentiality in peer review. Peer review is anonymous. When you write a paper and submit it to a journal, the editor sends it to three or more expert reviewers, that is, individuals who know the content very well. This keeps you honest (which is a good thing) and also helps to identify research or publications that don't necessarily yield anything new. The individuals who review the paper almost always know who wrote it; they are almost always provided with authorship information along with the paper. The reviewers need to keep their reviews confidential, that is, not share the contents of the paper or their reviews with other individuals. The more specialized a field, the more likely reviewers and authors are to know each other. Authors submitting papers to some journals are even permitted to *recommend* peer reviewers. Providing peer review for manuscripts when you are asked is one of the professional duties of a researcher.

Duplicate and Redundant Publication

If you have performed research and the time has come to write about it, try your level best to write about *all* of your results in a single paper. Slicing up your results into "least publishable units" so that you have separate papers for each of your experimental outcomes or endpoints—this is done typically to increase the number of publications you have—pushes the limits of ethical conduct. You did one experiment, so write one paper. Quaintly referred to as "salami publications" (because it is one study that is sliced up into bits for separate publications), they are typically recognized as such by your intended target audience and will somewhat dent your reputation. Sometimes publishing multiple papers about a single topic doesn't qualify as a salami approach. Often study endpoints or outcomes are explored further, or there might be some kind of depth or detail to a particular endpoint that is not reasonable to cover as part of a single report, or "side experiments" or observations were made that were not part of your original protocol or experimental design. Experiments also often engender subsequent studies that are more appropriately reported in separate publications. Sometimes advance or early reports are generated as letters in journals, abstracts at meetings or in presentations—that's not salami publishing.

Duplicate publications are equally unethical. These are papers that essentially duplicate the first paper you wrote, just slightly tweaked—perhaps to target a different audience—and then submitted for publication elsewhere. Ta dah! One study, two papers.

Instances of fraud, error, plagiarism, or other deliberate unethical manipulation of content have resulted in heightened vigilance by journals and detailed publication guidelines. The publish-or-perish attitude that prevails in academia may be responsible for the generation of more and more unacceptable manuscripts. We invite you to go visit a highly educational website called Retraction Watch (retractionwatch.org). They have a searchable database of all retracted papers and retraction notices. From January 1, 2019 to January 1, 2020, 380 original scientific papers were retracted (including by one Nobel prize winner who could not replicate her data). From January 1, 2000 to January 1, 2001, only 172 original scientific papers were retracted. The number of retracted papers in a single recent year more than doubled all of the papers published in a single year 20 years ago. Is unethical scientific publishing on the rise, or is just more of it being discovered? Obviously, the Internet has helped things along—finding information that leads to retraction is much easier now than it was previously. Many retractions are made by authors even before (or without) an alarm being raised.

Take-home

Ethical writing is based upon upstream ethical behavior. If studies are conducted unethically, or with unethical components, whether deliberately or ingenuously, the writing is doomed from the start. You cannot write ethically about unethical science. Ethical writing includes transparency and thoroughness, treating subjects, coauthors, and readers with dignity and respect, giving due credit when it is due, and being committed to the dissemination of scientific information, rather than writing simply to further one's career at any cost. Ethical scientific publishing similarly comes with a set of rules about authorship, plagiarism, dissemination, and disclosure. Learning about ethical scientific writing and publishing will help you become a better writer and a more informed critical reader of scientific literature.

Exercises

These can be performed individually in groups. Ideally, each exercise should be performed in a group, with the results exchanged for peer review. Creation of a peer-review rubric (e.g., for content, style, following directions, etc.) may facilitate review and discussion.

1. Design a very small study or experiment in your field.
   a. Use your topic and create a well-designed ethical protocol.
   b. Use your topic and create (deliberately) a protocol with ethical flaws.
   c. Exchange these experiments/studies with your peers and discuss the differences between the flawed and nonflawed designs, or, if not working in groups, summarize the differences between the studies— why are the ethical flaws ethical flaws?
2. Plagiarize something.
   a. Find a paper (or other published document) in your field and pla-giarize from it (one or two paragraphs) in some form (direct pla-giarism, incomplete paraphrasing, etc.). Each person in your group should plagiarize the same paragraph(s). Then take the same paragraph(s) and recommunicate the same information without plagiarizing.
   b. Exchange the plagiarized and nonplagiarized materials and discuss *how* they are plagiarized. Discuss the nonplagiarized paragraphs. Are the nonplagiarized paragraphs truly nonplagiarized? Are there any elements of plagiarism in them at all? If so, where and what kind?
   c. If you are not working in groups, perform the same exercise (i.e., write a plagiarized paragraph and a nonplagiarized paragraph from the same source) and summarize the differences between the two.
3. Criticize the experts.
   a. Science is rife with publishing misbehavior. Identify *one* such instance (you may have to look outside your field—many fields have examples of *historic* badness), acquire the document(s) in-volved (if accessible), create summary points, and discuss with your peers.
   b. If this is a solo project, identify the ethical publication transgression and write a summary of:
      i. The transgression and when it occurred
      ii. The journal (or other medium) involved
      iii. The individual(s)/companies/organization/journal (if the journal was the transgressor) involved
      iv. The fallout or backlash that occurred
      v. The consequences, punishment, or remediation requirements rendered upon the perpetrator(s). ("Damaged reputation" is a given.)

Tip: The ORI website has a summary of disciplinary actions taken against research fraud.

Another tip: if you type in the name of your field of study and "ethics" in Google, you should be able to get a topic in a fraction of a second.

And another tip: The searchable database in retractionwatch.org not only tells you *what* got retracted, it tells you *why* (hours of fun!)

4. You be the judge.

The scenario: like our protagonist at the start of the chapter, you are an undergraduate with the opportunity to work in a research laboratory—of a Big Name, no less. You observe the PI working (hands on) with human subjects without IRB approval—without even submitting a proposal to the IRB. Her reasons are rational: it is not grant-supported, the research is covered by IRB review and exemption at a different university where she is collaborating with colleagues, and the intervention she is conducting is not invasive; she's only interviewing patients who have been through behavioral counseling to see if their alcoholic tendencies have improved post-counseling. Is it okay? Are you mildly uncomfortable about it? seriously uncomfortable about it? What would you do about it? Would you report it? talk to someone about it? Who? Make a major whistleblowing effort? Quit working in the lab? What would you do and why?

   a. As an individual exercise, write an essay addressing your unease and how you would resolve it.

   b. As a group exercise, craft your own answers and not only discuss them but tally the approaches used by your discussion group.

5. Can you pass the test?

   a. There are a number of "plagiarism quizzes" online (some with tutorials you can take prior to the quiz). Choose one and take the test until you can pass it with a score of 100%. Two caveats: First, many academicians, libraries, and writing centers feel strongly about which tutorials to use and *not* use; ensure that you are using a tutorial endorsed by your professor (it might be on your own library site). Second: some alleged "plagiarism" sites are *not educational*, and in fact offer paper-writing services (significantly unethical, meriting not just censure but failure or possibly expulsion).

b. As an associated exercise, choose the question from the plagiarism quiz with which you struggled the most. Write a brief essay about the type of plagiarism covered in that question, and why you think you struggled with it. If you are working in groups, instead of an essay discuss the questions in the plagiarism test that caused you the most grief—keep in mind that there are plenty of people who disagree with some of the "rules" or examples in plagiarism quizzes—including your professors.

4

Critical Thinking

LIKELY   NOT LIKELY

© 2020 Max Beck

"Consumer Science" is one of the most popular courses in the university, with enrollment limited to 20, and you snagged a spot! Now you're up to the best part of the course, a project in which each of you is to find a "miraculous" product on the market, evaluate it, and present your results to the rest of the class. "Hm," you wonder, "what product should I . . . " and then up pops an ad on Facebook the second you log on: "LOCAL MOM MAKES PLASTIC SURGEONS FURIOUS WITH FIVE-DAY SOLUTION THAT'S AS GOOD AS A FACELIFT!" You immediately think: "Why, that's pretty miraculous—there's my project!" You go through a series of 14 consecutive web pages with newspaper articles and videos and news reports on them touting the miraculousness of this truly miraculous and best-kept secret cream called *Miraculorejuvaderm*, enter into a costly and impossible to cancel opt-out agreement, and order one month's worth of the stuff. The articles assure you it will only take five days to work and anyway the first month is free, after which your credit card will be automatically billed $75.95 monthly, which you can, of course, cancel any time. Bated breath is relieved when you receive your 0.75 ounce trial tube in the mail! Then, since you don't have any wrinkles yourself, you ask your mom to be your test subject. After the initial indignant refusal, she finally concedes when you tell her you'll get an F if you don't finish the project. "Besides," you tell her, "it's like a free face lift. It takes 20 years off your face."

"Andrea," says your mom, "seriously? I mean you *seriously* bought that?"

"Well," you say, "it was on the Internet. And there were consumer news articles and television reports about it." Your mother sighs and says maybe Consumer Science will be the most important course you ever take in college. You instruct your mother to put *Miraculorejuvaderm* on the left half of her face and to leave the right side of her face untouched, just like that amazing picture of the woman in the ad, so that you can document the difference.

What is an "opt-out" sales scam? While it can sometimes be legitimate (e.g., for magazine renewals) it is a common practice in online sales of "snake oil" products where you are offered a "free trial" (or a product at a substantially reduced price, for which you have to provide your credit card number) and are promised a full refund if you are not "100% satisfied." In reality, insurmountable and barely legal obstacles are put in the path of your inevitable cancellation, and in the meantime your credit card continues to be charged. The only thing that eventually works is to cancel the credit card.

Six weeks later, you and each of your peers give eloquent presentations in which you conclude that *Miraculorejuvaderm* doesn't work, gingko doesn't work, Miracle Brass Restorer doesn't work, the HCG diet doesn't work, Miracle Hair Restorer doesn't work, Miracle Spot Remover doesn't work, the Pyramid Energy Sleep Solution doesn't work, Muscle Mass Expander doesn't work, you can't actually learn a new language in 7 days, Lik-Nu spray-on car paint (as seen on TV!) doesn't work, the take-four-inches-off bathing suit diet doesn't work, the Fuller Eyelashes in Fourteen Days formula doesn't work, the Miracle Coconut Oil diet doesn't work, the Seven Days to Whiter Teeth system doesn't work, homeopathy doesn't work, you can't increase your IQ in two weeks, you can't learn how to draw in one month no matter how well you can draw "Sparky," raspberry ketone doesn't work (no matter *what* Dr. Oz says) and x-ray specs, disappointingly, also do not work. Collectively, the students are out a total of $1,644.58, and half of their parents aren't speaking to them. Your professor has an ear-to-ear grin and is reflecting on the fact that teaching is the most fun job in the world. (It is.)

What Is "Critical Thinking," Anyway? And How Is It Related to "Scientific Literacy"?

You get your syllabus for a new course, and there it is, the *de rigueur* nod to critical thinking: "students will develop critical thinking skills." You barely register the phrase as you move onto the next thing in the syllabus. Did you ever stop and actually wonder what "critical thinking" entails? Of *course* you know what it is. You think *critically*. That means, "um, here's a problem, and I'm going to . . . criticize it. I mean, I'm thinking about it." Can you define critical thinking? What do we mean by "critical thinking skills" in chemistry, or English, or philosophy, or psychology or physics or geology or medicine or . . . anything?

I found that I was fitted for nothing so well as for the study of Truth; as having a mind nimble and versatile enough to catch the resemblances of things (which is the chief point), and at the same time steady enough to fix and distinguish their subtler differences; as being gifted by nature with desire to seek, patience to doubt, fondness to meditate, slowness to assert, readiness to consider, carefulness to dispose and set in order; and as being a man that neither affects what is new nor admires what is old, and that hates every kind of imposture.

—Francis Bacon (1561–1626)

Critical thinking is thinking about your thinking while you're thinking in order to make your thinking better.

—Richard W. Paul (1948–)

According to the National Council for Excellence in Critical Thinking (as set forth by Michael Scriven and Richard Paul in 1987), critical thinking is defined as (emphasis ours):

the intellectually disciplined process of actively and skillfully conceptualizing, applying, analyzing, synthesizing, and/or evaluating information gathered from, or generated by, observation, experience, reflection, reasoning, or communication, as a guide to belief and action. In its exemplary form, it is based on universal intellectual values that transcend subject matter divisions: *clarity, accuracy, precision, consistency, relevance, sound evidence, good reasons, depth, breadth, and fairness.*

There are variations and expansions on this theme which you can explore on criticalthinking.org; check it out. Look at that definition again, carefully. Think about how you would apply this to your research, to the questions you ask (and how you answer them), and to the way you communicate your information. Are you being consistent? Are your sources reliable? Is your paper (or other writing submission) based on sound evidence? Critical thinking helps guide our thinking, research, *and writing* processes. It is one of the reasons we don't use Wikipedia or Big Ted's Web Page O' Hypertension as references in our research. It is why we don't jump to conclusions, why we fight the knee-jerk reactions we know that we ourselves are guilty of, why we exercise healthy skepticism, and why bias is so irritating to us. We read labels and editorials; we consider sources and context before drawing conclusions. Critical thinking makes us picky, thoughtful, thorough, fair-balanced, and nonjudgmental. It is why we do not think that Twinkies* are healthy just because they have no trans fats in them, or that candy is good for us because it has no cholesterol. If something you read or hear made you pause and ask additional questions, or if you're used to not taking things at face value, you're already a critical thinker.

Rubrics are a useful way to assess your (or others') critical thinking chops. Free downloadable rubrics on critical thinking are available from various sources. We recommend the American Association of Colleges and Universities (AACU) VALUE Critical Thinking Rubric or the Facione and Facione's Holistic Critical Thinking Scoring Rubric.

Not only do you need to be a critical thinker to write well (or do just about anything else), you need to be scientifically literate when you are communicating scientific information. *Scientific literacy* is the younger, neglected sibling of critical thinking. Tragically, scientific literacy is not faring very well in this country, with vast segments of the population believing that "intelligent design" is science, climate change (global warming) is a myth, and genetically modified organisms (GMOs) will kill you, your entire family, and your pets, among other things. What is scientific literacy? According to the National Academy of Sciences (NAS) (emphases ours):

Scientific literacy means that a person can *ask, find, or determine answers* to questions derived from curiosity about everyday experiences. It means that a person has the ability to *describe, explain, and predict natural phenomena.* Scientific literacy entails being able to read with understanding articles about science in the popular press and to engage in social conversation about the *validity of the conclusions.* Scientific literacy implies that a person can identify scientific issues underlying national and local decisions and *express positions that are scientifically and technologically informed.* A literate citizen should be able to *evaluate the quality of scientific information on the basis of its source* and the methods used to generate it. Scientific literacy also implies the capacity to pose and *evaluate arguments based on evidence* and to *apply conclusions* from such arguments appropriately.

Notice how NAS thoughtfully applies characterization of this ability to "citizens." "*Literate citizens*." At the risk of falling tragically upon the noncreative use of cliché, we are going to go ahead and say that you don't have to be a rocket scientist to be scientifically literate. If you go back and reread the definition of critical thinking, you can see that scientific literacy is an applied form of it. You cannot be scientifically literate if you can't think critically, but you can be a critical thinker without necessarily being scientifically literate. You would have, of course, the *capacity* to become scientifically literate, to hone those critical thinking skills and apply them to scientific concepts.

Time for a quiz. Think of a topic—any topic—that can remotely be thought about in a scientific way. (Pretty much everything can: Tower of Pisa; disposable diapers; Mount St. Helens; those weird spiders in your garden; your takeout Styrofoam containers; air; anything medical, physical, chemical, biological, psychological—animal, mineral, vegetable, or abstract). Picked something? Good. Now close your eyes—no cheating!—and think about it. Ask yourself a question about it ("Are disposable diapers really worse for the environment than cloth diapers?" "Does plastic really leach into the things you heat in the microwave?" "Is salt the biggest contribution to the development of hypertension?" etc.). Can you apply the tenets of critical thinking as defined earlier? Can you apply *some* of those tenets? Which ones? Write down your question and think about how you would go about researching and answering it. Are you approaching your question in a scientifically literate way? If you were doing research on the topic, would you have an idea about where to go for information? How would you begin to start thinking about it? Try it right now.

Now let's move into the Advanced Challenge. Pick a scientific topic about which you feel passionately annoyed (intelligent design, climate change, vitamins, meditation, etc.). What do you feel strongly about? For example, "If my sister doesn't stop trying to stuff coconut oil down my throat, I'm going to lose it"; "I can't listen to my neighbor's rants about how vaccines cause autism anymore"; "I'm pretty sure GMOs are safe to eat and I'm tired of seeing people getting hysterical over it"; "I am astonished at how much Ritalin is pushed on kids today." You *must* have a hot button (or several of them). Now *force yourself to look at the other side of the argument*, in a fair and balanced way. Can you do it? You may not like it, but critical thinking means being able to consider uncomfortable conclusions that may go against deeply held beliefs. Make a "pro" and "con" list. If you also want to be scientifically literate, you can't just have a "position"; you must have the means (i.e., sources, knowledge) to get to that position and a dispassionate understanding of any controversy around the position.

---

10 Tips for Critical Thinking

1. Don't believe everything you see, hear, or read.
2. Consider the *source* of the information carefully—including the *source* of the source.
3. Consider the *context* in which the information is being presented. Is information being provided out of context? Are you sure? How do you know? Are you suspicious?
4. Consider the *timing* of the information. Why did it come to light now? What else is happening that prompted it?
5. Consider and *evaluate* the *evidence* for a claim. Don't just accept or reject it.
6. What is your immediate reaction to what you're hearing? Look at the other side of the argument, even if it hurts.
7. Is the reasoning for the claim sound?
8. Do you have a knee-jerk reaction or a bias? (Of course you do.) See if you can figure out why, and try to deconstruct it logically.
9. Be honest and use integrity—in thought, word, and deed—no matter how passionately you feel about something or how badly you want to persuade others.
10. Listen—*really listen*—to others, with respect and an open mind.

Can you have an open mind and be skeptical at the same time? Sure. Have the open mind first, and then question everything you hear.

Basic Critical Thinking Skills: Do You Have Them?

You probably already have a lot of these skills. You couldn't have made it to or through college without developing some significant critical thinking abilities. Now you can focus on disciplining your thoughts and approach to communicating effectively in science.

But is critical thinking innate? Do you possess all of the critical thinking skills? Of course you don't. Do you have knee-jerk reactions? Of course you do. Are you biased? Of course you are—and so is everyone else. An important component of critical thinking, however, is *recognizing* that you have knee-jerk hot buttons, that you *are* biased, and that sometimes you *will* resort to dastardly forms of persuasion to achieve your ends. And when you're *challenged* on one of your knee-jerk topics, when the climate change denier starts citing data from the National Oceanic and Atmospheric Administration (NOAA—the ultimate authority) that appear to contradict climate change, instead of snorting derisively, you struggle to engage in a discussion whereby you can address the "evidence" provided and point out why it does not support your interlocutor's argument (in the case of the climate change denier, it is almost always because data are being taken out of context: cherry picked). The *really good* critical thinker will rejoice in the opportunity to exercise the powers of lucid argument.

Michael Shermer, a renowned skeptic, was asked at the end of a provocative 1997 interview with the *Detroit Free Press:* "Why should we believe anything you say?" He answered: "You shouldn't."

How Are Critical Thinking Skills Applied in Scientific Writing?

Remember that communicating science is half of "science." If one does research and then keeps it to oneself, we have a case of the proverbial tree falling in the forest—it doesn't matter if it makes a sound or not. Likewise, your communication about science is only as good as the science that precedes it. If the science was wrought recklessly, with no critical thinking, poor study design, lack of ethics, or an absent or flawed hypothesis or objective, or is lacking in any of the other components necessary for good science, it doesn't matter how well you write. If the science is flawed, the communication of it is *also* flawed, even before you start.

Your writing should reflect the thought processes you went through in your research, from hypothesis and observations to interpretation of those observations (or results, if the research was experimental). This is why working with collaborators and mentors is so important. You'll employ reasoning, assessment, evaluation, research, and observational skills in both planning and execution in *any* kind of scientific research.

Remember that last bit in the definition of critical thinking? *Clarity, accuracy, precision, consistency, relevance, sound evidence, good reasons, depth, breadth, and fairness.* Even in nonscientific courses your professors are going to require some or all of these in *any* kind of paper you write. Let's try to define these terms.

Clarity: Present your ideas, arguments, and conclusions in the clearest, simplest, most logical way possible. Do not use jargon or excessive verbiage. Do not use ambiguous language.

Accuracy: Check your data twice. And then check it again. Are your stats accurate? Are they reliable? Are you presenting your results *in context* (i.e., not cherry-picking?) If something is suspect, check it. Do not assume your software messed up; never lose sight of the GIGO (garbage in, garbage out) principle. This is also why peer review is so important—the kind of peer review you do internally *before* you submit a paper for publication. Revise, revise, revise.

Precision: Ensure that you are concise and thorough when writing, and provide the amount of detail necessary for complete understanding (and replication, if that's an option). Were you observing avifauna in Costa Rica, or observing rainforest avifauna in the canopy trees at La Selva Biological Station in San Pedro, Costa Rica, at an altitude of 75 meters?

> The difference between the *almost right* word and the *right* word is really a large matter. 'Tis the difference between the lightning bug and the lightning.
> — Mark Twain, *The Wit and Wisdom of Mark Twain*

Consistency: Are you referring to the same things in the latter half of your paper as in the first half? When rereading your paper, do you find that you switch back and forth between tenses, first and second person, pronouns, measurements (e.g., 500 mL vs. 0.5 L), wording in definitions, nomenclature and abbreviations? Are you using the same heading style, typeface and font style and size, margins, text justification, and figure labeling? Give your paper to others to review, and revise, revise, revise.

Relevance: Hopefully your research was relevant to begin with (i.e., fills a gap in knowledge), but when you write about it, are you including relevant information, that is, pertinent to the point you are making? There are two possible sins here: one of inclusion (including irrelevant information) and one of exclusion (leaving out relevant information).

Sound evidence: Obviously, this will also be part of your research design and findings. Whether or not you can disprove your null hypothesis, observe the things you were anticipating, observe nothing, get results or not, you need to be able to say why: "Here's my evidence." You need evidence in your research (or what's the point?), and you need to *provide all of that evidence clearly in your writing*. If you didn't see any birds while you were at the biological station but you heard them calling, you can't say "I assume they were there, since I heard them calling, and it sounded like they were scarlet macaws."

Good reasoning: There is a reason for the hypothesis you've developed or the observations you're making—"filling a gap in knowledge" or "learning how to perform bomb calorimetry" or "quantifying scarlet macaws in a one-hectare sampling area." There is a purpose for everything you're doing—what is that purpose? *Why* are you performing bomb calorimetry on leaves collected from a 1-meter square plot in the rain forest? Sound reasoning is, obviously, a critical component of both inductive and deductive reasoning; it not only needs to be included in your writing but it also has to be well expressed.

Depth: How much detail do you need? How much background? How much of an introduction? Depending upon what you're writing, you're going to need more or less detail. Short things, like meeting posters, abstracts, and lab write-ups need introductions or some kind of background, but you can't go into the same kind of depth, obviously, that you would in, say, a paper for a journal or a senior thesis. It is important to determine the level of depth you need: not too much and not too little. This is another area where peer review is critical. If you're not sure and you're writing a first draft, overdo it. Then pare it down in subsequent edits.

Breadth: How narrow should your focus be? A review paper may contain far more—or far less—breadth than, say, a clinical trial paper. Are you writing about the ecology of Costa Rica? the ecology of La Selva? the ecology of the avifauna community in La Selva? the ecology of *Ara macao* in La Selva? How long is your paper or lab report? 20 pages? 4 pages? If the latter, you need to save your real estate for the details that really matter for your purposes.

Probably the most famous "shortest paper" in journal history was a 1974 paper in the *Journal of Applied Behavior Analysis* by Dennis Upper, a clinical psychologist. The title of the paper was "The Unsuccessful Self-Treatment of a Case of Writer's Block." There was no text at all (*J Appl Behav Anal*. 1974;7(3):497; doi: 10.1901/jaba.1974.7-497a). Why wax eloquent when you can get right to the point?

Fairness: There are two sides to every story—often more. Often there are multiple hypotheses or observations that have been crafted to explain the same thing(s), and often conflicting results arise from similar research. All science, whether experimental or observational, has limitations. It's why experiments are designed so that as many variables as possible are controlled so that change in only one variable can be observed. But if the exact same experiment or observation is conducted in two different labs in exactly the same way, the fact that they're in different labs *itself* is a variable. Report and account for other rational hypotheses and research results. Discuss or at least mention the limitations to the study you have conducted. You cannot just ignore these.

How Are Scientific Literacy Skills Applied in Scientific Writing?

Likewise, the definition of scientific literacy can be considered in the context of writing:

> . . . *ask, find, or determine answers*
>
> . . . *describe, explain, and predict natural phenomena*
>
> . . . *read with understanding articles about science in the popular press and to engage in social conversation about the **validity of the conclusions***
>
> . . . *express positions that are scientifically and technologically informed*
>
> . . . *evaluate the quality of scientific information on the basis of its source and the methods used to generate it*
>
> . . . *evaluate arguments based on evidence and apply conclusions from such arguments*

Look at these components of the definition again, but from the standpoint of *writing*—not just thinking or designing experiments or constructing observations.

Ask, find, or determine answers: While these things are arguably important precursors to writing (not to mention essentially the entire point of science), you need to communicate to your readers how you asked your question(s), and where you found or determined your answer(s). Did you do a literature search for something? How did you conduct it? What lit database(s) did you use, what date range; did you use English language journals only?

Describe, explain, predict: You evaluate the quality of scientific information on the basis of its source when you are designing and performing an experiment, but you do it in a different way when you're writing. What papers are you using? what books? what websites? Are your sources credible? Are the sources used by your *sources* credible?

Express positions that are scientifically and technologically informed: These are not difficult skills to conceptualize from the standpoint of writing about science. By its nature, what you're writing about is (or should be) already scientifically and technologically informed—that's why you're writing about it. That being said, "informed" means that you have done your background research thoroughly—you know that you cannot write about your own science in a vacuum. Ensure that you are using references common to your discipline. Is it *just* journal articles or *mostly* journal articles? Or are books commonly used, or government websites such as NIH or NOAA? What about organization websites that may have the most recent guidelines for your field? Is there specific terminology in your field?

> Today's vocabulary quiz! Do you know what "ultracrepidarianism" means? You should. It's a very useful word!

Evaluate the quality of scientific information on the basis of its source: We warn you at several points throughout this book about illegitimate sources. These include helpful "educational" websites compiled by concerned citizens (or by teens in a remote country paid to produce "clicks per page"), predatory open-access journals that will publish anything (for a steep fee) without even bothering to read it, never mind have it peer reviewed; the newer advent of predatory scientific congresses where "esteemed scientists such as yourself" are invited to present basically anything; vanity publishing and online self-publishing; misinformation spread by more naïve journalists or simply your friends on social media. Much of the information at large on the Internet is wrong, biased, incomplete, or misinformed. Much of it may contain under-the-radar amplified misinformation (i.e., a legitimate publication that cites poor studies). Where are you getting your information? How do you know it is a legitimate source? Some things are (or should be) obvious: peer-reviewed papers in respected journals, standard textbooks in the field, government websites (usually), data repositories, and experts in the field. It doesn't hurt to check the reputation of the "expert," the reputation of the journal if you've never heard of it (especially if it's open access), the reputation of the website, and the reputation of the writer. Websites are probably the greatest source of peril, since they can come from anywhere. Suppose you're looking at a web page and are trying to decide if it's legit or not. Many people online are getting their "health facts" from ludicrous anti-science websites such as Mercola.com, FoodBabe.com, and NaturalNews.com. But what if you're reading something online that sounds lucidly written? No hyperbole, nothing dramatic or ominous, written by someone who appears to be "a doctor" at "a university." Start with due diligence and examine the website. You can Google the web page ("is HamstersGiveYouCancer.com a legitimate site?"), for starters. Critics and skeptics are quick to point out bogus websites, as is Snopes.com, though it's not always accurate. Your librarians are also a wonderful source for advice on this. We also discuss source legitimacy and offer some tips in Chapter 7.

What is *sponsored content*? Sponsored content is material provided or underwritten by a company and is paid content akin to product placement. Written to *inform* and (usually) not to *persuade* (that would be native advertising or advertorials), it provides readers with information germane to what's already on the page, thereby positioning the sponsor as an expert of sorts. Sponsored content is found on perfectly legitimate websites (and in print), typically magazines, newsletters, or organizations. It is considered to be a powerful marketing tool and is gaining in popularity (Facebook started allowing it in 2016). Does this mean that the content is not a legitimate source of information? Not necessarily. But good critical thinkers *consider the source of all of their information*. Nondisclosed sponsored content or even native advertising (advertorials) can appear as regular editorial content—by design. So do your homework. Never use a single source. And, while sponsored content might provide you with some clues for other research you can do, absolutely *never* cite it as a reference in a scientific paper.

Inductive and Deductive Reasoning

Deductive and inductive reasoning—or deduction and induction—are opposite forms of logic that allow you to come to a conclusion—a conclusion you can then test, if you're creating an experiment.

Deductive reasoning is a stepwise "if-then" process. If scarlet macaws eat fruits, nuts, and seeds, then it would be logical to assume that the scarlet macaws in Costa Rica eat the fruits, nuts, and seeds found in Costa Rica, and if you collected those bird droppings, you would expect to see evidence of that diet. If *Miraculorejuvaderm* is truly a miracle facelift in a tube, then, based on the television reports and newspaper articles and case studies on the *Miraculorejuvaderm* website along with the astonishing before and after photos, it stands to reason that if your mother wears the miracle cream on one-half of her face for 7 days, she'll look (at least half of her) like the amazing photo in the news report. Hence your experiment: if applied for 7 days, wrinkles will disappear. It can be *tested*.

"The famous "she's a witch" scene from *Monty Python and the Holy Grail* (© 1975, Python (Monty) Pictures).

Probably one of the most famous examples of deductive reasoning in popular culture is the "She's a Witch" scene in *Monty Python and the Holy Grail*, in which Sir Bedevere guides an assembled angry mob through a series of logical steps to see if they can determine if, in fact, the young woman they want to burn as a witch is indeed a witch. Do witches burn? Yes. What else burns? Wood. Does wood float? Yes. What else floats? A duck. So therefore . . . if she weighs the same as a duck, she is made of wood and is therefore a witch and can be burned. Sir Bedevere's flawless logic earned him a spot on King Arthur's Round Table. If you can pull off flawless logic like this, surely you will get that Sigma Xi scholarship for the lab equipment you need to analyze your bird droppings. Of course, this is also a fabulous example of a *false deduction* (logical fallacies, coming up!).

Inductive reasoning is more of a top-down process. Based upon your experience and observations, you craft an idea and then work backward to find the logical steps or principles that support that idea or concept. Induction is a *diagnostic* process (correspondingly, diagnosis in medicine is an inductive process). "If this is doing this, based on what I've seen, heard, and experienced, this must be the reason." If every witch you've seen burned readily, you conclude that witches are flammable. But are they? Why? How do you know? If you find only the seeds of a specific fruit in the droppings of a sample of 12 scarlet macaws, you conclude that's what they eat. Do they? All the photos and news articles and websites that you've read testify to the astonishing efficacy of *Miraculorejuvaderm*, so it clearly works, right? It works so well that it has even "angered doctors."

In Case You Didn't Know Department:

People in good moods are better at inductive reasoning and creative problem-solving.

—Peter Salovey, President, Yale University

The scientific process dictates that you prove that the reason you have identified is, in fact, the accurate explanation. Reporting that patients get better because you give them a pill is insufficient. Reporting that scarlet macaws eat one particular fruit is insufficient. Shrimp die when you increase salinity to 40 parts per million under experimental conditions. Indicator dyes turn your eluate red at a pH of 3.9. A hypothesis or question—communicated—has to exist before you explain a result.

Reflection, Assessment, Judgment, and Reassessment

The savvy connoisseur of the cinematographic classic and true critical thinker, however, may be questioning the validity of Sir Bedevere's deductions as he proceeds through his logical thought process. And in questioning those deductions, the conclusion at which he arrives is also questioned (although in his defense, in the movie he does go on to demonstrate that the accused young woman does, in fact, weigh the same as a duck).

It's important to remember that you can use logic (and statistics) to prove anything you want. Proper reasoning, however, does *not* start with "this is the conclusion I want"; it starts with "this is how and why I think it might work." An experiment or study specifically and carefully designed or conducted in such fashion as to prove something you want proved is flawed thinking, and flawed science. If that's your raw material, it's hard to write critically and lucidly about it without loading your oeuvre with enough caveats so as to render it useless. You look at 200 macaws, and only 4 of them have seeds in their droppings that do not occur in the other 196 samples of droppings. You can't assume they're anomalies and eliminate them from your results as outliers. If you give a drug to 200 patients and "only" 2 of them die, you can't assume that in reality it was something else that killed them and they're outliers that should not be reported. If *Miraculorejuvaderm* does not, in fact, provide your mother with a miracle facelift on one side of her face, you cannot assume that the tube you got happened to be a dud.

When crafting an hypothesis for an experiment or designing an observational study, your reasoning—deductive or inductive or both—goes through a series of organized steps: reflection, assessment, judgment, and then reassessment. No hypothesis is crafted before an observation is made, usually of a repeated event (every time you see a scarlet macaw, it seems to be eating possumwood fruits; every time you see a witch burnt at the stake, the witch burns). You start thinking about what you have observed. Eventually, you'll slide into "I wonder if . . ." and then you find yourself with an hypothesis. "I'm going to see if all they eat are possumwood fruits." "I'm going to see if it's only witches that burn." "I'm going to see if those snakes live only on that mountain." "I'm going to see if *other* species of shrimp also die at a salinity of 40 parts per million." "Did the drug kill those patients or did something else kill those patients?" You start off with "I wonder if," not "I know this is true and I'm going to prove it." This process is included in your writing—in more or less detail. A written piece in science with no rationale is untethered, doomed to wander all over the prosaic landscape without purpose or direction.

> I'd done research on daffodils
> But could not write it down
> Alas my hapless rationale had
> Wandered lonely as a cloud
>
> —Not Wordsworth
> —Not 1807

Then you decide what you're going to do to see how you're going to answer your question, and you set off to find out if you're right. You assess your idea, your hypothesis; you make your observations—whether they're in the lab, the field, or your kitchen. And you communicate this clearly in what you're writing.

Then you get some results: "Yeah, those birds *do* only eat possumwood, but only during the season before the fruits ripen," "*any* person burned at the stake seems to be flammable, including putative saints," "wow, there are some species of shrimp that actually live in hypersaline aquatic environments," "those snakes are everywhere, *including in my bathroom at the field station*," "the drug *did* kill those patients, but only the ones with very low—but within normal limits—white blood cell count." When you write about results, you are detailed and inclusive: this was the whole point of the study, right?

Then you think about your results (and remember, not getting the outcome you expected to get—a negative result—is still an important result and still needs to be reported). Is what you observed really addressing the question you originally wanted to answer? If you got results you didn't anticipate, why? Were you looking at the wrong variable, or not considering interactive variables, or only looking at something at one point in time when it was something that varied over time? Or did you do something wrong, mess up your experiment, or suffer some other consequence secondary to your scientific inexpertise, or accidentally incurred the wrath of the God of Entropy, who seems to delight in messing with scientists? These musings typically turn up in a "discussion" or "summary" component to whatever you're writing. You got some results—what do they mean?

So you. . . go back and check again. Repeat the experiment, make some more observations, maybe in a different place or different time or using a different population, or you collaborate with someone else who is doing similar research, or you up your N for more statistical power, or you double-check your stats and have someone else double-check them as well. You reassess. Reconsider. Repeat. Collaborate. Review. Replicate. The good scientist always questions and checks results. If you fail (and you will, and probably more than once), you are learning what it truly is to be a scientist. It's good for you.

> I have not failed. I've just found 10,000 ways that won't work.
>
> —Thomas Edison

Part of the review process of what you've done is to recognize where there is vagueness or ambiguity, either in the design of the experiment or study or in its conclusion. This is not only necessary in the science that you are performing yourself but in your assessment of *other* science related to what you're doing (e.g., review of the literature, replication of other experiments or observations, assessment of other results that support or conflict with your own). When reporting on the ecology of the scarlet macaw, indicating that they "eat fruit" is vague; stating that they eat unripened possumwood fruit is specific. Explaining how the tough beak of the macaw is adapted to breaking open and crushing hard fruits and nuts such as unripened possumwood fruits or macadamia nuts is ambiguous—their beaks can handle stuff like that, but do they *eat* those things or not? And the "news" website upon which you read all about *Miraculorejuvaderm* states with authority that 7-7-hydroxydextrolanolin has been scientifically proven to rebuild the subepidermal collagen reticular layer yet fails to mention what 7-7-hydroxydextrolanolin is or if it is, in fact, an ingredient in *Miraculorejuvaderm*, or what the "subepidermal collagen reticular layer" is or what it has to do with wrinkles. (We are making all of these things up.) Vague, ambiguous information, even presented authoritatively, is still vague and ambiguous.

What many scientists don't recognize is that *communicating results is also a form of interpretation*, a contribution to judgment and assessment. The results of an experiment or observational study can be subtly—or overtly—misinterpreted or misrepresented in the "communication half" of the research. This miscommunication can be ingenuous or unintended as a consequence of scientific (or writing) inexpertise, or it can be deliberate: an attempt to prove the point of an individual (or individuals) with an agenda, through the use of skew, spin, misused statistics, absence of context, cherry-picking, or distraction.

Worse, if this happens, the skew/spin/misused stats/distraction stuff can get *published* and then picked up by the naïve (or not necessarily naïve, if the source is authoritative) and reported in *subsequent* publications, amplifying and compounding the original malfeasance and cementing its place in "the literature." Remember that high iron content in spinach? That avalanche started in the 1800s with some innocent spinach research. . . that yielded inaccurate results and was amplified into the nutritional dogma it is largely still considered to be. Sure, the confusion was later deconstructed, but it is tough to un-slide an avalanche.

This is why critical thinking skills are an essential component of not only experimental design but experimental *reporting*. Eloquently reported bad science can sometimes slither through peer review and make it into the literature—you need to be able to recognize it and eschew its use in your own research and the reporting of it. This is also where collaboration and peer review are critical. If you have four people looking at your paper, one of them might know that one of your critical references has been discredited or that the lab itself whence come those data is one of questionable repute.

Logical Fallacies in Scientific Writing

Logical fallacies are reasoning pathways that take a wrong turn and produce glitches in the thought process. Sometimes this is conscious and deliberate—used for nefarious purposes of persuasion or to incite the masses into runaway mob mentality. Sometimes fallacies are accidental, secondary to insufficient philosophic undergirding. You've heard of some of them for sure, even if you can't define them: straw man, ad hominem, burden of proof, begging the question, circular reasoning, and so on. Deliberate use of logical fallacies can be used to confuse and misdirect, distract readers from the relevant, and set them off on a study of the irrelevant. Citing materials out of context is a frequent tactic for inflaming passions and can very effectively be combined with logical fallacies.

A fun (if oversimplified) way to review logical fallacies is on the Thou Shalt Not Commit Logical Fallacies website (www.yourlogicalfallacyis.com). Click on one of the cute icons to learn about some of the common fallacies, to help you identify (and avoid) "dodgy logic."

Given the strict rational constructs and processes that *good* science follows—which include sound reasoning—logical fallacies tend to occur less frequently in science than in, say, politics, "hot button" topics, and conspiracy theories. However, this does not mean they are *absent* in science, nor in the *writing* about science. Bias, spin, misinterpretation (deliberate or otherwise), flawed conclusions, and ethical rationalizations are unfortunately encountered in scientific writing all the time, if not as overtly as in, say, a political campaign. So what are they? What are these mysterious fallacies for which we need to be so vigilant, which we must take care to eschew?

Straw man. A diversionary tactic, a straw man argument is taking something an opponent (or perhaps another scientist's viewpoint) says, reducing it to an indefensible position by exaggerating or downplaying the claims, and then attacking it on those grounds. You would be, in essence, debating your own construct, while attributing it to another. "You're doing square *meter* biomass bomb calorimetry?! Soon it will be square *hectares* of biomass—you will eventually cause deforestation doing this, and the scarlet macaws will become extinct. I don't see how any credibility can be afforded a scientist who contributes to the extinction of an animal." In all likelihood, you have misrepresented the original argument. Don't do that.

© 2020 Erica Perez

Ad hominem. Literally "to the man," or "against the man." In this approach, one goes after the "man," not his idea. This is used to cast doubt on character (often abusively), in an attempt to divert attention away from the topic at hand. The ad hominem argument is irrelevant. "She's a 19-year-old undergraduate from Costa Rica—there's no way her study of macaws could be sufficiently informed and unbiased." Being a 19-year-old undergraduate from Costa Rica has no bearing on what kinds of seeds you find in bird droppings.

Burden of proof. If no one can prove you wrong, then you *must* be right. Right? It is a fallacy to claim that something exists just because its nonexistence cannot be proven. If that were true, we could make all the claims we wanted. You could claim that your scarlet macaws were in fact occult raptors, sneaking out at night and savaging the small mammals of the rain forest with their razor-sharp claws, their unforgiving beaks ripping out the still-warm innards of the hapless forest bunny rabbits who never saw them coming. When you are invariably laughed at by the ornithology community for making this claim, you can fold your arms indignantly and say, "Well, you can't *prove* that they're *not* vicious predators." Go ahead, write that paper.

Begging the question (circular reasoning). If you make an argument in which the conclusion is part of the premise, we have the proverbial snake eating its tail. Since Claim A is true, then Claim B is true. Since Claim B is true, Claim A is true. So Claim A is true. Scarlet macaws only eat possumwood fruits. Only possumwood seeds are found in scarlet macaw droppings because scarlet macaws only eat possumwood fruits. Can you think of examples of this kind of reasoning in the materials you've read? Sometimes it can be confusing to start a paper de novo to explain the research you've conducted. When you can show that your hypothesis is not incorrect, sometimes it can be a challenge to write about it without using your results or observations to go back and qualify your hypothesis or question. Again, this is when peer review is helpful. If someone reads your paper and keeps asking you "Why? This doesn't explain anything," and they're asking it every time you try to explain something—that is, around and around and around you go—you might be on the begging-the-question carousel.

Red herrings. Are you losing an argument or having one of your sacred claims challenged? Time for a distraction. Hitchcock was famous for the cinematic version of the red herring: distract the audience, lead them down a false path, get them to think that something else is happening, that someone else is the bad guy. Hitchcock would introduce a clue, in a deliberately (yet masterfully) poor attempt to make it subtle, but enough for viewers to say, "Aha. The butler did it." It's never the butler. It's always the innocent little blond-haired, blue-eyed tyke who barely even has a speaking part. If the 19-year-old Costa Rican upstart is trying to mess up your lab's reputation by claiming that scarlet macaws eat four kinds of fruit instead of only one, that's when it's time to say something like "This is why ornithology isn't taken seriously as a science," and watch with satisfaction as the conversation shifts from scarlet macaw nutritional preferences to heated discourse over why ornithology *is* taken seriously as a science. Seeds are forgotten. In science and scientific writing, this is one of the more readily obvious crimes. When you read the draft of your paper or a peer reads the draft of your paper and either of you say, "Wait a minute. What does this have to do with imperiling the legitimacy of ornithology?" Think about it: Did you add it because you didn't know what else to say, or were you getting defensive about negative results? Red herrings are rarely ingenuous. Maybe you can use it in a spoken argument when you're on defense, but it's too flagrant to get away with in science and scientific writing in general.

Correlation and causation. Or, if you want to impress your interlocutors, the Latin versions *cum hoc ergo propter hoc* (with this, therefore because of this—correlation) and *post hoc ergo propter hoc* (after this, therefore because of this—causation). Because two things occur at the same time does not mean (a) that one of them causes the other, or (b) that they are necessarily related at all. This can be a significant cause of science gone awry, all the way from making observations and establishing an hypothesis to interpreting data—and of course, writing about results, which is in of itself a form of interpretation. This crack in the Logic Wall can crop up anywhere, and when it does, it can contaminate anything that happens downstream in the process (even all the way through publication—it can even happen in peer review). Worse, this crack can crop up multiple times, functionally expanding and weakening that wall further. And even worse, this crack can have its genesis in the naïveté of inexpertise, or in the evil premeditated commission of a Logic Felony to further one's agenda. You cannot assume that those parrots have a beak that has evolved eloquently over time to be the unripe possumwood fruit-crushing machines that they are just because that's what they're using them for. Make that assumption, and next thing you know, you'll find out that they are, in fact, the most savage and feared winged nocturnal predator in the forest. "Oh, but come on," you protest. "That beak-fruit thing makes sense." Sure it does, as an inductive hypothesis, but not as an *assumption*. The parrots breed during a certain time of the year: clearly it's because of photoperiod? Or is it temperature? or availability of food? or scarcity of predators? or some combination of these? If correlation or causation is involved (finding that out is one of the goals of experimentation), it needs to be *identified*, not *assumed*. Correlation and/or causation may very well be involved; the fallacy is making the *assumption* that they are.

Tyler Vigen, a mathematician who went to Harvard Law School (so we know he is a legitimate scientist) has used Actual Math to determine a number of significant correlations. For example, doctorates in computer science are correlated with total revenue generated by arcades. Look, you just can't argue with a correlation coefficient (r) of 0.99. Are you getting your doctorate in civil engineering? This is very tightly correlated ($r = 0.99$) with the per capita consumption of mozzarella cheese in the United States. A math doctorate? Tightly correlated at $r = 0.95$ with the amount of uranium stored at U.S. nuclear power plants. Visit Tyler Vigen's website (www.tylervigen.com) for some serious education in statistics. It's science!

Appealing to emotion, the mob, authority, or nature. If you're not getting anywhere with your claim, if you find you're under constant challenge or attack, it's time to amp up the firepower. *Miraculorejuvaderm*, wonder elixir that it is, appeals to your *emotion*, your fears and vanity: you're not getting any younger, my friend. Face it. Look at the ("simulated") example. The *mob*: "Are you going to let a bunch of doctors and "so-called health experts" keep you from trying something proven to be effective? You know they are keeping these secrets from you to protect their own interests, and why hasn't the media reported this?!" And then, paradoxically, *authority*: "Doctors have shown. Clinically proven. Developed by a plastic surgeon. Proven results." And then *nature*: "*Miraculorejuvaderm* is all-natural, composed of harmless natural compounds to which you are already exposed on a daily basis (but, of course, compounded into a highly effective facelift-in-a-tube). It doesn't have any *chemicals* in it. If it's *natural*, it's good for you; it's safe." Okay, well, let's look at this two ways: *Miraculorejuvaderm* consists of mostly water, mixed in with a little oil. 100% natural. You put it on your skin like the weak hand lotion it essentially is. It *is* safe. On the other hand, it *does* have chemicals in it. Water, H_2O, is a molecule. Made out of atoms. It's a chemical, and perfectly natural, so perfectly safe. Strychnine, $C_{21}H_{22}N_2O_2$, is *also* a molecule made out of atoms. It's a chemical, perfectly natural . . . you can see where we're going with this. As the critical thinker you are, it dawns on you that "natural" doesn't mean "safe," "better," or "it works." When Andrea starts reading about *Miraculorejuvaderm*, she is guided through multiple links containing case studies, testimonials, "newspaper articles," "news reports," and commentary from physicians—including the Man Himself, the mysterious Scandinavian doctor who found a unique way to erase time itself in only 14 daily applications of a special unguent. There's the authority. Sure, in tiny pale grey type on the bottom of the page there are disclaimers about how the information on the page is "not intended to be taken as fact" and that these results are "not typical" or results are intended for "illustrative purposes only," but when you watch a couple of those local not-at-all staged "Action News" segments wherein a breathless commentator reports on her astonishing experience with a product she absolutely didn't believe in to start with—you think, well, maybe I ought to give this a chance. But fake evidence is not real evidence. If a "local mom" discovers something, particularly if it's something that "angers doctors," read that pale, tiny type on the bottom of the page. All those "weird tricks" ("by following this one weird trick . . .") are not going to suddenly provide you with an epiphany about whiter teeth or less belly fat; they're going to route you through several pages of bogus testimony, videos through which you cannot fast-forward, and "data" where you will ultimately be encouraged to buy something. The old cliché "if it sounds too good to be true, it usually is" stands. If the facts are lame, beware the appeal to emotion ("you're old"), the mob ("she's a witch!"), authority ("Scandinavian doctor"), or nature ("100% natural!").

---

10 Tips for Recognizing "Appeal" Fallacies

1. Appeal to the mob: incites righteous indignation in an us/them fashion. "Best kept secret by doctors," "Yet this information has been suppressed in the media," etc.
2. Appeal to emotion: designed to elicit an emotional response with regard to vanity ("You're old and fat") or hot buttons ("This animal shelter kills 40 animals per day, LIKE and SHARE to get the word out!").
3. Appeal to nature: touting the benefits of a "natural" product (typically nutritional or cosmetic) or the evils of things that are unnatural ("GMOs will kill us all").
4. Appeal to authority: makes a case using an "expert" for support ("famous doctor used by Hollywood stars," "leading researchers") and use of legit authority for cherry-picking (e.g., "NOAA proves there's no climate change").
5. Appeal to force or fear: persuasion using a perceived threat ("Cyclotrons can cause a black hole").
6. Appeal to anger: a form of a mob fallacy used as an agenda ("the pharmaceutical industry only exists to make money and doesn't care about health").
7. Appeal to novelty: assumption that anything newer is better ("new and improved!").
8. Appeal to tradition or common practice: assumption that anything that's been done the same way for a long time is better ("why does my child have to learn Common Core?!").
9. Appeal to belief or popularity: playing on personal beliefs as alternatives to science; common in evolution and vaccine arguments. If so many people believe something, it must be true.
10. Appeal to ignorance: playing upon the lack of knowledge; if there's no contrary evidence, it must be true (see "burden of proof").

Loaded questions. A loaded question comes preloaded with a presumption of guilt. There is no way to commit this fallacy in an ingenuous matter, and better examples of it are seen in politics than in science. Lab A is defending its Scarlet-Macaws-Only-Eat-One-Thing research, and then when spunky and inconvenient upstart B comes along and says, "No, they eat four things," Lab A asks, "Why did you try three different analysis techniques? Because the first two only showed the presence of a single seed type in bird droppings?" Suddenly the young upstart finds herself defending herself from an accusation. This is dangerous. If you focus on trying to defend yourself from an accusation, you will sound guilty no matter how innocent you are. The appropriate response is to lob the attempt back to them and ask why they're asking such an odd question. While feuds in scientific writing may exist occasionally as epistolary jousting in Letters to the Editor or editorials in journals, these are rarely things you see in formal scientific writing or science. The simple skill of recognizing it tends to suffice in scientific writing. If it is actually present in something you're reading, whatever it is, it's tainted.

Ambiguity. Find yourself on defense? Time to get creative with semantics or context. Remember, if you're under attack, adverbs are your friends: *always, never, often, sometimes, frequently* . . . such lovely, useful words, into whose dark and ambiguous recesses one can hide when challenged on an unanticipated outcome: "Well, I said '*sometimes*,' not '*always*.'" The good and effective scientific writer, of course, eschews ambiguity in the first place. "Frequently" is perfectly useful if it's qualified. "Never" and "always" aren't ambiguous at all but can get you into trouble in science. By its nature, science evolves, and everything is subject to the Second Law of Thermodynamics. Of course scientific "laws" are *also* science and in of themselves have the potential to evolve.

Ambiguity, of course, can be an issue in any kind of writing, not just scientific writing, and it is not uncommon to get called on it by a professor in any class you take. However, in scientific writing, recall that *precision* is crucial. If your writing contains elements of ambiguity, it is by definition not precise. This is another sterling example of the value of peer review. You want to iron out all of those "what-do-you-means" before you submit a paper to a journal.

Black or white. Also referred to as the "false dilemma," black-or-white reasoning assumes that there is *only* one choice between two things: yes/no, up/down, out/in, black/white. If there is one thing you should know about science by now, it is *all* grey.

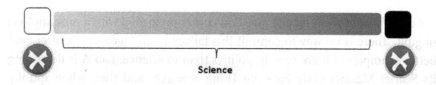

Science

Black/white might be useful in dichotomous keys or decision-point algorithms, but there is always going to be an exception or caveat or qualification—even if it hasn't been identified yet. If your parrot doesn't eat fruit, it must be a predator. It either eats this fruit or that fruit. This cream works or your money back. If, as you do your reading and research, you find any definitive statements about "this or that," consider it askance. Is it conceptually possible? It could be, particularly if one of the choices is broad; for example, it's either a scarlet macaw or "some other" bird. Given your luck, of course, as soon as you publish that, scarlet macaws will be determined to be a kind of winged salamander.

Weasel-wording. Weasel-wording is a lovely coalescence of logical fallacies into a single flowing narrative that sounds so good, so authoritative, so legitimate, so right, so logical—but, upon closer inspection, means little to nothing. What does "clinically proven" mean? "Clinically proven to reduce wrinkles." What does that mean, what is the context, what are the semantics, who is the authority, why is it important (if it is), how is it true, where are the data, what is the evidence, what is the source, is there an agenda, are the arguments valid . . . *are* there even any arguments? How do you measure "reduction"? Most individuals—including many scientists—will not properly question this claim. Clinicians who know what "clinically" means *may*, however, recall that nothing is definitively *proven* in clinical trials. Trials show that this drug keeps more people alive than placebo, but that's not proof. Side effects from Drug A kill fewer people than side effects from Drug B. That's not proof. All that shows is that side effects from Drug A kill fewer people than side effects from Drug B. *Miraculorejuvaderm* slathered liberally upon one's skin will indeed show a change in that skin: It will get moist, so it will probably look better than when it is dry and flaky. This does not mean that it's "clinically proven" to turn back time.

As a critical thinker and writer, it is your duty to qualify your claims and write what you mean. Newer writers may be writing ambiguously, for example, using buzzwords (or buzz-phrases) that are commonly iterated in their field yet are still meaningless ("it has long been known that," "it is generally believed that," "clinically proven," etc.), and desperate writers might be trying to make less-than-exciting data seem exciting ("correct within an order of magnitude," "obtainable data were assessed for," etc.). You may not even be aware that you're doing this, and even peers might skate over that "common prose." Oftentimes, however, weasel-wording is a sin of commission. With more and more experience, you'll notice it, and notice it if you're doing it yourself.

Take-home

Critical thinking is the single most important skill you can possess as a combined sci-entist, researcher, and writer. Our message is:

Don't believe everything you see, hear, or read. Ever.

Recognizing flawed thinking, design, or arguments—either in the science itself or the communication of it—will help you be a better writer and save you a lot of time up-front. Once you have completed a paper, it's more challenging to go back through it and identify the critical flaws and fix them. Sometimes that fix might require a sig-nificant rewrite. Think before you write. Create an outline. Is it logical? Flesh out the outline. Is it still logical? Are you seeing any flaws creep into it as you expand your thinking? *Be a lucid skeptic in thought and—literally—word.*

Exercises

1. Find your own "*Miraculorejuvaderm*" online. Perform your own "Consumer Science" assignment (this can be done alone or in groups). Analyze it from the standpoint of (or use the AACU or Facione and Facione's Critical Thinking Rubrics):
   a. Its claims
   b. Its evidence/data to support those claims
   c. The source of the information (and the sources of those sources)
   d. Legitimacy of sources
   e. Authority
   f. Bias/agenda
   g. Fallacy
2. Now that you've reviewed the logical fallacies, which is the one you tend to commit the most? It probably occurred to you when you were reading about it. We all have one or more of these to greater or lesser extents. Pick your nemesis and write two pages about it, complete with examples, for group discussion or as a submitted writing assignment.
3. What do we mean when we say "writing is the other half of science"? Explain this in general and then using an example from your own field. Why is this important in the context of critical thinking?

4. What is the difference between critical thinking and scientific literacy? *Is* there a difference?

5. Discuss the single biggest issue in your field that is suffering as a consequence of the impaired scientific literacy of the laity. What are the challenges contributing to that impaired scientific literacy (e.g., complicated nature of the science, emotion, religion, politics, lack of proper education, poor research, amplification of bias or misinformation in publication, ethics, etc.), and what is one way those challenges might be addressed?

5
Communicating Basic Statistics

WHY DID I DO RESEARCH THAT INVOLVES DATA?!

© 2020 Kris Mukai

You've spent *years* happily squishing about in the mud, collecting your savage bloodworms. They're *fascinating*, with their eversible probosci, how and what they eat, where they live, how fast they travel through the mud, their anatomy and physiology, how and why they're used as bait, why they're called "bloodworms," and how they're related both to their own close relatives and to other worms. You've covered a lot of geography studying them and their little fangs. You've squelched about in hip boots up to your thighs in the sulfurous sucking mud in Maine swamps, so caught up with your beloved polychaetes that you didn't notice that the tide was coming in—*fast*, the way it is wont to do in the upper latitudes. You have breathtaking scanning electron micrographs of their pretty, fluffy appendages. You have reams and reams and notebook after notebook full of notes. Verily, you are a proud emerging expert on the genus *Glycera*.

You march your notes, triumphant, into your advisor's office. She ruffles through your notes and then looks up at you over the top of her glasses, and says "Okay, how many? How much? Compared to what? What's the relationship of this to that? Does it change? When? How? How related are they to that, and to each other?" Ominous music seeps in from somewhere; you trudge back to your cubicle and look down at your traitorous notes. A minute ago, they were awesome notes. Now, they're just pages after pages after pages of—the ominous background music comes to a crescendo—*data*. "No," you whisper in horror, "*No . . . I'm a field scientist . . .*"

Yes . . . it's time to start comparing stuff to other stuff.

That's what "statistics" is: comparing stuff to other stuff. Sometimes it's comparing the same stuff to each other, under different conditions. That's basically it.

We're not going to teach you how to do statistics in this book. You're probably going to have to take one statistics class, if not several. But if you're writing a thesis, dissertation, journal article, or some other kind of scientific document, eventually you're going to have to report on or write about statistics. Depending on the scientific field in which you find yourself, you might have to use stats in lab reports. Writing about (and interpreting) statistics is a critical skill, in both science and writing.

What Are Statistics? Do We Really Need Them?

Afraid so. If you have a mess of data, they aren't going to make much sense, nor will you be able to use them to make your point, answer your research question, or convince others that you've really done anything. (Notice, incidentally, that we refer to "data" as "they" and "them." The word "data" is *plural*, no matter what *The New York Times* says.) Don't be afraid—again, all you're doing, when you get right down to it, is comparing stuff to other stuff. Think of your precious worms. You wanted to see what the ecologic differences were between worms in lower latitudes and higher latitudes. Stuff was different: water temperature, water salinity. Muddier mud, sandier mud. More worms lived in certain conditions as compared with others, bigger worms here, smaller worms there.

Is the mere mention of "statistics" giving you a creeping sense of terror? Don't worry! That's normal, especially if you're just starting out in it. There are some great "user-friendly" references that you can acquire to help and guide you. There is no shame in acquiring these books.

Rumsey, DJ. *Statistics for Dummies*. (Wiley)

Kranzler JH. *Statistics for the Terrified*. (Pearson)

Lang, T. *How to Report Statistics in Medicine*. (ACP). (This book is specifically about *communicating* statistics. While it is medical content-focused, the emphasis is on writing about stats, not learning stats per se.)

Before you can address the noble scientific question "why?" you have to be able to say, "Here's what I've got; these are my observations." In other words, "stuff." Compared with other stuff. Stuff in context. Significant stuff—or nonsignificant stuff that was supposed to have been significant. Persuasive stuff. Comparative stuff. Summed up stuff. Stuff to support your arguments. Different scientific fields will require varying amounts of statistics. If you're working on a degree in ecology or epidemiology, get yourself a programmable calculator before the first day of class. Math or engineering or chemistry or physics may not require as much stats, but it is still a skill you need. A complete education in a scientific discipline is in all likelihood going to contain statistics, and as a practicing scientist, you're going to need how to read about and understand them, interpret them, and communicate them, even if you don't perform statistical analyses with any degree of frequency.

In science, data and statistics are the ultimate support for scientific argumentation. If you submit an article to the *Journal of Physical Chemistry* reporting that a fabulous new molecule you just created peaked at a wavelength of "roughly 70 cm^{-1}, with some other minor peaks" without including a graph, the reviewers will tick the "reject" box without a second thought—a verdict for which you waited four months. You cannot engage in intellectual scientific discourse (socially or academically) and say something like "there is no evidence for" or "there is a significant amount of evidence for" without backing that up.

What Are We Trying to Prove With Statistics?

Well, when you're writing about science, it's to make a point. You want to say to your audience, "Look. We were wondering if this drug really works, so here's the information we have to support our contention. And we did, see? Here are our results." Before you even start collecting your data, you ask your question: "Does this drug work better than other drugs?" "Are there more worms in colder water or warmer water?" "Is wavenumber absorbance really higher in the synthetic molecule as compared with the naturally occurring molecule?" After you ask your question, you decide what kind of information you're going to need—and how much—to answer that question: "We need at least 200 people taking the study drug and 65 taking the comparator drug;" "I need to make sure I count the number of worms in at least 25 different spots in both the colder and warmer locations;" "I need to repeat that spectral analysis at least 10 times with each molecule." You're armed with this information *before you start*. (How you find out these numbers before you start is something you learn in statistics or is provided to you by statisticians.) Since you know this beforehand, you collect enough data to begin with. You *don't* want to under-collect and sit dismally in your cubicle and say, "Oh *man*. I can't make any conclusive statements about *anything* with this." As with ethics in study design, upstream preparation is going to determine whether or not you can make legit conclusions.

> "Statistics is the grammar of science."
>
> —Karl Pearson (upon whom you can blame a great many fundamental statistical concepts)

Before you start, you may also decide that you want to evaluate more than one thing (e.g., salinity *and* temperature for your worms, drug efficacy *and* patient-reported outcomes for your drug, spectral analysis *and* electrochemical properties of your molecule). May as well: if you're squelching through sulfurous mud, handling very annoyed worms with large venom-delivering fangs, and trying to dodge rapidly rolling incoming tides, you may as well get your money's worth, so to speak. If you're collecting information for more than one thing, these things are conventionally referred to as *primary, secondary, tertiary,* etc. objectives (or outcomes, depending on the kind of research you're doing—if what you're doing is experimental). It is quite possible that you have more than one primary (or secondary, etc.) objective. Primary and secondary objectives (or endpoints) are particularly common in clinical trials. Imagine that you are testing a drug for heartburn. You might be comparing its efficacy against a placebo or a comparator drug as a *primary objective* (efficacy is almost always the primary objective or endpoint in a clinical trial), but the primary objective would be something specific, such as "percent esophageal healing as compared with x," whereas a *secondary objective* might be something like "patient-reported outcomes on subjective relief."

A corollary to an outcome—something you're assessing—is the *variable*. An *independent variable* varies either by itself (e.g., age, sex) or as an experimental manipulation (e.g., salinity, temperature, length, absorbance, etc.). You're assessing presumed *cause* (independent variable) and presumed *effect* (dependent variable). A well-controlled experiment assesses one or more independent variable(s). If the impact of the independent variable is of primary interest, then a well-designed experiment will examine a single independent variable. However, if interactions *between* independent variables are of interest, then a well-designed experiment can examine two or more independent variables. The experimental design should encompass the goals of the inquiry. *Dependent variables* vary as a consequence of changes in the independent variable (e.g., patients get better on an experimental drug; dissolved oxygen goes up in colder water), among other things. In statistics, you'll learn how to analyze single-variable data (e.g., dissolved oxygen in seawater), and you'll also learn multivariate analysis (multiple variables, including variables that may have an effect on each other, e.g., dissolved oxygen *and* water temperature, or dry ice *and* temperature).

How Am I Going to Be Writing About Statistics?

In a statistics class (and yes, some of your science classes), you'll have to learn all kinds of unreasonable math-requiring calculations, rife with italicized things and Greek letters and a bunch of superscripted and subscripted ephemera. It's good discipline. Then, you get a stats calculator, or a spreadsheet program that will import your data, and crank out your analysis of variance (ANOVA) results with the push of a button. There are a number of ways you can compare data. What you decide to compare and how will determine the statistics you use. Once you are faced with the perilous task of analyzing your data, then your stats class will make sense: *you can apply what you have learned.* This is when statistics will not only make sense, but you'll be thankful that you have learned it and will be keen on using it (usually). Performing a *t*-test or an ANOVA or a regression for the sake of having to learn how to do them is one thing, but applying those skills and writing about your research is another.

Now that you have your stuff and have *compared* your stuff, you have to communicate it lucidly. If you are writing about your data—for a lab report, a paper, a presentation, a meeting poster, a thesis or dissertation—there are conventions that are followed.

10 Tips for Organizing Your Statistics When Reporting Results

1. Describe the purpose or rationale of the study or experiment.
2. Explain why the study or experiment is important (e.g., what gap in knowledge it was designed to address).
3. Identify your outcomes or goals (all of them, in the event you have primary, secondary, or more).
4. Describe what's being assessed (including all variables, if there are multiple variables).
5. State how these outcomes are going to be measured.
6. State how these measurements are going to be analyzed (which tests?).
7. Provide raw data (or summarized raw data—this depends on the length of what you're writing and the type of document you're writing).
8. Provide statistical analysis of data.
9. Present results lucidly (tables? graphs?).
10. Explain results.

See how easy it is? If you are writing the standard introduction-methods-results-and-discussion (IMRAD) paper, part of this falls under "methods," and part of it falls under "results" and part of it falls under "discussion," depending upon how the study was designed. Depending on the genre and level of detail of the document you're writing, you may even be including information about the brand of scales you're using and the publisher of the software used for your calculations (this would be in your thesis but probably not on a meeting poster). If you are writing an IRDAM-style paper (where methods come at the end), the content is the same, but the organization of the paper is slightly different (more on IMRAD and IRDAM later). And more and more as of late, journal publications don't even contain full Methods sections—they're available as online supplements. In this case, you would be following the dictates of the journal.

If your science is experimental, it needs to be reproducible. You make a landmark discovery like, say, cold fusion in a jar of water, and you *publish your data*, instead of, say, hypothetically, calling a press conference to brag about it (hint: don't do this. This actually happened in 1989 and was met with much derision from the physics community). If you perform well-executed experiments and analyses and report them comprehensively, other scientists should be able to reproduce your efforts and come up with the same or similar results ("Holy cats—the bloodworms in Maine *are* more vicious than the ones in South Carolina").

Let's have a look at a fine publication from the end of 2011 from the *British Medical Journal*. These were some authors, clearly, who wanted to put applied statistics to some truly practical use. The paper certainly caught *our* attention, since *our* publications were so boring, by comparison. The *British Medical Journal* gets playful around Christmastime, typically publishing one paper that is real science but also entertaining:

Stanaway FF, et al. How fast does the Grim Reaper walk? Receiver operating characteristics curve analysis in healthy men aged 70 and over. *BMJ*. 2011;343:d7679.

If this isn't a useful application of statistical analysis, we don't know what is. Let's face it; who *wouldn't* want to know how fast the Grim Reaper walks? Let's have a look at their tidy abstract and how it addresses each part of a solid data presentation and analysis.

How fast does the Grim Reaper walk? Receiver operating characteristics curve analysis in healthy men aged 70 and over. (Stanaway FF et al. BMJ. 2011;343:d7679.)	
ABSTRACT	
OBJECTIVE: To determine the speed at which the Grim Reaper (or Death) walks.	Objective/goals
DESIGN: Population based prospective study. **SETTING:** Older community dwelling men living in Sydney, Australia. **PARTICIPANTS:** 1705 men aged 70 or more participating in CHAMP (Concord Health and Ageing in Men Project). **MAIN OUTCOME MEASURES:** Walking speed (m/s) and mortality.	What's being assessed
[Main outcome measures continued] Receiver operating characteristics curve analysis was used to calculate the area under the curve for walking speed and determine the walking speed of the Grim Reaper.	How outcomes are going to be measured
[Main outcome measures continued] The optimal walking speed was estimated using the Youden index (sensitivity + specificity-1), a common summary measure of the receiver operating characteristics curve, and represents the maximum potential effectiveness of a marker.	How measurements are going to be analyzed
RESULTS: The mean walking speed was 0.88 (range 0.15–1.60) m/s.	Data
[Results continued] The highest Youden index (0.293) was observed at a walking speed of 0.82 m/s (2 miles (about 3 km) per hour), corresponding to a sensitivity of 63% and a specificity of 70% for mortality.	Data analysis

[Results continued] Survival analysis showed that older men who walked faster than 0.82 m/s were 1.23 times less likely to die (95% confidence interval 1.10 to 1.37) than those who walked slower (P = 0.0003). A sensitivity of 1.0 was obtained when a walking speed of 1.36 m/s (3 miles (about 5 km) per hour) or greater was used, indicating that no men with walking speeds of 1.36 m/s or greater had contact with Death.	
CONCLUSION: The Grim Reaper's preferred walking speed is 0.82 m/s (2 miles (about 3 km) per hour) under working conditions. As none of the men in the study with walking speeds of 1.36 m/s (3 miles (about 5 km) per hour) or greater had contact with Death, this seems to be the Grim Reaper's most likely maximum speed; for those wishing to avoid their allotted fate, this would be the advised walking speed.	Results/conclusion

Verily, a truly practical use of statistics. Now you, gentle reader, know the truth: to outrun Death, you're going to have to trot along at least 0.83 meters per second.

How the Fourth Estate Is Going to Ruin Your Statistics

"SCIENCE PROVES THAT THE GRIM REAPER IS A REAL ENTITY," screams the front page of *USA Today*, "AND IT TRAVELS AT TWO MILES PER HOUR." While the metabolic researchers at the Centers for Disease Control and Prevention are hopping around with joy thinking that this might get people to exercise more, they are also, of course, snorting their Diet Cokes because they're laughing so hard. This brings us to the topic of *generalizability*. This paper does not, in fact, prove that the Grim Reaper is real, nor that it picks off the slowpokes. This paper does not *prove* anything. This paper shows that, based upon very specifically defined preestablished parameters, men over the age of 70, in a particular study, who are not institutionalized for any reason, in Sydney, Australia, tend to die at a reliably lower frequency if they walk at least 0.83 meters per second, as compared with men over the age of 70, in the same study, who are also not institutionalized for any reason, in Sydney, Australia, who walk slower than 0.83 meters per second. The authors want to be playful, and perhaps get noticed, so they toss in the "Grim Reaper" bit, and, to their delight, go viral. Had the title of their paper been "Walking speed and survival analysis in community-dwelling men aged 70 or older in the Concord Health and Ageing in Men project," nine people would have read their paper. Can you generalize those results and apply them to the obese teen couch potato epidemic in the United States? No, you can not, even if the local tabloids feel they would be remiss if they did not alert the public that ACCORDING TO SCIENTISTS, PORTLY SLOWPOKES ARE GOING TO DIE. It's the truth: the Grim Reaper is stalking video game-playing couch potatoes and will smite your aged dog if he waddles along at 0.75 meters per second.

CLIP OF NEWS ANCHOR (PERKILY): . . . new study showing that drinking a glass of red wine is just as good as spending an hour at the gym.

JOHN OLIVER (AGHAST): *What?!* That last one. . . no! No! *No!* That last one doesn't even sound like science! It's more like something your sassy aunt would wear on a t-shirt!

—From: *Scientific Studies: Last Week Tonight with John Oliver* (May 8, 2016)

Mind you, some statistics is designed—and is in fact useful—because it *is* designed to be generalizable, or at least somewhat so, within certain limitations. However, a common mistake is to take *any* scientific result, no matter how small, as long as it has a statistic attached to it, and loudly trumpet "here's proof that." This would be approximately 98% of everything you see on social media, including on, unfortunately, many science sites. Sometimes information from a *single case report* (particularly if it's health related) turns into "this-causes-cancer" or "this-prevents-cancer." As we discussed in the preceding chapter, it's generally safe to not believe everything (probably most of everything) you read online.

Correlation and Causation: Generalizability's First Cousin

"Listen," says Correlation and Causation at the family picnic when speaking to Generalizability, "that paper was totally *awesome*. 'Slow walking causes death.' *Love* it." Generalizability laughs bashfully, punches his cousin in the arm affectionately, and goes off for seconds on the macaroni salad. In science, we are often looking at variables, and the data you collect show you how those variables are *related* and if they're *really related*, or *kind of related*, or *not related at all*. Science, methodical and pure, never *proves* anything. Sure, if a bunch of scientists collect *enough* similar (replicated) data over many years, you can get to the point where you can say "smoking causes cancer," but what the data are showing you, in fact, is that one is *more likely* to get cancer if one smokes, as compared with people who *don't* smoke. You can't write up your dissertation on bloodworms and say they're more vicious in Maine because the colder water makes them crankier. You *can* say that there is a higher correlation of viciousness in Maine bloodworms as compared with South Carolina bloodworms as measured by depth of puncture wounds and degree of wound necrosis in graduate lab assistants collecting the animals. Another danger of correlation and causation is that a *different* variable might be at work, one that might not even be in consideration. Maine bloodworms might be crankier because of the tidal amplitude, for example.

© 2020 Maritsa Patrinos

The takeaway is this: you need to take care while you're writing. Even if you yourself are not *interpreting* the data, when you *write* about data, that in of itself is a form of interpretation, and can reflect the bias or inexpertise (such as unfamiliarity with the language of statistics) of the writer. It is therefore always helpful—for this and other reasons—to have other people read your paper (or whatever it is that you're writing), preferably individuals who know the topic, preferably other scientists, collaborators, or peers. Hearing someone say "are you sure about that? what did you mean?" is much better than hearing someone say "looks fine to me."

What Do You *Mean,* "Inexpertise?"

Yours. *No* one knows more about the anatomy and physiology of the parietal cell than you do (that's the cell in the stomach that produces hydrochloric acid)—no one even argues with that. But if you design an elegant experiment that shows that there is an elevated risk of developing esophageal cancer when the parietal cells go amok and you write that parietal cells indirectly cause cancer or that when you tested certain acid-reducing compounds you got insignificant results, your paper will end up in the editorial trash bin. Being the reigning parietal cell expert in North America doesn't mean you're an expert at *writing* about it. Are you sure about "insignificant"? We know, of course, that we can't say that parietal cells (and the hypersecretion of acid from them) cause cancer. We can talk about hypersecretion of acid and increased risk, but that's how we characterize it. *None* of the acid-reducing compounds you tested worked? Those results are significant indeed—not at all "insignificant." That is very important *information*. If something works, that's important to know; if something *doesn't* work, that's *also* important to know. What your data might have shown was that the efficacy result of the compound you tested as compared with a placebo compound for acid reduction was statistically *non*significant (i.e., your study compound did not perform in a statistically significantly better way over placebo). Remember, there are no *in*significant experimental results. Even if you don't get the results you anticipated, that's telling you something. It might mean that there truly is no difference between experimental and control variables. Or it might mean the experiment was not well designed, or that something went wrong during the experiment. All of this is information. It might be frustrating information; it might not be the information you want, but it's information all the same (i.e., significant). *Nonsignificant* means you aren't getting a statistically significant difference between the two conditions of the variable you're measuring—for whatever reason. Fanged worms in Maine and South Carolina, as it turns out, are equally vicious, when the viciousness data are analyzed statistically. There might be interesting statistically significant differences in some other variable (e.g., size) which might be a secondary objective. But when you find out that northern and southern fanged worms are equally vicious— that's a piece of information. It's even a *publishable* piece of information. It's not *in*significant. The neophyte scientific writer may be well versed in the viciousness of bloodworms or the alleged carcinogenicity of acid hypersecretion from parietal cells, but using "insignificant" when she means to say "nonsignificant" will make reviewers go "pfft, this is a neo." "Insignificant" is not a statistical term.

So does "nonsignificant" mean the results have to be exactly the same in the two things you're comparing? Rarely. It can happen, but if it does, it is more a consequence of chance than anything else, particularly if you're talking about continuous variables (say, length of worms), as compared with discontinuous (or categorical or discrete) variables (say, male or female worms). Suppose the average length of the worms in colder water is longer than the average length of the worms in warmer water. How much of a difference does there have to be between those means for that difference to be "significant?" Well—that's for you to find out in your stats class. But even your senior thesis committee isn't going to be able to look at those two mean values and tell whether the difference is significant or not. You'll need to clue them in with . . . the magical *P*-value. You'll find out all about *P* in your stats classes. The reason we're not getting into *P*—apart from what it tells your readers and how to write about it—is because it is defined as "the probability of obtaining by chance a result at least as extreme as that observed, even when the null hypothesis is true and no real difference exists; if it is $\leq$ .05 the sample results are usually deemed statistically significant and the null hypothesis rejected." Pay attention to the "$\leq$ .05" part. Your readers will skim over the colder-water-worms-averaged-37-cm-in-length-and-the-warmer-water-worms-averaged-32-cm-in-length part of the results and they'll hit "*P* < .001" and say "Oh, okay. They're statistically significantly different in length in colder than in warmer waters." Not, mind you, that this is necessarily *meaningful*—it's only a statistical result. The *P*-value cutoff for "significance" is generally (and somewhat arbitrarily) considered to be .05. As a writer, you need to remember that a *P-value does not "prove" or "disprove" anything*; it's just a piece of information that is calculated in somewhat standard fashion in statistics and is widely recognized, even by those not particularly conversant with statistics. The Food and Drug Administration (FDA) wants to see a *P* of .049 or less if you're submitting clinical trial data. A *P* of $\leq$ .049 is okay; a *P* of $\geq$ .05 is not—you can see the arbitrariness. Statistical significance is calculated different ways for different things (have fun in that stats class and pay attention) and is heavily influenced by the number (N) of data points being assessed. Remember under-collecting your worms? Find out in advance how many you need to have enough *statistical power* to generate useful comparisons. We're saying this for a reason. If someone has handed you some data—maybe even already awesomely graphed!—and is reporting a *P* of < .0001 but the N is only 3, it's time to get suspicious. Something isn't right. We'll talk about this shortly in our discussion of statistical weaseltry.

Another important measure is the *confidence interval*, which, simply put, is a range. If your mean worm length is 37 centimeters but the 95% confidence interval (CI; the range of lengths into which 95% of the worms you collect will fall) is 2 cm to 94 cm, that is doing some other things to your data (although it is providing some interesting information on variability). If your N = 235 and your *P* < .001 with a [95% CI 35.7–39.1], that's pretty good. If N = 4, and

$P < .0001$ with a [95% CI 1.2–184.8], your readers are going to get suspicious. You ought to be too.

> "I've seen grad students finish up an empirical experiment and groan to find that $p = .052$. Depressed, they head for the pub. I've seen the same grad students extend their experiment just long enough for statistical variation to swing in their favor to obtain $p = .049$. Happy, they head for the pub."
> —Carl Anderson (VP, Head of Data and Analytics at WeWork)

Another thing we're not going to get into in any detail is the flak that the almighty P-value is taking. The "significance cutoff" of $P < .05$ can originally be traced to Ronald Fisher, recognized as the father of modern statistics. At no time did Fisher ever say "a P below .05 is significant, above .05 is not," but he *did* suggest that $P < .05$ was a convenient point at which a result definitely due to a treatment was likely. "Convenient" is hardly dogma, and Fisher provided the caveat that using this "convenient" cutoff wouldn't necessarily be accurate for small effects or small amounts of data. A P-value greater than .05 probably indicated that you needed more data. Poor Fisher knew not what havoc he had wrought . . . somehow, unintentionally, $P < .05$ evolved into dogma. In other words, people are locked into P and focus on it as a significant/nonsignificant result.

P is only a number. You can have five years of valuable data, with significant and rational interactions and effects—a P-value of .05 does not invalidate your results: that's not common sense, and that's not how science works. It can be a challenge to write lucidly about statistical results if they have not been lucidly derived or analyzed. In a 2014 paper in the *Journal of Pharmacology and Experimental Therapeutics*, Harvey Motulsky wrote:

> . . . investigators fool themselves due to a poor understanding of statistical concepts. In particular, investigators often make these mistakes: 1) P-hacking, which is when you reanalyze a data set in many different ways, or perhaps reanalyze with additional replicates, until you get the result you want; 2) overemphasis on *P*-values rather than on the actual size of the observed effect; 3) overuse of statistical hypothesis testing, and being seduced by the word "significant"; and 4) over-reliance on standard errors, which are often misunderstood. (*J Pharmacol Exp Ther.* 2014;351:200–205)

As a scientist, you'll become more well-versed in these areas and better able to recognize the sins committed (deliberately or ingenuously) by yourself or others. Understand that these are things that you also need to recognize as you read other resources—your writing is as good as the sources you use.

In your stats class, you will learn more than you ever want to know about statistical terms and values such as *P*, N, CI, variables, and significance. Here, we will show you how to write them (and read them!) correctly, since they are often poorly expressed and can lead to significant misunderstandings. These values are frequently expressed incorrectly, including by those who ought to know better. You will hear the argument "this is the way we've always done them" and "it doesn't matter as long as you're consistent." It makes our ears bleed, and we hear it from scientists and colleagues all the time.

---

10 Tips for Writing About Statistics

1. *P* and *p*. *P* is italicized. Always. One of the things many individuals writing about science do wrong is randomly use upper case or lower case *P*s, often priding themselves on their consistency. But they mean different things. If you're talking about *P* for your *entire sample* of something—worms, molecules analyzed, patients—you use the upper case *P*. If you're talking about a *subsample within the entire group*, use lowercase *p*. If you collected a total of 550 worms for your experiment, that's your entire sample, and you use *P*. If 290 of them were from Maine and 260 of them were from South Carolina, you use *p* for each of those subsample statistical results. Ideally, these samples should be representative of total populations and subpopulations. Sadly, you'll find that many journals and style guides simply emphasize consistent use of upper- (or lower-) case *P*, regardless of sample population.

2. *P* < .01. *P* is written in a very specific way. *P* can never equal 1, so there is no zero to the left of the decimal point. This error abounds in scientific writing—*abounds*! As with "*P*," though, you will more likely than not see *P*-values with a zero to the left of the decimal.

3. .05, significance and nonsignificance. Remember that a *P*-value of .05 or higher is considered to be nonsignificant (usually). A *P*-value of .049 and lower is considered to be significant. Remember to use "*non*significant" and not "*in*significant." Keep in mind the caveats about "significance."

4. N and n: Use the same rules as for *P* and *p*, although note that they are not italicized. 550 worms is your N (total population), with Maine *Glycera* n = 290, and South Carolina *Glycera* n = 260 ("n" is your subpopulation). *N and n do not refer to the same group*, but as with *P* and *p* and zeroes to the left of the decimal in *P*-values, this is frequently and wantonly disregarded, including in many scientific journals.

5. Student's *t*-test. You'll learn about all of the different kinds of *t*-tests (note that "*t*" is italicized and lower case) and what they're used for in your stats class. Student's *t*-test is not called that because you're a *student*. "*Student*" is a

name. Names are proper nouns. Proper nouns are capitalized. Here is some fabulous trivia you can use to wow folks at your next party: "Student" was a pseudonym for one William Sealy Gosset, a chemist/statistician working for the Guinness brewery. Gosset developed a new statistical technique for monitoring the quality of the beer, but was only allowed to use a pseudonym when he published his work on t. Thanks to beer, we have Student's t-test. Gosset, using the pseudonym "Student," published "The probable error of a mean" in 1908 in the journal *Biometrika* thus wreaking semantic havoc on statistics for generations to come.

6. Correlation and causation. *Reflect on what you're saying.* It is very easy to write in a way that implies causation—without it sounding to you that you're doing it. Even if you're sure you're not, read it out loud and think about it strictly from the standpoint of correlation and causation.

7. Generalization. Your results apply to *your specific experiment or study*, and not necessarily to all populations as a whole. Generalizing to the total population can be done if your experiment was well designed and your sampling was representative of the total population (use a high N, and ensure that variability is representative).

8. General style. When writing about statistics, you may be required to hew to specific *styles*, such as ACS (American Chemical Society) style or AMA (American Medical Association) style or APA (American Psychological Association) style (or others). It is useful to check before you start writing something lengthy. Useful hint: if you are mandated to use a particular style, find a paper in that journal (e.g., *Journal of the American Chemical Society* if you have to use ACS style, or *JAMA* if you have to use AMA style, etc.).

9. *True* authority. Style notwithstanding, math is math, chemistry is chemistry, physics is physics. There are right ways to write *everything* (e.g., statistical notation, physical and chemical constants, mathematical formulae, etc.) that *do not vary* from academic field to academic field—or shouldn't, anyway. Consult the International Union of Pure and Applied Chemistry (IUPAC) Physical Chemistry Division's *Quantities, Units and Symbols in Physical Chemistry* (the "IUPAC green book"). When you took chemistry and physics, did "T" and "t" mean the same thing? No they did not. T_{max} is wrong; t_{max} is right.

10. *Real-life* authority. You are rarely going to be the one who calls the shots on how you write your stats: your professors, editors, or peer reviewers, or PIs may end up calling the shots. They may require you to do heretical things like use nonitalicized p-values, or all upper-case (or lower-case) p-values, wanton zeroes to the left of the decimal place, commas and no commas, thin spaces, Tmaxes and T_{max}es and T_{max}es. . . "as long as it's consistent." But at least you'll know the truth.

Remember, data are important pieces of information, but they almost always have to be taken in context. Raw data are useful, but when you're comparing a bunch of raw data, your readers are going to be looking for a P-value (or other indicators of significance that you'll get into in a stats class) and an N at the very least, and often the CIs (or any other indicator of variability) and measures of central tendency (such as mean, median, or mode) as well. We'll get into visual communication later, but if you're looking at graphed data, you at least want to see the P and N (or p and n) on the graph somewhere. If they're missing, always wonder why. CIs are useful on a graph, but they can also clutter a graph. Sometimes the CIs are a critical component of discussion (particularly if you're focusing on variation), and to make your point, they need to be on the graph. There's no "rule" about what to include; the research should dictate context.

What Are Objective and Subjective Data?

An objective exam in school is a multiple-choice test. A subjective one is an essay exam. Objective data are solid, measurable yes/no, right/wrong, a-b-c-or-d things ("57% of the participants were female"). Subjective data, on the other hand, are prone to interpretation. *Looks* pink, *feels* better or worse, worms *seem* way more vicious up here than down there, etc. Subjective data aren't exactly quantifiable in the same way objective data are, although they can be characterized that way, using survey categories and scales. If you analyze or compare subjective data, you should make every effort to quantify them. Surely you've taken surveys or answered questions like "on a scale of 1 to 10 ... " That's subjective (i.e., your perception of what is measured [pain, quality, enjoyment, etc.]). The scale-of-1-to-10 thing is a Likert-*like* scale. A Likert scale is actually a specific assessment tool used in psychology and psychiatry, but its name has been coöpted and used everywhere. Looking at your worms? How pink are they on a scale of 1 to 5? That is a subjective assessment. You are probably even practicing subjective assessment of objective data, by eyeballing things before you do your statistical assessment—that's why sometimes we get surprised ("whoa! No *way* that's nonsignificant!"). Subjective analysis is very important in some clinical trials (or other studies evaluating the efficacy of therapeutic interventions). Many medical complaints have subjective components to them—usually pain. That's why we go to doctors, right? "This *hurts*." Some drugs have therapeutic effects that aren't necessarily "sensed" per se (e.g., anti-osteoporosis drugs or chemotherapy, which actually makes you feel *worse*). But others do. This can only be described subjectively: this drug eases pain; this drug lets me sleep; this drug makes me less depressed; this drug helps my heartburn. Where there is a subjective component to a medical complaint, researchers would do well to look at patient-reported outcomes (PROs; or "self-reported outcomes"). If you run a clinical trial and have fabulous statistical significance—$P < .0001$!—with your insomnia drug, showing that time asleep has *definitely* increased as shown by polysomnography, but your trial population is saying "cheap generic sleeping pills work much better," it would be marketing suicide to continue dropping millions into trying to sell that drug. So, from the standpoint of medicine, at least, we have statistical significance, but it must be tempered with *clinical relevance*. All the statistical significance in the world doesn't matter if it ultimately doesn't make the patient feel better or doesn't make the patient well.

Subjective data are frequently acquired using scales, surveys, and questionnaires. Some epidemiologic data are subjective, such as the widely used subjective survey assessment tool called the Visual Analog Scale (VAS). It's useful. It's an on-a-scale-of-1-to-10 Likert-like assessment tool that can be used with preverbal (e.g., children), postverbal (e.g., aphasic), or non-English speaking individuals. Does it hurt? How much? Figure 5.1 is a standard VAS used in hospitals (developed for pediatrics):

Figure 5.1

Figure 5.2 shows a more modern one:

Figure 5.2

. . . although they probably don't use it in pediatrics.

Can you quantify subjective data enough to perform statistical voodoo on them? Yes. Notice the scale has numbers. You can assess the responses provided to each number individually, or you can group them (0–3, 4–6, 7–9, 10–12). They may not be objective data that you can measure with a sphygmomanometer or a blood chemistry assay, but you can say "in a study of 2220 patients, 87% reported a decrease from 10–12 on the VAS for pain to 0–3 following treatment with the experimental antinociceptive compound." They're PROs, and it's extremely useful information. If you're the single observer of your worms, and you have your own color scale, and you're eyeballing the "1 to 5" pinkishness category (using your own color scale) into which they fall, those are also data, if not hard, discrete data. Results from these kinds of studies can undergo the same kind of statistically rigorous analysis as objective data.

The reverse, incidentally, is true. You may have statistically nonsignificant results for this drug you've spent years and years and millions and millions of dollars on researching, but the patients love it and swear by it, and don't care one bit about statistical significance. Of course, you'll have the FDA hurdle to clear: if they don't see statistical significance, they won't be happy. What if subjective improvement is a placebo effect? This is another thing to consider when writing about results: they might be statistically significant, but are they *relevant*? Is there some subjective component that may not be reflected by statistical significance or nonsignificance? Or are they statistically nonsignificant but subjectively significant? Is the writer ethically and honor-bound to report this information, if this information is collected? The answer is *yes*.

Don't forget that FDA is only one example of a target. Depending upon what you're writing, you're going to want to provide all your information but in different ways. Are you working on a project that is ultimately targeted to the public? They aren't going to want to see (or care about or even understand) *P*-values or statistical significance. They want to see "4 out of 5 doctors . . ." FDA wants to see $P < .05$. The National Science Foundation, from whom you are trying to get a grant, wants to see that your data are unique and that there is a necessity to collect more. Your thesis committee wants to see *all* of your stats, correctly displayed and interpreted (don't think they're not going to check). The students you're TAing in Intro Ecology, who are slogging miserably through by-hand calculations of Student's *t*-tests, are going to be delighted by the idea that applied stats is useful for the brewing of awesome beer. The journal to which you are submitting a manuscript wants to see the complete experimental protocol, analysis, data, and conclusions. Your Thursday seminar group is going to be interested in a summary of the wound analysis data generated from the fanged attacks of vicious bloodworms. Pictures will be cool, but back them up with some graphs.

Eschewing Statistical Weaseltry

The first edition, published by WW Norton in 1954. It has sold well over a million copies since then, and is the all-time best-selling book on statistics.

In 1954, Darrell Huff, a mere freelance writer with no training in statistics, wrote a book that has since become the best-selling book in the field of statistics. This is telling, since the title of the book is called *How to Lie with Statistics*. In the front of the book, he quotes Benjamin Disraeli, also frequently quoted in many statistics books: "There are three kinds of lies: lies, damned lies, and statistics" (this quote was attributed to Disraeli by Mark Twain). Statisticians themselves cheerfully discuss how this is done. For obvious reasons, statistics lends itself well to jollity:

"Did you hear the one about the statistician?" "Probably."

"82.5% of all statistics are made up on the spot."

"He uses statistics as a drunken man uses lamp-posts . . . for support rather than illumination." (Andrew Lang, 1894–1912)

"Statistics is the art of never having to say you're wrong."

To wit, if you're creative, you can use statistics to support *anything*.

No, no! you protest, of *course* I won't lie with statistics! Well, there is deliberate prevarication, and there is accidental prevarication. Recall our discussion earlier of bias—a bias every scientist has, a bias you may not know you have, or statistical or scientific reporting inexpertise, or subtle pressure from a collaborator, peer, or mentor, an agenda, however innocent—all of these things may color your writing, even if your data are sound and you are the most ethical and vigilant of communicators. *What* you report and *how* you report it—right down to your style of writing—provide their own influences, subtle or otherwise.

If in your own writing style subtle bias creeps in, particularly if you are new at the whole scientific writing thing, and you are objective and unpossessed of any agenda, this is not weaseltry. If you are, of course, writing about your own research, or writing about an in-depth topic that has been assigned to you, it is likely that you're going to be excited about it; care must be taken not to report data as if you are writing a persuasive narrative. If your results are persuasive, it must be the data that are doing the persuading, not you.

There are, however, plenty of ways to practice *deliberate* weaseltry, subtle or otherwise. Subtle weaseltry, is, of course, the most effective (weaseltry in scientific writing extends beyond the "creative use" of statistics as well). You can *leave data out,* for example. Hypothetically, not that this would ever happen, a pharmaceutical company might elect not to report all of their safety data, if some of it was negative. (Actually, this *did* happen.) This would be a bad choice, since it could result in patient deaths, lawsuits, fines, and removal of the drug from the market.

You can *imply* (but never come right out and say, of course) that there is correlation with causality, banking on your target audience's inability to pick up on it. If you suggest that Death cruises along at 0.82 meters per second, and this has implications for your fat dog, and our friends in the Fourth Estate get a hold of it, the takeaway will be that if your dog doesn't walk at least 0.83 meters per second it is doing to *die*. The Grim Reaper will glide up behind your beloved beagle and hiss "you call that fetching a Frisbee? I think not." The data sit there forlornly in your receiver operating characteristic curve, innocent of all claims or implications of canine mortality. That part comes in with the writing.

Taking data out of context is a favorite of those with an agenda. This is deliberate manipulation, picking and choosing data to support one's message. Unfortunately, the data chosen are typically *real* data, often from recognized authorities. A climate change denier, for example, might cite data from the National Oceanic and Atmospheric Administration (NOAA). They're real data, and they may refer to cooling global temperatures, but if those data are used in a NOAA discussion of long-term global *warming* trends, the "cooling" part may be selectively used for an anti-global warming agenda. When challenged, the writer can say "Hey. These are NOAA data. I didn't make them up." Staid NOAA reports tend not to go viral, but impassioned, agenda-driven, misleading communications often do, and the general population rarely check the facts, or don't check the facts *enough*. And even if facts *are* checked, individual out-of-context statistics will still check out as "legitimate."

> Try it yourself! Visit NOAA's Climate.gov website and click on the "maps and data" tab. You can plot your own time period to cherry-pick information to support your contention that either climate change (warming) is occurring or not occurring! Start with a 100-year plot, and then zoom in on a period when temperatures are increasing or decreasing. Instant argument—and you're getting the facts straight from NOAA, so you're right.

Again, there is *statistical significance* and there is *relevance*. Something shown statistically to be "significant" may not be relevant in an applied sense. This is particularly true in clinical science. A clinical trial may show, for example, that a depression drug significantly improved depressive symptoms based on an assortment of assessment tools, but if the patients don't report feeling any better, then what is the statistical significance worth? *Emphasizing statistical significance over applied relevance* is weaselly. If an experiment has been thoughtfully designed to exclude any assessment of subjective results, you can get away with writing about it, but if there are subjective data and those data suggest that the statistical significance is irrelevant, subjective results might be downplayed. In a similar vein, nonsignificant results may be characterized as "trending toward significance";

a feeble attempt to say "well, they were *almost* significant." "Trending toward significance" will not impress FDA if they're looking for $P < .05$. In some types of statistical analysis, noting a trend toward significance may legitimately provide some valuable information, but qualifying truly nonsignificant results as "trending toward significance" is weaselly. Massaging the significance/nonsignificance thing is called "spin," and it is, unfortunately, not uncommon in the literature (deliberate or not, but usually the former) (see Chapter 3 for more on spin).

Manipulating your data to improve chances of statistical significance is also weaselly. You can slice your data up into different groups or hack your *P*s, or just eliminate those pesky outliers. There are ways you can adjust for outliers in legitimate statistical ways, and in many cases outliers *can* be excluded. But you cannot just "decide not to include them." Are you using survey data? What kinds of questions were asked and how were they asked? Were they leading questions? Does the survey question read "do you believe scientists are manipulating global warming data?" or does the survey question read "what is your position on global warming?" When the global warming survey is sent out, to whom is it distributed? A specific, biased target, known to be hostile to global warming, or to a cross-section of enough respondents representing diverse views? Are the survey questions going to engender honest results ("I always wash my hands before leaving the restroom," "I always wear my seatbelt," "I have never experimented with marijuana," "I always use a condom," etc.)? When you're reporting statistics, you're relaying important information. Rationalization has no place in reporting your data. A priori study design intended to deliver the results you want is unethical, and the study results are meaningless. Also, watch out for absolutes like "always," and "never." Science just doesn't work that way.

> "When you get into statistical analysis, you don't really expect to achieve fame. Or to become an Internet meme. Or be parodied by *The Onion*—or be the subject of a cartoon in *The New Yorker*. I guess I'm kind of an outlier there."
>
> —Nate Silver (statistician)

Your results were statistically nonsignificant. Okay, fine. That's still information. It might not be the outcome you wanted, but just because it isn't, don't try to make it look like it *is* significant. A popular way to do this is using the *weaselly y-axis approach*. If you stretch out that *y*-axis (i.e., reduce the upper limit to lower than 100%), you can make the difference between the two things you're comparing look bigger than it is. Adjacent bars in a bar graph on a scale of 0% to 100% are going to look a lot more similar in height than, say, on a stretched out *y*-axis range of 0% to 30%, and *hugely* different on a graph ranging from 20% to 30%. The upper limit of your y-axis should comfortably accommodate the bar with the greatest value (e.g., if the greatest value is 32%, the upper limit of the y-axis can be 40% or 50%, unless you're keeping your graphs the same size throughout your document). Your *P* could be .15, but the graph can make the results look *supremely* statistically significant. If you want to hedge your bets, leave the *P*-value off the graph too, and if the N in your "large population study" was only 8, you may as well leave that off too. This is particularly effective weaseltry if you have created this graph for a slide presentation. The presenter can whip by that slide, leaving the unsuspecting individuals in the audience with a snapshot image of something they'll recall later as "whatever it was, there was a pretty big difference." You are less likely to get away with this in print, since the savvy reader will recognize the weaseltry immediately, and that's even if you get it past the journal editors or your professor or thesis director—which you won't. These are writing issues. You have data, and they have to be communicated. How you communicate them correctly and ethically requires education, practice, and discipline (and, of course, peer review).

When considering measures of central tendency, clarify your means, medians, and modes. Huff uses the example of salaries (in which in his 1954 book the big boss adorably earns $45K/year; see figure 5.3). The average (*mean*) US net wage for 2016 was $46,640.94 according to the Social Security Administration. Well, that's livable. No economic crisis there. Of course, that value doesn't reflect the *median* (the salary occurring exactly in the *middle* of the entire range of salaries) for 2016, which was $30,533.31. That's less livable, as incomes go. And then there's the *mode* (the salary occurring *most frequently*), which, as you might imagine, is well below $46,640.94. The "mean net wage," as a single statistic, also does not take into account marital status, number of children, race, geography, skilled versus unskilled labor, and property, investments, total compensation, or other assets. The "1%" are not in that mathematical mean salary category. Depending on the point you want to make, you might want to focus on the same data in a different way. Is it wrong? No. Biased? Probably. What do you do about it? Report the mean, median, *and* mode—if these are the kinds of measures of central tendency germane to your research.

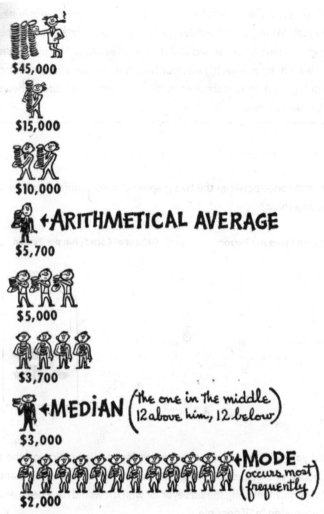

$45,000

$15,000

$10,000

+ARITHMETICAL AVERAGE
$5,700

$5,000

$3,700

+MEDIAN (the one in the middle, 12 above him, 12 below)
$3,000

+MODE (occurs most frequently)
$2,000

Figure 5.3 *How to Lie With Statistics* (Darrell Huff; 1954).

Take-home

Statistics are a powerful and necessary tool in science, and you need to be able to communicate about them accurately and ethically. You need to understand your statistics before you start your research, and design your projects accordingly—with input from others. Be confident and educated about how you analyze your data, and understand the conventions of reporting the kinds of statistics you use in your fields;

clinical science is different from chemistry, which is different from physics, which is different from biology. In fact, often these disciplines have their *own* statistics classes. It is easy (and a good thing) to be excited about your research, but be level-headed when you write about it. Be prepared, take your stats class, know what you're doing before you start doing it, take great care when writing up your results, and eschew statistical weaseltry in all its forms.

Exercises

1. What's the difference between the two graphs? Is there a difference? Do they mean the same thing?

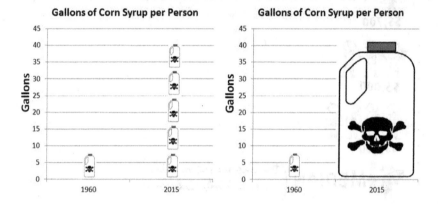

2. How would you write survey questions assessing public opinion on the following? Choose one of the following and write a nonbiased version and a "leading" version, identifying the surveyor for which it would have been developed. Answer option "g" if possible.
   a. Teaching evolution in elementary schools.
   b. Stem cell research.
   c. Genetically modified foods.
   d. Perceived safety of cyclotrons or nuclear energy.
   e. Complementary and alternative medicine, such as homeopathy, nutritional supplements, chiropractic, etc.
   f. Alternatively, pick a topic within your field about which passions run high in the general populace, and how you'd craft your surveys (nonbiased and "leading").
   g. Can you remember a survey in which you've been asked to participate (whether you did or not) that was biased? How so?

3. Within your field, find a paper (or analyze a paper provided to you by your professor) that contains flaws such as correlation/causality, generalizability, or some other crime. What is it? Explain the lapse, whether or not you feel it is deliberate or ingenuous, and how you'd fix it.

4. When reporting or writing about your own data, how would you decide if correlation and causation were *truly* related? Obviously, you'd have to make a statistical decision about this before you started writing, but how would you write about it?

5. As mentioned previously, some of the most entertaining quotes in all of science (and beyond science as well) are about statistics. Pick three of your favorites and explain what they mean. Alternatively, this would be a really fun discussion group or seminar activity.

   a. "Definition of statistics: The science of producing unreliable facts from reliable figures." —Evan Esar (humorist)

   b. "Smoking is one of the leading causes of statistics." —Fletcher Knebel (author)

   c. "Facts are stubborn things, but statistics are more pliable." —Mark Twain (humorist)

   d. "There are two kinds of statistics, the kind you look up and the kind you make up." —Rex Stout (writer)

   e. "If your experiment needs statistics, you ought to have done a better experiment." —Ernest Rutherford (physicist)

   f. "It is the mark of a truly intelligent person to be moved by statistics." — George Bernard Shaw (playwright)

   g. "Statistics: the mathematical theory of ignorance." (ouch) —Morris Kline (mathematician—double ouch)

   h. "Statistics are the triumph of the quantitative method, and the quantitative method is the victory of sterility and death." —Hilaire Belloc (turn of the [last] century writer and historian, unlikely to have heard of *p*-hacking)

   i. Find your own, related to your own field. This should take you approximately 60 seconds online.

6

Visual Communication

© 2020 Maritsa Patrinos

"Check out fab roadcuts," you tweet, tossing in a hashtag for "roadcuts" and links to your Facebook, Instagram, and Snapchat pages. Within moments, you've shared the geologic find of your young lifetime with your geology buddies and those friends of yours on Instagram who indulge your obsession with rocks (wear the Geek badge with pride). Now your friends can see the cut, strata, folds and inclusions, and, up close, see the wonderful Devonian brachiopods you've uncovered. You're the envy of your peers, especially Ted, who's been lording his brachiopod collection over everyone in the department.

We have become a *visual* culture. We tend to communicate more in images than words. This is convenient for scientists, because many scientific disciplines are very visual. In science, as elsewhere, not only are pictures worth a thousand words, they say things that words can't even say themselves. In Chapter 1 we introduced you to the famous Watson and Crick paper announcing their model for the structure of the DNA. We omitted the illustration, the famous and by now iconic double helix. The paper's impact relied in great part on that diagram, which went on to make history. If you do an experiment, and you have groups of dogs, cats, and college students fed water (control), decaffeinated coffee (experimental condition #1), and espresso (experimental condition #2), you're going to get pretty interesting results. Put them in a graph (see Figure 6.1).

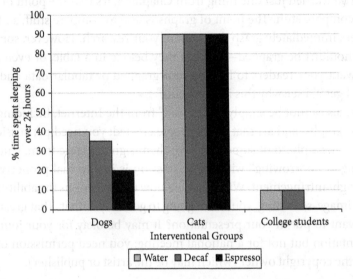

Figure 6.1

Science is a very visual discipline. When you study scientific writing, you're learning how to communicate ideas and information clearly and effectively. An important part of that communication is the appropriate use of visual information.

When Should I Add Visuals?

You should use visuals whenever they *contribute to or illuminate content*. What will make your reader understand, *truly* understand, what your data mean? Your data are the star of your visual. There is no such thing in science as a graph for the sake a graph, or gratuitous art—or, heaven forbid, clip art. Your graphs and charts and tables should be as *clear* and *efficient* as possible: you use them to make a point. Superfluous embellishments such as 3D graphs, beveled edges, graded colors, or shadowing clutter, obscure, and even distort data, as visual communication guru Edward Tufte (1983) eloquently argued in his classic text *The Visual Display of Quantitative Information*.

Don't distort your data or present it out of context. It is easy to graph data in such a way that *P* looks significant when it isn't. You can accomplish this by adjusting the scale of the *y*-axes, omitting confidence intervals (error bars) when they really need to be there, or graphing cherry-picked data from your study that make it look good, interesting, or significant when it isn't. This is misrepresentation, and it's weaselly if it's deliberate, particularly if the data are transiently presented (i.e., in a PowerPoint presentation).

If you learned just one thing from Chapter 5, it's that the point of statistics is to compare stuff. The point of graphs is *also* to compare stuff, so that your readers immediately grasp the point of your research. However, some things just shouldn't be graphed—because they belong in a table, or even in prose. You want your readers to look at your graphed or tabular data and say "oh, okay, I get it. I see why this is important."

Visuals cannot be simply "borrowed" from the Internet, including photos, tables, graphs, and art that other people created: you need permission. Also, if you take a photo of someone, have them sign a release form. Cutting and pasting or "borrowing" without permission is intellectual property theft or copyright infringement. With software now that affords us the ability to do re-verse image-searching, you're not going to get away with it. That great cartoon you want to put in your presentation? It may be okay for your journal club presentation but not for a national meeting: you need permission or license from the copyright owner (which may be the artist or publisher).

Ensure that your images are correctly labeled, captioned, and credited (where necessary). Label the *x*- and *y*-axes in charts, identify units of measurement, and explain all abbreviations in the legend. The legend might also be a good place to indicate some key statistical parameters, such as *P*-values or confidence intervals.

If you are submitting to a journal, ensure that visuals are allowed, how many are allowed, and if the journal publishes color images. If you're struggling with a word count for a paper or meeting abstract or some other document and trying to get away with including more information, it might be that it can be added in figure legends and table subscripts, but journals may also impose a printed page limit along with a word limit. So check. Rarely are images actually embedded in the text of an article for a journal submission. They often occur at the end of the article or in a separate file, with only a reference to where they go in the article (and, more frequently, may only occur in the online *supplement* of the article, and not the article itself). Referencing articles also differs significantly in style from journal to journal. You may be using "Figure 1" or "**Fig 1**" or "see figure 1" or "fig 1," or any of a host of other styles. Ensure from the get-go that you are referencing your figures correctly and that you are providing legends correctly and containing the information required for the journal. What does the journal use in tables? "P?" "*P?*" is there a space around the operators in the *P*-value ($<$, $>$, $=$)? How do tables and figures get footnoted? With numbers, letters (upper case? lower case?), or symbols ($*$, $\S$, $\dagger$, $\ddagger$)? Journals will also have technical specs such dots per inch (dpi, or how many dots of color appear on a linear inch) for print or pixels per inch (ppi, an electronic version of a dot) for electronic (print publications and photographs typically require a much higher resolution), image size (in dimensions or MB), or file format (e.g., JPG, PNG, TIFF, etc.).

How Do I Create a Visual?

There are some general principles you need to consider before you start working with images.

Colors, Patterns, and Shading

If you know what you're writing is going to contain a lot of graphics, *check to make sure first if you can use color.* The journal to which you might be submitting may not allow it, the book for which you are writing a chapter might not allow it, or it might not be allowed for your dissertation or thesis. This is very important. If you are crafting a paper that centers around micrographs of some beautiful cells that are fluorescing a brilliant green, will it "work" in black and white? What if you have chromatography, or tracings or wavelengths in contrasting colors? What if you have color-mapped brain imagery—can you show the enhanced glucose uptake without the different colors in the brain scan? What if you're writing about bird plumage? *Check first.* If you cannot make your point without the use of color, you must ensure that you can find a publication route where it can be used (perhaps as an online supplement) or ensure that you can capture the information adequately with prose.

Not all color is created equal. If a journal or printer accepts "full color," that refers to the basic four-color printing ink system: cyan, magenta, yellow, and black (CMYK), which, combined, creates full color. Greyscale images are not technically black and white (which is very stark) but rather images reproduced in shades of grey. If you absolutely don't need color to make your point, greyscale can often provide the information you are trying to convey. Half-tone images look like they're composed of little dots (see Figure 6.2).

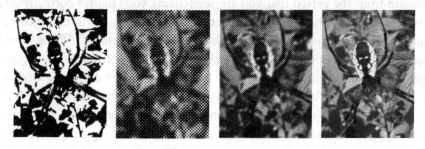

Figure 6.2. *Argiope aurantia*, the garden spider, rendered in (left to right) high-contrast black and white, two varieties of halftone, and greyscale.

Use color wherever it conveys information and clarifies meaning (e.g., elucidates differences in data of any sort, creates meaningful contrasts, or is needed for accuracy). If you are limited to black and white (by publishers or printing costs), you can accomplish many of the same things you would do if you could publish in color. You can use shades of grey or patterns to differentiate between entries in a graph, such as dashed versus solid lines, open circles versus solid squares, or shades of grey to differentiate between bars in a bar graph (see Figure 6.3).

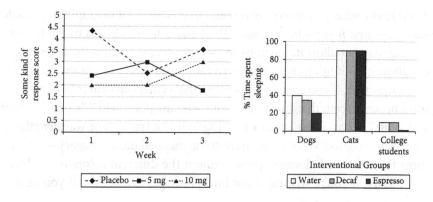

Figure 6.3

Composition

Another important concept to keep in mind is *composition*. If you are taking a photograph yourself or making your own illustrations, you have more control over this than if you are acquiring images from elsewhere. Effective images contribute to the content; if you are speaking of a particular ligament in the knee joint, an image of that immediate area would be better than an image of the leg, thigh and calf inclusive. If you are photographing something— either directly or through a microscope lens, try to isolate that which you're photographing against a neutral background. Get the seaweed out of the way of the mussels you're photographing and clear the bench of any equipment not being discussed in your photograph. Try to find a cell in your field of view that is not overlapping others and is not near a bubble or an errant piece of lint. Keep it as simple and clean and demonstrative as possible.

Electronic and Print Images

Images prepared for print need a much higher image resolution than those used in electronic media. This is important to keep in mind if you are creating something that starts out as electronic but later (or additionally) is rendered as a print piece. If you are getting an image from elsewhere and it is destined for print, remember to acquire a *high res* version of the image! With lower-resolution images, you will lose detail. That is fine for a bar graph, but not if you're trying to show delicate staining patterns in a mouse cerebellum or crystal patterns in a geologic polarized photomicrograph. Good resolution for print pieces should be at least usually 300 dpi, preferably higher, and usually a minimum of 200 ppi for digital images (although these "cutoffs" vary, depending upon the image). A great digital image at 100 ppi will not look so crisp in print.

Let's say you have the most fabulous high-res photos of the stress fractures in the steel samples you've generated in your physics lab, and they look great in the assignment you've written, but what happens when your prof, duly impressed, suggests you present a poster on Research Day? Next thing you know, you enlarge the photo for the poster and your good mood is ruined: enlarging all of those tiny little dots from 150 dpi to 400 dpi turns your glorious high-res photo into a fuzzy, impressionist rendering. Loss of resolution may also mean loss of the most important information embedded in the image. So, plan ahead. Don't make that poster the evening before Research Day.

What Kinds of Images Are We Using in Scientific Writing Projects?

Your choice of visual depends, as always, on your rhetorical goal. What do you need your audience to focus on? What's the take-home message? If you have raw data (e.g., age, sex, geologic period, etc.), put them in a table. Experimental or observational results that measure and compare stuff? Graph them. Atmospheric observations? Dissections? Rock formations? Stress fractures? Photographs. Processes? Put those in decision trees, algorithms, or diagrams (see Table 6.1).

While it's certainly true that, depending upon what you're writing about, there is going to have to be a fair amount of narrative description, an image can go a long way toward helping your audience understand what you're saying and saving you space. The whole "picture is worth a thousand words" thing is absolutely true. Do you have a four-page paper due? Is the journal article you're writing limited to six final single-spaced pages or have a cap of

Table 6.1. Common Types of Visuals and Their Uses

Visual	Goal
Tables	To present exact values; to organize raw data or data that do not fit simple patterns; to classify information
Graphs	To show trends or patterns in data
Line graphs	To summarize trends, compare continuous variables, relate data to constants, emphasize patterns over specific measurements, show change over time
Bar graphs	To clarify differences or draw comparisons
Pie graphs	To show percentages out of a whole, compare, contrast
Histograms	To display the distribution of data by frequency (note that a "histogram" is not a "bar graph"; these terms are frequently interchanged—incorrectly)
Scatter plots	To display individual data points that, taken together, indicate a trend (or not)
Diagrams (e.g., flowcharts, algorithms, decision trees)	To illustrate complex relationships, configurations, processes, decision trees, or interactions, causation, or hierarchies
Maps	To illustrate geographic relationships or trends that involve location and distance
Photographs	To show a realistic representation of an object
Drawings or illustrations	To explain mechanisms; to describe objects and processes; show detail, provide specific labeling; to show complex interactions

2500 words (three single-spaced pages)? Does your lab write-up have to fit onto the preprinted template you've been given? Page, word, and character limits can be the bane of scientists trying to communicate. And *what* are you writing, anyway? Sure, there are papers, but there are also lab write-ups, you may be called upon to make presentations, meeting posters, bulletins or newsletters. Review papers are very different from IMRAD papers, which are very different from meeting abstracts. Are you allowed to have images in your meeting abstract? In your assignment at school?

Tables

What goes in a table? Lists of things, raw data, discrete categories of data. Tables can be gigantically huge (check the back of your trigonometry or chemistry books) or really small. Remember Mendel's Punnett squares? (Figure 6.4) Also, please note the "Punnett" is a proper noun). As with any visual display of data—the simpler the better. Use only the minimum amount of vertical and horizontal lines that you can get away with to delimit data (tip: you can safely eliminate all vertical lines and then work backwards to see where you *really* need them—e.g., if columns are too close to each other).

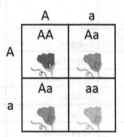

Figure 6.4. Reginald Punnett and Gregor Mendel: Ruining Peas 4 Lyfe.

After our intrepid geologist finishes gloating over his Devonian brachiopod find, someone somewhere in a position of educational authority is going to make him communicate about it beyond his Instagram page. We are going to assume that our geologist has been thorough in his collection of information *before leaving the site.* This isn't experimental, but again, considering things in context is what's important in science, and the details are what provide the context. You do not want a lab notebook full of data that are useless without context (heads up on that one). So what goes in his tables?

He hit a gold mine of fossils in a hidden part of a roadcut through the Oriskany Sandstone formation. He saw all manner of brachiopods, clams, snails, cephalopods, trilobites, crinoid fragments, and, upon closer inspection with a loupe, he even found some conodonts. These treasures varied in number and size; some were the actual fossils, and others were molds. He could identify some of them, and others he could just classify as "bivalve" or "brachiopod." This, you must agree, certainly qualifies as a "list of stuff." How can this information be organized into a table (see Table 6.2)?

Here he's made a rough table, scrawling away in his lab notebook as he sifts through his treasures, into which he has organized his fossils with columns to indicate how many there were total, and of the total, how many were full fossils, partial fossils, or molds. Assuming there isn't much variability in size, he also has a column for "size." If there were extreme variability in size, he would probably require a different kind of (or additional) table. This is a very common type of table encountered in IMRAD-style papers: it displays basic facts (e.g., genus, size, whole/partial fossils, etc.). Continuous data, such as size, etc. can be expressed as means or "packaged" into discrete ranges (e.g., "1–3 mm," "4–6 mm," etc.).

Table 6.2. Table Full of Stuff

Fossil	Number	Rough meas.	Whole fossil	Pieces of fossil	Imprint, mold or cast
Brachiopods	42	3–6mm	9	21	12
Rensselaeria	18	Appr 4mm	10	2	6
Leptostrophia	19	Appr. 5–6mm	10	0	9
unidentified	5	3–6mm	0	5	0
gastropods	18	2–6mm	3	11	4
Loxonema	12	Appr 2mm	5	5	2
Platyceras	5	5–6mm	1	4	0
unidentified	1	3mm	0	1	0
bivalves	16	4mm-ish	13	3	0
Actinopteria	14	"	13	1	0
unidentified	2	2mm	0	2	0
cephalopods	5	7mm	0	0	5
Michelinoceras	5	"	0	0	5
trilobites	7	5mm	6	1	0
Dalmanites	6	5mm	6	0	0
unidentified	1	7mm	0	1	0
Crinoid parts	31	3–42mm	0	31	0
Conodonts	Est 10–15/cm2	Appr .75–1mm	all	0	0

When creating a table, start with a very basic grid shape, as our geologist has done in his lab notebook or on a blank page on his laptop. You can make the table look better later. While you're filling in the table, though, you want it to be as clear and usable as possible—use big enough cells to capture all the information. You want to be able to fill in the table clearly. Once your rough table is complete—that is, filled with all of your data—you can tidy it up. Almost all word processing programs have table templates that can speed the process up for you tremendously.

Our geologist has been hunched over in his lab, tirelessly tallying, identifying, and measuring his finds, either from photographs of the roadcut, or from actual samples he has collected. Now his tidy, publishable table looks like Table 6.3.

Notice how much cleaner and visually appealing this table is: formatting has become consistent; words and values are line-centered; scientific genera have been italicized; groups have been clearly delineated with shading, indenting, and bold type; labels have been simplified. Notice that

Table 6.3.

Organism	N	Appr. size (mm)	Full	Partial	Mold
Brachipods	42	3–6	9	21	12
Rensselaeria	18	4	10	2	6
Leptostrophia	19	5–6	10	0	9
unidentified	5	3–6	0	5	0
Gastropods	18	2–6	3	11	4
Loxonema	12	2	5	5	2
Platyceras	5	5–6	1	4	0
unidentified	1	3	0	1	0
Bivalves	16	4	13	3	0
Actinopteria	14	4	13	1	0
unidentified	2	2	0	2	0
Cephalopods	5	7	0	0	5
Michelinoceras	5	7	0	0	5
Trilobites	7	5	6	1	0
Dalmanites	6	5	6	0	0
unidentified	1	7	0	1	0
Crinoid fragments	31	3–42	0	31	0
Conodonts	10–15/cm^2	0.75–1	all	0	0

this table will readily reproduce in black and white (or, preferably, grey-scale) and be as clear as the original. The table should be accompanied by a legend that has just the right amount of information: "Devonian fossils from the Oriskany sandstone formation, Sussex, NJ" works as a legend. It's not necessary to add in the exact latitude and longitude of the site or that these fossils were extracted from a roadcut or how they were extracted or the exact period during the Devonian or any other details like these (although you will see this in some papers). Extra information, if important, will be in the text. Conversely, the details about number, size, full or partial fossils, etc. will not be in the text but rather in the table. At the appropriate place in the text, the author will refer to the table. The caveat here, of course, is that your destination for what you're writing may require more or less detail: more detail in a thesis, dissertation, or government document, less detail in an abstract or meeting poster or four-page lab write-up. And no matter what kind of legend you create, your editor or professor will probably help you clarify it anyway. You will see in some scientific papers or meeting abstracts extraordinarily long legends accompanying tables. These are useful if very specific detail is needed for the table (e.g., information about statistical methods, or physiologic pathways, etc.) *or* the author(s) are trying to save word count in their paper by including some of the information in the legend.

10 Tips for Creating Tables

1. Do not use any horizontal lines unless needed for clear delineation of sections. Alternating lines (or sections) of tables can be created in separate colors or shades for clarity (many table templates will do this for you). Keep in mind that if you're submitting to a journal, they'll have their own style to which you must hew.
2. Do not use any vertical lines (but see aforementioned exception).
3. Use alternating colors, indentations, or font styles (e.g., italics, bold face, etc.) to delineate groupings and subgroupings.
4. Be consistent with line centering and column centering—within the table and throughout your document.
5. Make sure your column headers are short and comprehensive (e.g., "*P*" will suffice; you don't need "*P*-value").
6. Leave enough space between columns so that the text is intelligible.

7. Do not use multiple colors, patterns, textures, or typefaces unless critically necessary. If your table is really that complicated, consider making two tables—or can any of it be graphed?

8. If the table contains footnotes, ensure that they are written correctly according to the style you are following (e.g., APA, ACS, AMA, journal style, etc.).

9. Make sure the table is intuitive; if you have to explain it to someone, it doesn't work. *Use the conventions of your field.*

10. Group only like values in each column unless they are significantly related to each other (e.g., don't combine color and height, but N and % are okay).

Graphs

Graphs are the scientific mainstay of visual presentation of data being compared. While tables are a useful device for *listing* things, graphs are used, broadly speaking, to *compare* things, reveal complex relationships between data sets, and, in general, extract the maximum amount of meaningful information from your experimental or observational results. They are not always needed: take our geologist, for example. He probably *could* have graphed those data, but the result would not have been meaningful because he was not really showing trends or relationships. If your readers wonder why you've graphed something, you shouldn't have graphed it. Another reason why peer review is so helpful!

Most word processing and spreadsheet programs have some kind of graphing software built into them. For discipline-specific graphing, there is typically software available that provides more sophisticated graphing tools. Many statistics software programs (such as SAS and SPSS) also have graphing functions. You may be importing graphs from other sources (get permission) or creating graphs from scratch using your own data. A great deal of scientific equipment makes its own graphs; some examples are seismographs or chromatographs or electrocardiographs generated by tracing software or things like line graphs or histograms or scatter plots automatically generated by data collected by a computer. What makes for a good graph (see Figure 6.5)?

What kind of graph should you use? The first thing you should consider are the conventions of your field. There are specific ways, within your scientific discipline, to graph certain things: regressions, scatter plots, odds and risk ratios, radar graphs. The type of data you're trying to graph will dictate the type of graph you use.

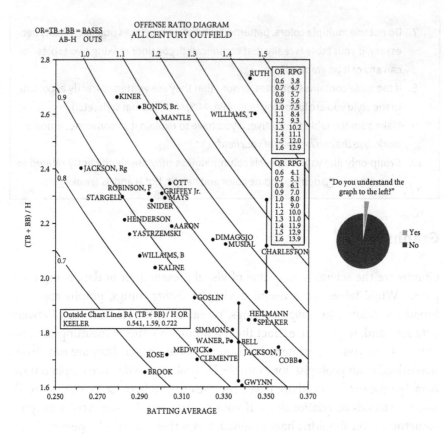

Figure 6.5. When one asks one's friends if anyone has a "really confusing graph," there's nothing like a buddy who's an engineer *and* a Sabermetrics fanatic. He cheerfully provided the graph on the left (which has something to do with baseball). This is an example of a *ratio graph* (offense ratio, in this case). The graph to the right is an example of a *pie graph* or *pie chart*.

Bar Graphs

Bar graphs (also called *column graphs*, or, incorrectly, *histograms*) are very commonly used in science to provide a snapshot of comparative data, either of a single group of variables over time or of multiple groups of variables. Our dogs-cats-students on espresso graph earlier in the chapter is an example of a bar graph, which compared dogs-cats-college students as groups and water-decaf-espresso as groups. Ensure that you label both the *x*- and *y*-axes. If necessary (depending on the kind of document you're writing), you can add things to the graph, like an N, *P*-values, or confidence intervals. Do you need all that information on the graph? Again, it depends on what you're writing. An editor or professor may even provide the specifications up front. What you *don't* want to do is make it 3D in an hilarious typeface with gratuitous shading and textured fill—not unless you want to give your readers seizures. Gratuitously fabulousized graphs are unnecessary (see Figure 6.6).

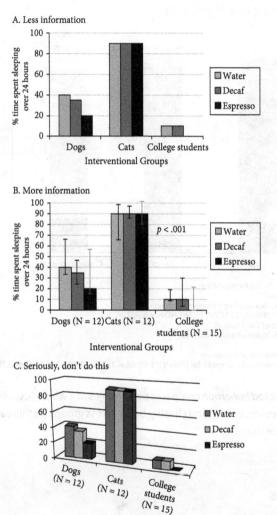

Figure 6.6. What kind of bar graph do you need? A. Make your graphs as minimally complicated as possible. Do you really need confidence intervals (error bars)? B. If you're submitting to a journal, chances are your graph will need all the bells and whistles: confidence intervals, *p*-values, N, and any other host of statistical information necessary to interpret the graph. C. Don't do this.

Bar graphs can be simple or multiple (Figure 6.6 is multiple), horizontal or vertical (there is no clear agreement as to when to use one over the other, so it's often a matter of preference), stacked (or composite) or not, which allows for multiple comparisons between variable components. What if you're trying to compare multiple variables across groups? You might want to consider a stacked bar graph (see Figure 6.7).

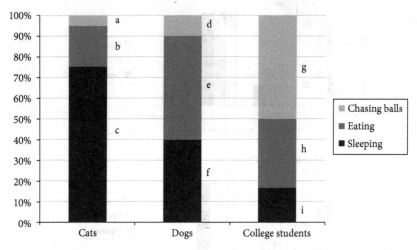

^aOnly if you need the ball for something
^bIf you call this "food"
^cPreferably on your face covering your nose and mouth
^dOMG CAN YOU THROW IT NOW? NOW? HOW ABOUT NOW?
^eWhat are you eating? Can I have it?
^fTrying to train me to stay off your bed is futile
^gSeasonal
^hHuge sale on ramen over at the safeway
ⁱDude I'm trying to sleep and your spotify list [IMPOLITE CHARACTERIZATION YOUR ROOMMATE'S SPOTIFY LIST]

Figure 6.7. A *stacked bar graph* compares percentages of a whole (usually) for categories (in this case, sleeping, eating, and chasing balls) across groups (in this case, cats, dogs, and college students).

Line Graphs

Line graphs are useful for comparing data over time: how has the earth's temperature increased over time, how has teen obesity in the population been increasing, what are rates of smoking over time? Note that these data can also be expressed in bar graphs, with a single bar representing a year or range of years. Figure 6.8 is a graph from the Centers for Disease Control and

Prevention (CDC) tracking the incidence of gonorrhea in the United States over about a 70 year period of time.

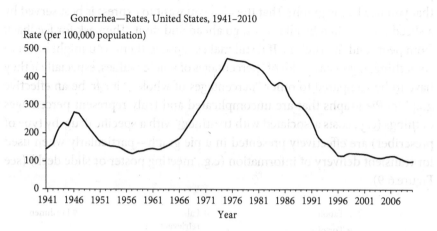

Gonorrhea—Rates, United States, 1941–2010

Figure 6.8. In this figure the rate (number per 100,000 individuals—this is a standard way of reporting epidemiologic data) of gonorrhea peaks a bit in the mid-1940s (can you think of why?) and then ... what was up with those fabulous '70s, anyway?

Line graphs can have single or multiple lines in a single graph. When you have *many* independent variables to plot, it makes sense to actually plot them separately and present them side by side, rather than trying to fit them all onto a single graph. While, say, a specific adverse event of two doses of an experimental drug with their respective placebos can be graphed with solid (experimental) and dashed (placebo) lines, if you have so many variables that the *y*-axis has to be compressed, consider separate graphs.

Pie Graphs

Pie graphs are used to show percentages out of a whole in order to compare relationships among component parts. The fact that they always add up to 100% is a double-edged sword: on the one hand, they are easy to understand at a glance; on the other, they may be too simplistic and often inadequate in representing how data sets compare. In fact, some professional organizations such as the American Medical Association (AMA) go as far as to discourage the use of pie graphs of any kind in their publications, suggesting that they are better left for lay audiences. If you have something that can be put in a pie graph, you'd be better off putting it as a summary in the text or, if you must, in a table. We will not tell you *not* to use pie graphs, but we do urge you to pay extra consideration to your needs and purpose. Are you absolutely certain that you need a pie graph? That the idea you want to express is best served by a sliced circle? Then by all means, go ahead and do it. Then ask for feedback from peers and instructors. If it still makes sense to them, you might be on to something. If you have a lot of "percentages of whole" values, especially if they have to be compared to other "percentages of wholes," it can be an effective graphic. Pie graphs that are uncomplicated and truly represent percentages of things (e.g., costs associated with treatment with a specific drug, by type of prescriber) are effectively presented in a pie graph—particularly when used for transient delivery of information (e.g., meeting poster or slide deck) (see Figure 6.9).

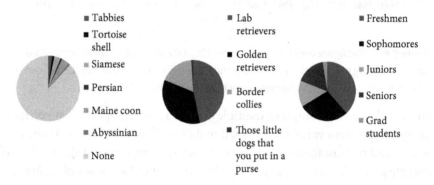

Figure 6.9. These pie graphs afford a quick visual breakdown of the individual populations that are chasing balls. How else could you graph this? Would it be as effective?

Histograms

Histograms are similar to bar graphs but are used to show the distribution (often grouped) of numerical data for a single variable in terms of frequency or intensity. Unlike bar graphs, there is no space between each data "column," and the columns are typically minuscule in width (hair's breadth in many equipment-generated cases). Histograms can be used to estimate probability distribution, such as those seen in the normal distribution graphs (e.g., the "bell curve").

Scatter Plots

Scatter plots are good for representing a number of individual data points and how close they are to (or how far they deviate from) a trend (or curve). Be sure to use simple symbols to represent your dots! Many scatter plots have a large volume of data and give a visual presentation of clustering of data points—and how they trend (or don't trend). Trend within multiple data points is typically determined by a regression analysis (have fun in stats!). Scatter plots are usually two-dimensional (see Figure 6.10), but there are some 3D (multiple axes) equipment-generated plots that contain thousands of points.

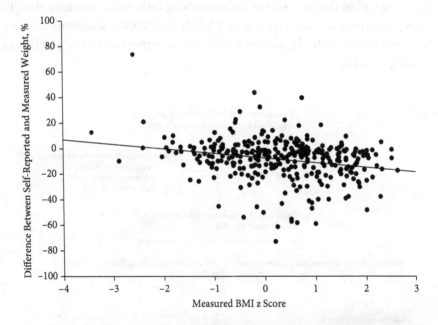

Figure 6.10. A scatterplot from Centers for Disease Control and Prevention, plotting self-reported body mass index (BMI, based on height and weight) by children ages 6 to 11. Each data point represents one individual. The trend is assessed with a linear regression (which yields the *z* score on the *x*-axis), shown as the line that is "fitted" to the scattered data points. Negative values (check the *y*-axis) represent an underestimation of weight. (Beck J, et al. Accuracy of self-reported height and weight in children aged 6 to 11 years. *Prev Chronic Dis.* 2012;9:120021)

Diagrams, Including Flowcharts, Algorithms, and Decision Trees

Use these to show processes, relationships, mechanisms, and hierarchies. Diagrams are probably one of the most accessible types of visuals to you, as you can create them with Microsoft Word and even open-source software. You need to walk a fine line here between overtly simple diagrams (which are better avoided and explained in text) and overly complicated diagrams (which will be difficult to interpret). As usual, avoid any extraneous details such as unnecessary shading or patterns that may make the image unclear, 3D images, or decorations. Use arrowheads only if their direction has a specific meaning within the process you are describing (otherwise, use only straight lines to connect a visual element to a label). CONSORT diagrams are very common in clinical trial papers and often a requirement of journal publishers (see Figure 6.11).

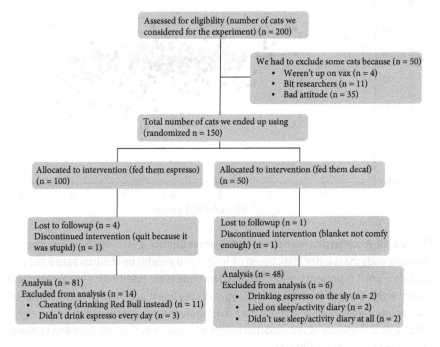

Figure 6.11. If you were doing a legit scientific study on the effects of espresso on cats, dogs, and college students, your study disposition might be a CONSORT-y one like this, not that study dispositions typically contain results, as this one does.

Images

Photographs and Drawings

When your topic calls for a more accurate visual representation of an object, you can't go wrong with either of these. Drawings are preferred for representations of microscopic entities that are not easily captured by photography because of their size (e.g., cellular, atomic, or molecular processes). Unless you create the drawing or the photograph yourself, make sure you obtain permission from and credit its creator. If you use photographs of human subjects, make sure you get them to sign a release or consent form. We have already emphasized the importance of resolution and color in photography. Never lose sight of your final medium (print? electronic? slides? meeting poster?). What kind of resolution do you need?

Maps

Maps not only show the geography of a place but can reveal patterns in the distribution of data (population, disease, climate, soil composition, water resources, etc.). Use them whenever the geographic location is relevant for your topic—and *always* include the typical map conventions: orientation (e.g., "north"), the scale of your map, and legends for the interpretation of colors or patterns in your map. Our geologist, for example, would be well served by a map of the area wherein the Oriskany sandstone formation is located. What are the different kinds of maps our geology student could use?

Writing Accompanying Text for a Visual

All your visuals must be properly labeled and captioned, as well as referred to in the text. Making visuals stand out ensures that even a busy reader such as our tweeting geologist can grasp the importance and meaning of the results even though he might not have sufficient time to engage in the whole text. It is true: most readers will zoom in on the visuals before they get around to the text. By convention, label tables "Table" and number them sequentially and everything else "Figure," also numbered sequentially (and separately from tables). If your paper only has one table or one figure, do not number them at all (just label it "Table" or "Figure"). Place the title of the figure immediately after the figure number (and don't repeat the title *inside* the graph).

The title of your figure or table should be succinct, specific, descriptive, and summarize the content. In general, use title case—that is, capitalize all major words in a title, but check with your instructor or the journal guidelines first. As a rule, table titles go on top of the table, whereas figure caption titles go under the figure; however, you should check the journal's publishing guidelines to see exactly what conventions they follow. Do not include the type of figure in the title:

> *Avoid*: Figure 1. Stacked Bar Graph Indicating the Distribution of Teenage Births by Country, 1978–1979
> *Better*: Figure 1. Distribution of Teenage Births by Country, 1978–1979

There is one exception to this rule: when you are using an advanced imaging technique that requires identification so that the reader can correctly interpret what they are seeing—for example, electron micrographs, fluorescein angiogram, autoradiograph, etc. In this case, write "Figure 1. Electron Micrograph Showing the Structure of Streptococcal Strains."

Captions (or legends; used interchangeably) are always printed below or next to the figure, and should (depending upon the type of figure) (a) identify the figure; (b) describe the figure; (c) explain abbreviations, labels, symbols, scales, units of measurements, and other variables when necessary; (d) provide statistical calculations based on which trends were compiled; (e) include footnotes for information too cumbersome to include in the figure; and (f) credit, when necessary (e.g., "used by permission from" or "courtesy of," etc.).

Even if your figure is now independently intelligible (more or less), thoroughly labeled, titled, and captioned, you still need to mention it in text—either in a descriptive sentence ("Table 1 shows . . .") or as a parenthetical notation at the end of a sentence giving an explanatory account of the data ("see Table 1" or just "Table 1"). Some disciplines prefer one style over the other (AMA, for example, prefers the latter), so make sure you check the conventions of style in your field and journal requirements if you're creating a manuscript for submission. Avoid writing "the figure above" or "below" since where figures are placed in text is often unpredictable, especially during the publishing process. Refer to the figure or table number. When you're preparing manuscripts, the publisher is likely to request that you indicate approximately where in the manuscript the image or table should be placed.

10 Tips for Graphs

1. Remember the "GIGO" rule: garbage in, garbage out. Your graph is only as good as the data you use to generate it. So check your data. Sometimes it's actually when you see your data graphed that you realize something is amiss.
2. Pick the right kind of graph for the data you are trying to display (see Table 6.1 for some general guidelines).
3. Ensure that all the labels you need for the graph to be truly useful are on the graph, *but*...
4. ...don't use gratuitous labeling (i.e., putting every label possible on the graph. Do you really need those confidence intervals?).
5. Ensure that the legend is complete enough but not overlong.
6. Follow directions for graph style (e.g., are the footnotes "1, 2, 3," "a, b, c," "*, †, ‡," or something else?).
7. By convention, plot independent variables on the x-axis and dependent variables on the y-axis.
8. Do not expand or compress the y-axis to make nonsignificant data look significant. This is unethical.
9. Provide the statistical method used to generate the data and the statistic that summarizes the relationship between the dependent and independent variables, such as a correlation or regression coefficient, in the figure or legend.
10. Keep. The. Graph. *Simple*. You don't need 3D, beveled edges, shadows, separated pie pieces, or any other graphing fabulousness options the graphing software may offer.

Document Design

Formatting a scientific paper does not have to be a complicated affair; again, keep it simple, and again, follow the dictates of your professor or journal, or hew to the editorial style peculiar to your scientific discipline. Without specific dictates, a good rule of thumb is to use a simple, common typeface such as Times New Roman with a 12-point font, or Arial 11 point. There was a trend toward using san serif typefaces (typefaces that have no "tags" on the letters; Times New Roman is a serif font, whereas Arial is sans serif), although presence or absence of serifs (and how big the serifs might be) fluctuates in fashion. There is no dogma for typeface or font. Follow necessary directions and if none are provided, keep it simple and clean.

Unless otherwise directed, use one-inch margins and double-space the entire document. Provide a separate title page, a separate abstract page, and a separate references page (again, this may vary depending on the destination of your document). Do not use fancy or large fonts for the title page—as always, keep embellishments to a minimum. On the title page, center the title and provide information about yourself underneath: name, degree if applicable, place of work, date, and, if necessary, word count for the paper (which should not include front and back matter such as title page, abstract, and references). If you are submitting a paper to a journal, you may have an exact template to follow that includes other requirements, such as author disclosures or which author did what for the study (e.g., collected data, performed statistics, etc.).

Paginate your document (typically in the upper right corner) and include a header with your name, date, and an abbreviated version of the manuscript title in the upper left corner. Use headings, bullets, and other design elements consistently (stick to a style sheet), and, if possible, place figures and tables as close to the place in the text where they are mentioned. If that is not possible or proves to be difficult, move them all to the end of the paper and use figure placeholders in the text. Try to keep tables together on one page as much as possible. Avoid splitting them onto two or more pages. This is difficult if you're using double spacing (if the paper is published, everything will be single-spaced and look normal, but drafts are almost always double-spaced). A "double-space" dictate means everything is double-spaced: tables, captions, references, and other elements in the document.

Finally, *pay attention.* A serious pet peeve of professors is that all of this document formatting information is typically in the syllabus or included with the assignment—it's general information for *every assignment you submit.* Journals may punt a manuscript back without reading it if you have not followed formatting directions. Remember: it's better to do it right from the get-go than to have to go back and reformat later. So: think before you start.

Take-home

Visuals are an integral part of science and a powerful way to support your observations and arguments. A few simple but effective rules govern the use of visuals in sciences: economy, simplicity, efficiency, comprehensibility. Get familiar with the visual conventions in your field in order to learn how to plan for your own visuals. Put some work into the titles and captions of your figures: they are well worth it and your readers will thank you, too. Above all, make sure your visuals have a well-defined purpose and showcase the data.

Exercises

1. You have examined mortality rates for males and females with diabetes in individual states of the United States. Should you use a table, a graph (what type?), or some other type of figure? Explain your choice.

2. You have conducted a review of literature to identify the prevalence of autism spectrum disorders throughout the United States from 2000 to 2010 according to sex and age of children diagnosed. What is the best way to present the data? a table? a graph (what type)?

3. You have set out to study sediment quality in near coastal waters in the Gulf of Mexico in order to assess impact of the most recent hurricane. What sort of figures do you envision to produce for this paper? How can your audience best understand your data?

4. Find an example of a good visual and one of a less effective visual while perusing the literature in your field. What makes it good and what makes it weak, and why? How can you improve the weaker visual?

5. We are providing you with a series of tables documenting the leading sources of death in the United States in 1980 and 2007, respectively, according to several variables such as sex and race (source: CDC). Based on these tables, create meaningful and interesting graphs that reveal the patterns hidden in the data, if any. Explain your graphical choices in a brief paragraph. Work individually or in groups, and compare your work. Which representation was the most effective and why? Which data did you choose to emphasize and which did you choose to omit or downplay? Why? What other type of information would you need to make the most sense out of these data?

Table 1. Leading causes of death and numbers of deaths, overall population: United States, 1980 and 2007

	1980		2007	
Rank	Cause of death	Deaths	Cause of death	Deaths
	All causes	*1,989,841*	*All causes*	*2,423,712*
1	Diseases of heart	761,085	Diseases of heart	616,067
2	Malignant neoplasms	416,509	Malignant neoplasms	562,875
3	Cerebrovascular diseases	170,225	Cerebrovascular diseases	135,952
4	Unintentional injuries	105,718	Chronic lower respiratory diseases	127,924
5	Chronic obstructive pulmonary diseases	56,050	Unintentional injuries	123,706
6	Pneumonia and influenza	54,619	Alzheimer's disease	74,632
7	Diabetes mellitus	34,851	Diabetes mellitus	71,382
8	Chronic liver disease and cirrhosis	30,583	Influenza and pneumonia	52,717
9	Atherosclerosis	29,449	Nephritis, nephrotic syndrome and nephrosis	46,448
10	Suicide	26,869	Septicemia	34,828

Table 2. Leading causes of death and numbers of deaths, males: United States, 1980 and 2007

	1980		2007	
Rank	Cause of death	Deaths	Cause of death	Deaths
	All causes	*1,075,078*	*All causes*	*1,203,968*
1	Diseases of heart	405,661	Diseases of heart	309,821
2	Malignant neoplasms	225,948	Malignant neoplasms	292,857
3	Unintentional injuries	74,180	Unintentional injuries	79,827
4	Cerebrovascular diseases	69,973	Chronic lower respiratory diseases	61,235
5	Chronic obstructive pulmonary diseases	38,625	Cerebrovascular diseases	54,111
6	Pneumonia and influenza	27,574	Diabetes mellitus	35,478
7	Suicide	20,505	Suicide	27,269
8	Chronic liver disease and cirrhosis	19,768	Influenza and pneumonia	24,071
9	Homicide	18,779	Nephritis, nephrotic syndrome and nephrosis	22,616
10	Diabetes mellitus	14,325	Alzheimer's disease	21,800

Table 3. Leading causes of death and numbers of deaths, females: United States, 1980 and 2007

	1980		2007	
Rank	Cause of death	Deaths	Cause of death	Deaths
	All causes	*914,763*	*All causes*	*1,219,744*
1	Diseases of heart	355,424	Diseases of heart	306,246
2	Malignant neoplasms	190,561	Malignant neoplasms	270,018
3	Cerebrovascular diseases	100,252	Cerebrovascular diseases	81,841
4	Unintentional injuries	31,538	Chronic lower respiratory diseases	66,689
5	Pneumonia and influenza	27,045	Alzheimer's disease	52,832
6	Diabetes mellitus	20,526	Unintentional injuries	43,879
7	Atherosclerosis	17,848	Diabetes mellitus	35,904
8	Chronic obstructive pulmonary diseases	17,425	Influenza and pneumonia	28,646
9	Chronic liver disease and cirrhosis	10,815	Nephritis, nephrotic syndrome and nephrosis	23,832
10	Certain conditions originating in the perinatal period	9,815	Septicemia	18,989

Table 4. Leading causes of death and numbers of deaths, whites: United States, 1980 and 2007

	1980		2007	
Rank	Cause of death	Deaths	Cause of death	Deaths
	All causes	*1,738,607*	*All causes*	*2,074,151*
1	Diseases of heart	683,347	Diseases of heart	531,636
2	Malignant neoplasms	368,162	Malignant neoplasms	483,939
3	Cerebrovascular diseases	148,734	Chronic lower respiratory diseases	118,081
4	Unintentional injuries	90,122	Cerebrovascular diseases	114,695
5	Chronic obstructive pulmonary diseases	52,375	Unintentional injuries	106,252
6	Pneumonia and influenza	48,369	Alzheimer's disease	68,933
7	Diabetes mellitus	28,868	Diabetes mellitus	56,390
8	Atherosclerosis	27,069	Influenza and pneumonia	45,947
9	Chronic liver disease and cirrhosis	25,240	Nephritis, nephrotic syndrome and nephrosis	36,871
10	Suicide	24,829	Suicide	31,348

Table 5. Leading causes of death and numbers of deaths, black or African American: United States, 1980 and 2007

	1980		2007	
Rank	Cause of death	Deaths	Cause of death	Deaths
	All causes	*233,135*	*All causes*	*289,585*
1	Diseases of heart	72,956	Diseases of heart	71,209
2	Malignant neoplasms	45,037	Malignant neoplasms	64,049
3	Cerebrovascular diseases	20,135	Cerebrovascular diseases	17,085
4	Unintentional injuries	13,480	Unintentional injuries	13,559
5	Homicide	10,172	Diabetes mellitus	12,459
6	Certain conditions originating in the perinatal period	6,961	Homicide	8,870
7	Pneumonia and influenza	5,648	Nephritis, nephrotic syndrome and nephrosis	8,392
8	Diabetes mellitus	5,544	Chronic lower respiratory diseases	7,901
9	Chronic liver disease and cirrhosis	4,790	Human immunodeficiency virus (HIV) disease	6,470
10	Nephritis, nephrotic syndrome, and nephrosis	3,416	Septicemia	6,297

7

Research and Documentation

© 2020 Kris Mukai

Biotechnology: that's pretty sure the program you want to be in. Bending science to your will, manipulating nature to achieve progress—progress that can save the world! Challenged with "pick a topic, any topic" in your intro to biotech class, you decide to explore genetically modified foods. Hot topic. Maybe you can create your own Frankenfood right there in the lab! Your professor suggests perhaps you should start with one of the *steps* in the construction of Frankenfood and try to replicate an extant process using the basic lab equipment available. Isolation and purification of a protein—a growth hormone, perhaps? Hormones are proteins. Proteins are made out of amino acids. Amino acids are made out of carbon, hydrogen, oxygen, and nitrogen, with the occasional sulfur atom mixed in for seasoning. How do you start? Easy. You take carbon, hydrogen, oxygen, and nitrogen, put them in a test tube, and shake it up. When you tell this to the graduate TA running your lab, though, he gives you "the look" and suggests you do a little background research first.

While this isn't as much fun as creating organic molecules from scratch on the lab bench, you recognize the utility of background research. So, as any conscientious student, you fire up Wikipedia to get started.

Does that sound familiar? And as you ponder that: Is that what your supervisor expects to find in your reference list?

We thought so.

While we do not discount the value and many uses of Wikipedia as a casual source of information, we strongly discourage you from using it. If you must, use it only as a starting point for concepts that need to be verified, double-checked, and entirely replaced with more credible sources of information. Otherwise, your quest to create Frankenfoods or isolate a protein will not get very far. Wikipedia information is only as good as its editors, who are not financially compensated or incentivized by fame, fortune, or tenure; sources can be—and often are—unverified, missing, biased, or even fake. Even that stalwart of curious minds everywhere, Google, relies on popularity algorithms that should have nothing whatsoever to do with finding reliable scientific information.

The research references you have to produce in support of a *scientific* paper have to meet a much higher standard than Wikipedia, howstuffworks.com, and random Google search results. If you write "so far there have been no studies on the significance of N-terminal purity in protein synthesis" because they didn't show up in the first 20 Google hits, you had better be *very* sure that no such studies exist. If you cite two lesser studies in support of one idea, but neglect to mention other more significant or well-known studies, readers are going to call you on that. If you cherry-pick the sources that support your ideas, philosophy, agenda, or research, you're going to get—justifiably—called on that too. And finally, and most importantly, if you reference the *wrong* types of sources (i.e., non-peer reviewed), no one in science will take you seriously. You have to be precise, thorough, and rigorous at all times, exercise good judgment when identifying legitimate sources, and respect the scientists that came before you and on whose research you rely. As Newton famously once said of his predecessors, "If I have seen further, it is by standing on the shoulders of giants."

The Role of References in Scientific Writing

There are important reasons you should put a lot of effort into your references:

Authority

Your references are a sign of your *ethos*, a rhetorical term used since antiquity to denote authority and credibility—or, in other words, mastery over your subject. You are what you cite—and the way you cite can also make or break your paper. Your references are a sign that you have read widely and that you have judiciously considered the relevant literature already existing on your topic. They show your scholarship, maturity, hard work, and critical judgment.

Continuity

You cite because you want to place your work in the context of pre-existing research. You start the research on your topic way before you even start doing your actual, original research—to learn from the experience (both good and bad) of others, avoid duplicating results, and place your research in the most appropriate and useful context of the field. While there is justification (and often a need) in your wanting to replicate or verify the intriguing results of a previously conducted study, there is little more irritating than finding out that your original research hypothesis has in fact already been covered by a variety of studies and has already become well accepted or even disproven. To provide a rationale for your study, you need to provide context; to provide context, you need to be thorough about your research; to show your thoroughness, you need cite that pre-existing work as support for your rationale. Where is there a gap in knowledge that your research can fill? Which of your methods have been validated by previous research? Showing the readers that you have carefully given this some thought and research helps position your work within your field.

Reproducibility

You acknowledge the work of others out of courtesy, to provide acknowledgement, show respect, and behave ethically—of course—but also to allow your reader to locate your sources independently and thus have a chance to learn *at least* as much as you have from the work of others. Any reader should be able to corroborate everything in your citations and bibliography. Without complete citation, there is a hole in the reproducibility of your work. A good reference is sometimes worth its weight in gold from the standpoint of your own efforts to decide where to start in your research. If you find that an early step in your study design has been well researched and validated, you may be able to start your research a little farther down the line than you had originally intended. For example, if you find out in multiply iterated published studies that *E. coli* is an extraordinarily effective vehicle for replicating protein via fermentation, then you probably do not need an early stage in your research wherein you try to identify the most effective vector for protein amplification.

Surprisingly, sloppy use of references is more common than you think. One scientific editor in fact revealed that he often found it difficult to track down references in scientific papers, having to resort to a veritable scavenger hunt to find some of them. He discovered that references had the "wrong year, wrong volume, or wrong page," and sometimes wrong author(s), title, or, worse, wrong content (Ole Bjørn Rekdal, Academic urban legends, in *Social Studies of Science*, 2014). Every time information gets entered into a document (including references), there is a window for error, which can easily get amplified via cutting and pasting. While often unintended, such errors might occasionally be heinously deliberate moves. Let's say you read a helpful article that makes reference to another study that sounds quite promising. What would *you* think if you tried to locate that reference and were not able to because the title, author, journal, volume, issue, or page numbers were wrong—or worse, you found it, but it said something completely different than what was represented in the article that cited it? Would that cast serious doubts about the veracity of the original article? It should. Rest assured: professors as well as reviewers *will* check your references—and when they do, make sure they can find them in good order based on the bibliographic information you provided.

References in Scientific Writing: Sources and Legitimacy

Types of Sources

When you write a scientific paper in any field, you should rely on other *scientific* and *peer-reviewed* articles, books, or reports. This means credentialed authors published in reliable scientific journals, issued by established academic presses or provided by reputable research institutions, organizations, agencies, associations, or councils. What all these sources should have in common is, in rhetorical terms, *ethos*: authority conferred by a respected editorial board, blind peer-review, a principled editorial process, and a research and/or education-driven mission (as opposed to for-profit). Depending on your topic, you will use a mix of *primary, secondary, and tertiary* sources. Learning the differences between these is important as your professor or supervisor may direct you to use only one type of sources for a project.

Primary sources: report original research: journal articles; conference proceedings; interviews; lab notes; patents; technical reports; theses and dissertations

Secondary sources: summarize existing state of knowledge at the time of publication: review articles synthesizing primary research, monographs; textbooks; treatises

Tertiary sources: report condensed information or overview, not original ideas—used for general reference: compilations; handbooks and textbooks; dictionaries; encyclopedias

Using *solely* sources vetted by the scientific community is a basic requirement of gathering essential background information for a scientific paper. It is a principle as simple as it is difficult to follow—after all, the siren call of Google or Wikipedia is *so* tempting! As with all rules, there can be exceptions—for example, you may want to use popular sources (e.g., news sites, reportage, etc.) to describe a popular cultural phenomenon that provides a springboard for your argument. Thus, if you were writing about the controversy surrounding the studies showing a link between vaccines and autism, you might be allowed to cite recent popular press coverage of the issue if you feel it is relevant to your paper, or cite the occasional data from an unscientific survey of public opinion (as long as it is identified as such). When in doubt about the authority or legitimacy of a source, check with your instructor, editor, or librarian first.

Legitimacy

How do you know a source is authoritative? Who is an "expert?" You might have an intuitive answer to that, particularly as you become educated in your research area, but the staggering glut of information out there can blur the boundaries between legitimate and dubious sources. While the Internet is certainly the main vehicle for literature search, both scientific and nonscientific, and while it's true that the Internet has made resources more readily and conveniently accessible, it is also true that it has introduced a lot of unwanted noise into the resource identification process. With solid critical thinking skills and continued exposure to literature searching techniques in your science classes, you can educate yourself on how to tell a good source from a questionable one and learn what qualifies a source as valid and in your field. Doing background research on cloning? You'll be wanting to check out the Office of Biotechnology Activities at the NIH, not Clones R Us.

Websites

If you find yourself on a website that is drowning in dubious ads and pop-ups and peddles fairly obvious conspiracy theories, your alarms will blare. However, someone with sufficient programming skills, an eye for design, and decent writing skills, can craft an authoritative-looking website that has nevertheless a nonscientific agenda or distorts science. You will need to dig deeper to figure out whether or not you can trust that site.

---

10 Tips for Identifying a Website as Unreliable

1. It is trying to sell you something (look at ".com" vs. ".org," ".edu," and ".gov").
2. It has ads for other products ("refi now! Local Mom Makes Docs Furious!").
3. It may be poorly designed: bad layout, neon colors, flashing icons, gratuitous clip art or images that appear to be scientific but which are not.
4. It contains poor, awkward, or inconsistent writing, riddled with grammar errors and typos.
5. "Voice" is inconsistent or unscientific.
6. Links take you to questionable pages (with clear bias or ads).
7. Using the reverse lookup ("link:[web URL]") yields some important clues (if 27 conspiracy theory sites link back to it, take that into consideration).
8. Claims are unsubstantiated and references are not provided ("scientists have clearly demonstrated that . . .").
9. Authors are not identified, or background information is not provided for the authors.
10. Page is hosted by a for-profit entity (which is either not identified, or not identified as "for profit"). The "About" page can provide further clues regarding ideology, financing, etc.

References in scientific writing bear a striking resemblance to references you are asked to provide on your résumé: who writes them matters a whole lot. And just as your potential employer will call your references to verify that they do, in fact, know and endorse you, your readers will check your references to make sure they are, in fact, appropriate and authoritative and say what you claim they say. Avoid sourcing your "hard facts" from second-hand reportage of the kind usually found in science writing, press releases, or pop science newspaper articles, because you are always taking the risk that the writer or reporter misunderstood or misreported an original study, either blatantly or subtly—which may in turn lead to your own misinterpretation of data, thereby compounding the original error. That is why we don't recommend including references even from established websites such as WebMD— a large portion of the writers there are not medical professionals, and even when they are, they address a variety of audiences, including lay audiences, often skipping significant details in the process. That detail might not be crucial for lay audiences, but it likely is for your purposes. A good rule of thumb in your assessment of sources would be to exclusively consider sources whose target audience is other scientists.

Journals

Your university library gives you access to databases devoted to scientific subjects and indexing all the major journals in your field. In general, most databases will index journals that are legitimate sources of scientific information—peer reviewed, with an editorial review board that includes experts in the field, and published by an established and reputable publishing house. These kinds of sources are usually behind a paywall, and your institution is paying good money to offer you access to them. As an exception to that, a lot of federally funded research should be available for free from various databases (notably, PubMed).

Impact Factor and h-Index

There are many ways of calculating the "value" of a publication; the most common ones are the *impact factor* (IF) for journals and the *h-index* for authors.

For example, the 2017 IF of a journal is calculated as follows:

X = the number of times articles published in 2015 and 2016 were cited in indexed journals during 2017

Y = number of citable items published in that journal in 2015 and 2016 (excludes editorials, letters, etc.)

IF = X/Y

Thus, if a journal has an impact factor of 1.2 in 2017, then its papers published in 2015 and 2016 were cited an average of 1.2 times in 2017.

The higher the IF, one may argue, the more prestigious the journal. Some of the most prestigious science journals (*Science, Nature*) hover in a celestially high IF range (37–40), while your run-of-the-mill, respectable second- or third-tier journal will have an IF under 5. By contrast, the *h-index* is a type of "personal" IF, measuring the productivity and impact of a particular author.

Both the IF and the h-index have been criticized for good reasons. For example, what if that multitude of citations comes from indexed journals that have a very low impact and are considered of marginal importance in the field? And who decides what's "marginal," anyway? What if something is cited for the sole reason of being refuted, disavowed, criticized, or paraded as an example of what *not* to do? What if one's h-index is bolstered by self-references (a scientist keeps citing her own works as often as possible)? These (and many others) are factors a simple calculation cannot take into account. However, while the accuracy and value of the IF and the h-index may be questionable, these measurements are still considered to be barometers of "relevance" by editors, publishers, journals, colleagues, libraries, scientists, and promotion and tenure committees, which is why you should be aware of what they mean, especially as you try to bolster your references section. Citations also serve an important social function in science; the higher the impact factor of the journal in which one publishes, the higher one's personal h-index and prestige. That is why publishing in high-impact journals such as *Science* or *Nature* is akin to striking the proverbial gold.

Open-Access Journals

How about sources that look acceptable and legitimate and are published as "open access" such as those under the aegis of PLoS (Public Library of Science)? Many such sources could be legitimate, but you should always consult both your professor and librarian for advice. Open-access journals arose as a reaction to the traditional publishing house model, which is slow and inefficient for researchers and expensive for the users (a single article behind a paywall may cost anywhere between $20 to $70, roughly). Offering the articles for free on the web substantially diminishes traditional print and subscription-related costs and provides faster access to new research. Its numerous advantages notwithstanding, open-source publishing can also allow for predatory editing practices (e.g., charging high fees to print articles regardless of quality). There has been a recent explosion of journals in a variety of scientific fields whose peer review process is essentially nonexistent and that take advantage of the academic pressure to "publish or perish" by adopting a publish-for-fee model (thus bypassing actual editorial work or printing costs). There are ways to figure out which publications (and publishers) to avoid, but opinions on the topic are, at the time of our writing, still in flux. We recommend that you consult your librarian and additional sites promoting integrity in research and publishing, such as Committee on Publication Ethics, the Directory of Open Access Journals, or the Open Access Scholarly Publishers Association. Simply entering the name of the journal into Google along with "reputation" or "predatory" should turn up red flags as well, if there are any. There are a number of free and paid predatory journal databases and whitelists. In the meantime, we can offer you some general advice on how to parse the wheat from the chaff:

---

10 Tips for Identifying Reliable Sources

1. Identify the *authors*. See if the authors' credentials are available for review, and if the authors (or main author) is qualified to weigh in on the topic. What else did they publish? Ensure that they have expertise in the subject and are associated with an educational institution or some other reputable organization. Check to see if contact information is available for the author(s).

2. Identify the *publisher* or *publication*. (Note: Publishers like Elsevier or Taylor & Francis may publish hundreds of scientific journals.) Publishers should be well known and a reliable source of information. See if the publisher is associated with a reputable professional (e.g., American Medical Writers Association) or scientific organization (e.g., the American Society for Microbiology), who sits on their editorial board, and how long they have been in business.

3. Identify the *audience*. Is the source written for a lay or a professional/scientific audience? Always choose scholarly over popular sources. Pieces written for popular audiences often oversimplify and sometimes even distort the original scientific research upon which they are based.

4. Determine the *scope*. How in-depth (or not) is the source? Ensure that it meets your needs and is *useful* to you. Read the source it in its entirety before you decide to use it. Journals that are too broad or vague in scope (e.g., *American International Journal of Research*) should probably raise your suspicions.

5. Ensure that the source is relatively *recent* and covers the latest knowledge about your topic. In science, timeliness is paramount. Information is becoming obsolete at lightning speed; in many scientific fields even 6-month old articles may be outdated—some are outdated before they even go to press. Do your best to find the most recent relevant sources for your project. (Your instructor may also impose date limits on your bibliography—check with her.) Remember, if you are using a website as a reference, the day you access it and use the information from it is the date that is permanently affixed to it in your bibliography (website access dates must be included in a bibliographic entry).

6. Make sure your source is *objective*. Your source should be free of bias of any kind and be possessed of no agenda beyond "education." Read the source critically and decide whether the author has employed judicious methods

and tried to eliminate bias. Also look for disclosure of competing interests, organizations funding the research, and other potential affiliations with commercial, political, religious, or other interests which might *potentially* indicate bias.

7. Make sure the source has its own proper *documentation*. The references cited in the paper must be legitimate (e.g., authoritative, objective, written for the right kind of audiences), and do not forget that your sources are only as good as, well, *their* sources. The sources you use should be based on scientific, objective, bias-free research written by vetted, legitimate authors with proper credentials in peer-reviewed publications.

8. Determine the *location* of your source. Remember that your path to your resources gives you a clue to their legitimacy; your level of trust should not be the same for PubMed, Google, Wikipedia, the library, and recommendations from friends, peers, professors, and librarians.

9. Your source needs to be *relevant*. Stick to what you need; use resources relevant to your *particular* topic, not *related* topics.

10. Your source should be *accessible* to other scientists. People who read your work should be able to find all of your resources, provided they have access to a good university library.

Let's go back to Wikipedia as an example. Can you tell why most instructors may have a problem with it as a source of information? Here are some top reasons:

- Wikipedia authors are often anonymous, so credentials cannot be verified.
- Entries can be edited by anyone, so pages can (and often do) contain errors or can be "hijacked" or "pranked."
- The audience is usually a lay audience.
- It is impossible to ascertain bias in the current Wikipedia operation model (although some efforts are made in that regard).
- While most articles have sources, they are not comprehensive, and not all of them are legitimate.

There are many legitimate institutional and governmental sites that offer solid secondary references that are peer reviewed, objective, based on quality research, and written by scientists for a general to specialized audience. In fact, sometimes you can only find the most recent data you need on such websites: guidelines published on association websites, environmental reports, agricultural forecasts, or continuous epidemiologic data such as those available on the National Health and Nutrition Examination Survey on the Centers for Disease Prevention and Control website. In general, the savvy scientist will exercise healthy skepticism when dealing with Internet sources and will apply judicious criteria like those outlined here to see whether a site passes muster.

© 2020 Kris Mukai

Source Formats

Print

We wouldn't be surprised to learn that 99% of your search for sources begins and ends online—from Google to PDF, so to speak. While the journal publishing model is definitely evolving, and few journals worth their salt are *only* available in paper format these days, there are still a few sources that in the old days you could only find in print but which are now more and more frequently found in electronic format as well:

- *Books—monographs, treatises, and others.* These may or may not be available electronically but are primarily intended for print.
- *Textbooks, handbooks.* They can be a useful source of "background" or general information on your topic. They are best used to reference general phenomena, processes, and procedures that are considered basic knowledge in your field. Advanced, higher-level textbooks are also a good source of, well, basic advanced, higher-level information—when these books are used, individual chapters are usually cited by title (as you'd do with a scientific paper), with the book information included. Many textbooks—particularly basic or introductory texts—are available in e-format or may contain an electronic form of adjunctive materials (e.g., online or in an included CD-ROM).
- *Journals.* Nothing can stop you from using the print version of a published journal article. NOTHING. Additionally, older issues of journals are often not available online—the journal may only archive back to, say, 1995, but the critical "landmark" paper you need dates to 1988. If you are working in an area of science where "the classics" are important, such as zoology or geology, you may find yourself trotting off to a museum to get the obscure 1903 paper on the type specimen of the saltwater shrimp you're working on.

Electronic

Most journals usually have electronic versions, enabling your convenient access, and some will *only* have digital versions. Journals that decide to go exclusively digital still have to make sure that the integrity of the peer-review process is maintained and the rigors of scholarly publishing are respected. Note that not all electronic journals are open-access journals (though many are). Making articles accessible online has numerous benefits:

- it ensures a broader distribution and exposure of the articles;
- it enables faster publication time (which is why you would see next to some articles the notation *Epub ahead of print*, meaning the article has cleared all peer review and editorial hurdles, is correctly formatted and ready for print, and is immediately accessible before the actual, physical print date—sometimes weeks or months in advance);

- it allows for convenient features such as in-text linkage to references, sources, etc.;
- it allows for publication of extra material, data, and figures that are not usually accommodated by print due to space and cost restrictions.

The other main type of electronic sources, websites, should be used with extreme caution, for all the reasons we have been outlining: they are rarely peer reviewed, may have commercial interests or other hidden agendas, are not always written by credentialed scientists, can be dated, and may be poorly documented. Use our criteria and exercise your critical thinking skills to figure out if you can trust the information on a site to as a resource for your scientific writing. In some cases, the web pages of respectable organizations, institutions, or governmental agencies may be acceptable (or even necessary) for your purposes as they may meet all or most of our criteria. In the end—it's a judgment call, and when in doubt you should *always* check with your instructor, mentor, or librarian to verify that a website would be acceptable as a source under particular circumstances.

When considering "electronic" sources, keep in mind that virtually *everything* has the potential to be electronic: books, manuals, articles, photographs, videos, recordings, newspapers, encyclopedias, and yes, even tweets.

Other

Sometimes references can come from unexpected sources: email or personal communication, interviews, press releases or press conferences, internal company memos or reports, newspapers, newsletters, government documents, archives, manuscripts, public records, surveys, special collections, microform sets (increasingly digitized these days), or even multimedia sources such as film. Again, you need to apply judicious criteria as to the credibility of the source and the way you intend to use it. Are you going to use a newspaper article reporting on the cultural controversies that for a while enveloped the use of Gardasil, the first vaccine against HPV (human papillomavirus), a known factor in cervical cancer? It depends—are you using it to provide information on the pharmacokinetic properties of the vaccine or are you using it to describe the unanticipated public backlash associated with the marketing this vaccine? If the former—think again. If the latter—perfectly okay.

Finding References

Databases

A vast majority of your sources will be indexed in one of the major databases to which your university or organization subscribes (if your organization does not subscribe to any, you should be prepared to incur no small amount of expense paying for papers, unless they're available online for free). The following are some of the most common databases used in scientific fields.

- PubMed. PubMed is hosted by the US National Library of Medicine (part of the National Institutes of Health) and provides access to well over 20 million citations in the natural sciences from around the world. PubMed is a resource tool that provides links to articles, books, databases, and informational websites. It links to full text articles when possible. It is the largest database of medical literature in the world, indexing over 9,500 journals since 1965, and the successor to *Index Medicus*, a print database started in 1965.
- ERIC (Education Resources Information Center) is hosted by the US Department of Education and indexes more than 2000 journals as well as tens of thousands of research reports available on microfiche from 1966 forward. ERIC is useful for searching the literature on "education," although this is broadly interpreted in a sociologic sense and contains much scientific content.
- EBSCO ("Elton B. Stevens Company," as it was originally named in 1944) Host Research Databases cover a variety of useful resources such as Academic Search Premier, one of the largest scholarly, multidisciplinary databases, and CINAHL (Cumulative Index to Nursing and Allied Health Literature), covering nursing and allied disciplines.
- ProQuest offers an aggregation of many different databases, such as Health & Medical Complete, ProQuest Medical Library, ProQuest Science Journals, ProQuest Newspapers, and ProQuest Research Library. Most articles are full-text.
- Dissertations Abstracts Online or Digital Dissertations—also available from ProQuest, offering access to American doctoral dissertations (more than 1.5 million of them since 1861, and counting!). Sometimes you can find hidden treasures within doctoral dissertations that were never published as articles or monographs.
- InfoTrac—available from Thomson/Gale, offers access to InfoTrac One file, which indexes 8,000 academic journals, general magazines, newspapers, and

more (more than half of those are available in full text), and Expanded Academic ASAP (3,500 academic journals).

- LexisNexis is a database known for its legal coverage but which also covers topics such as the environment or statistics (Lexis Nexis Statistical).
- PsycINFO covers about 2,000 journals from 1872 to the present, mainly in psychology and related fields, including anthropology, medicine, pharmacology, physiology, psychiatry, and sociology.
- Ovid covers basic science, nursing and health professions, behavioral sciences, as well as humanities and technology. Although PubMed has a more friendly interface, Ovid may offer you access to more databases—Medline, for one, and Embase, produced by Elsevier Sciences, which has a broader scope, covering more European and foreign-language journals.
- Web of Science (part of Institute for Scientific Information in Philadelphia) indexes over 8,500 journals internationally in all fields of the hard sciences, social sciences, and arts and humanities. This database is particularly apt in identifying journals with the best impact factors.
- Google Scholar provides wide access to scholarly literature (not just science) where you can access "articles, theses, books, abstracts and court opinions, from academic publishers, professional societies, online repositories, universities and other web sites." Sometimes if your keyword combination isn't finding you what you need on Pubmed, the same keyword combination on Google Scholar might find exactly what you need—on Pubmed.

This is not a complete list of databases. Your librarian or instructor may direct you to yet other similar sources that may be particularly appropriate for your field of research. Make sure you master the particular search algorithms or subject headings of the database you are most likely to use. They are not all the same, and do not all play by the same search rules. A session with your librarian will be invaluable in offering you the key to what may otherwise seem like the tedious, frustrating, and gargantuan effort. You can use keywords to search for what you need, but sometimes these databases do not work like your typical Internet search, so make sure you understand how you can get the most of them before you embark on a lengthy research process. Some databases such as PubMed have extremely good tutorials providing an excellent introduction on using the database, and you are well advised to take advantage of such resources.

In most cases, the database search will return bibliographic information and abstracts. The title of an article should be good enough to get you to want to read the abstract, and the abstract should be informative enough for you to want to get the full text of the article. Never—and we feel we should emphasize this—NEVER should you use just the abstract as a source for your paper. The abstracts may often mislead you or give you a partial or distorted picture of the results. Often the information in an abstract can be misinterpreted out of context, and it is not uncommon that data in an abstract do not match data in the paper. If *only* the abstract to a paper is available, there are specific ways it can be referenced, depending upon the writing style you use (e.g., AMA, APA, ACS, etc.). Note that a *meeting abstract* is a completely different kind of abstract, with its own referencing conventions. If your library does not provide full-text coverage for the article, book (or book chapter), periodical, or other literature you need, you can always resort to interlibrary loan, or ask your librarian if there is another way to obtain the resource you need.

Journal Websites

In case you have a better idea what you want, your instructor directs you, or you simply want to zoom in on a few top-tier journals in your field, you may opt to go directly to journal's website and search this way. As with most databases, you can usually search by author, keyword(s), or year, or you can simply browse through current and archived issues. Most individuals who have expertise in a particular field will read the table of contents (TOC) of their flagship journals each time those comes out. In fact, many journals offer a service whereby you can sign up to be emailed the TOC of the journal each time it is published (or even before), thus enabling you to keep abreast of the topics of research *du jour*. If you belong to a professional society or association, your membership fees typically include one or more of the journals that organization publishes.

Associations, Organizations, and Other Internet Resources

Many professional organizations, in addition to sponsoring journals, may also offer on their websites resources such as bibliographies (sometimes annotated), press releases (don't rely just on those: always use the original source), fact sheets, diagnostic or therapeutic guidelines, and a variety of other resources that are usually trustworthy. There is an organization for anything under the sun—from Bald-Headed Men of America to the Restless Legs

Syndrome Foundation and the Brotherhood of the Knights of Vine (vintners and grape growers). In fact, there is an annually updated *Encyclopedia of Associations* from Gale describing more than 22,000 nonprofit national organizations; we're sure you'll find one that is of particular interest to you among those. If you don't have access to that encyclopedia, you can try a web directory such as CQ Press's *Washington Information Directory* of government agencies, Congress and its committees and subcommittees, and private and nongovernmental organizations in the Washington, DC area. Barring that, faculty and librarians at your institutions are veritable repositories of knowledge in their particular areas of expertise, and you would be wise to ask them first what associations are relevant in your field.

Print and Electronic Media

Sometimes it's time to hit the book stacks in order to find that elusive source or get a better idea what you're in for; your best bet in this case is to rely, yet again, on help from a librarian. And sometimes it's useful to follow the right people—and organizations—on social media websites like Twitter or blogs to see up-to-date announcements on various new articles and research on topics that interest you.

> Never cite a tweet or a blog as a reliable source, though you can use them as tools that empower you to be up to date with the latest research in your field. Always verify their sources if the information in the tweet or blog is germane to your research. Mind you, nontraditional media such as blogs, tweets, YouTube videos, Reddit forums, or Instagram pictures are a valuable source of fascinating information—especially if sociality is part of your research (e.g., you study the controversy over a scientific topic). These sources can even be cited; many writing style guides offer guidelines as to how to cite social media.

Automatic Alerts

Just as you can get TOC updates from journals, you can set up Google to keep an eye out for certain keywords. If the keyword is unique enough (a specific molecule, taxonomic classification, mineral, process, etc.), or you're keeping tabs on a scientist with a unique enough name ("Kylander" is safe, "Kyle" is not—unless you want hundreds of hits a day), Google can scan the overpowering onslaught of hourly e-information and drop you an email when

something crops up. Don't forget to *always* track the original source of information (i.e., the original study that may be referenced in a news source).

Documentation, References, Citations

Documentation, references, and citations are terms that are often used interchangeably, and we admit there's considerable conceptual overlap among
them. Let's attempt a little clean-up via definitions:

- **Documentation** is, generally speaking, the process of providing evidence
  by acknowledging sources.
- **References** refer to the specific bibliographic information that allows readers
  to find and verify your sources; most commonly, you identify these on a separate "References" or "Works Cited" page. Needless to say, documentation is
  as reliable as the sources identified, so always make sure you pick the right
  types of references—authoritative, peer reviewed, objective, and timely.
- **Citations** occur in-text, when you provide either a direct quote or a paraphrase of an original source together with abbreviated bibliographic information about that source. Remember that, unlike in the humanities or
  even the social sciences, direct quotes are almost never used in scientific
  writing (with possible exceptions for memorable, hard-to-paraphrase,
  historical, or otherwise important passages). Always summarize the gist
  of the idea and indicate the source depending on the documentation style
  you're using, which can be a superscript or bracketed number pointing
  to a full bibliographic entry in the references, or in some cases an in-text
  citation (i.e., author, year, page).

Example of a citation (in text: superscript number at the end of the sentence):

This study concluded that the effect of clindamycin on inflammatory lesions and
the effect of tretinoin on noninflammatory lesions were maximized by combining
clindamycin and tretinoin.[3]

Example of a reference (end of paper, AMA style):

3. Jarratt MT, Brundage T. Efficacy and safety of clindamycin-tretinoin
   gel versus clindamycin or tretinoin alone in acne vulgaris: A randomized, double-blind, vehicle-controlled study. *J Drugs Dermatol.*
   2012;11(3):318-326.

What Kinds of Things Are Documented?

How scrupulous do you need to be about citing? Do you need to cite your lectures? your lab coordinator? your textbook? Is there an end in sight?

You[1] betcha![2] That's[3] how you prove[4-8] that[9-11] you've done[12] a thorough[13-19] job[20] on your assignment.[21-33] We kid, of course. You do *not* need to document *everything*. In an astronomy paper, no one expects you to provide proof that the sun is at the center of our solar system—Copernicus did a pretty good job several centuries ago and at this point it is accepted fact. We have to build new knowledge on the assumption that some theories are the accepted truth—either proven beyond a shadow of a doubt or accepted as common knowledge (at least within specific disciplines). So, when you cite, follow these rules:

You need to cite	You don't need to cite
• Data and arguments from published research (yours or others'; yes, you need to cite yourself if you published—there is such a thing as self-plagiarism) • Recent concepts, theories, or arguments that have not yet gained wide acceptance in the scientific community or are still controversial in some way • Any concepts, theories, and arguments that have clearly been attributed to one scientist or scientific team	• Commonly accepted matters of fact, such as basic natural principles or occurrences (e.g., the vertebral column has 33 vertebrae) • Well-defined and uncontroversial terms (e.g., the notion of acceleration or nuclear reaction) • Generally accepted theories and concepts that do not solely belong to one scientist or team

Your citation habits should not be limited to material printed in the form of books or articles. If you use information from atypical sources, you should document them:

- any other forms of print or online media (newspapers, magazines, circulars, newsletters, official reports and documents; websites, emails, "tweets," blogs, etc.)
- any forms of visual media (movies, TV shows or news, videos—whether on the Internet or not, radio shows, music, etc.)
- any visuals (photos, graphs, tables, other) that you did not create (we offer more detailed guidance on documenting images in Chapter 6)
- software (such as statistical packages or specific data capture or analysis software)

- any types of legal documents (patents, motions, court transcripts, judgments, contracts, etc.)
- any type of business documents (reports, memos, proposals, technical specifications, etc.)
- personal communication (e.g., interviews, phone conversations, emails, and any other forms of correspondence)

The style manual you use for your discipline will have specific rules for dealing with these—and other—types of documentation. It is your responsibility to use the correct documentation format for each of these citation types.

Another distinction you should make is between primary and secondary citation. If you find a source cited in another (respectable) source but you can't locate the original because it is out of print or not available in English, then some citation manuals (APA, for example) allow you to include a secondary citation: "Bonfrere, as cited in Nicholson (2010), found that . . . ." Note that the *difficulty* of locating a source is not a legitimate reason to use a secondary citation, and you should do your best to avoid secondary citations. Research shows that failure to consult original studies leads to propagation of scientific myths—such as that eating spinach will make you strong because of all that iron it contains. Again, citing sources correctly is extremely important—and can have serious practical consequences, as in decades of erroneous dietary advice.

> The "spinach is rich in iron" story is complicated and infinitely more fascinating than it might seem. Read the whole saga in Rekdal OB. Academic urban legends. *Social Studies of Science*, 2014;44(4):638–654.

When and Why Documenting Is Necessary

Expert researchers can take a single glance at your references and be able to tell whether, in fact, you've done your homework or note. Incomplete or inappropriate referencing is a sign of inexperience, naiveté, or laziness. Your references show how well informed you are and help establish your ethos. A poor list of references, which looks and reads as an afterthought, may indicate lack of preparation and seriousness and may prevent you from getting published—or getting that A. You need to document to acknowledge your fellow scientists but also to build your own argument and authority.

---

10 Reasons Why You Should Document

1. *Give credit when credit is due.* It's the *fair* thing to do and it establishes your place among the rank and file of like-minded scientists.
2. *Avoid plagiarism.* It would be unethical not to recognize where some of your ideas come from. Plagiarism comes with a hefty price—and puts a permanent blemish on your academic (and professional) record.
3. *Build your own library of data on a topic.* Keeping tabs of the research on your topics of interest is smart and allows you to revisit and reconsider what's been done and what remains to be done.
4. *Show the research pathway that led you to a particular finding, conclusion or opinion.* Documenting allows others to see how you arrived at your particular conclusion or how you narrowed down a topic.
5. *Validate and confirm hypotheses and statements.* By documenting some of your statements you show that you haven't just pulled them out of thin air. All those scientists before you? They have your back.
6. *Place your work in a broader research context.* Thorough referencing also places your own research within a specific school of thought, trend, or body of research.
7. *Build a strong argument for the rationale for your research.* Documenting sources can highlight the gaps in past research that can be filled by your own research.
8. *Enable readers to find similar (or different) studies on the topic you researched.*
9. *Enable readers and reviewers to verify your research.* What is science without verification? Providing documentation validates the scientific process—in this case, of making sure you know your stuff.
10. *Showcase your scientific rigor.* Documentation is a badge of academic honor—display it proudly, as it shows you have, indeed, done your due diligence.

Documenting is something you do for *yourself* (e.g., to keep track of your topic of research), for your *readers* (e.g., to help them locate more information), and for the sake of the *scientific process* (e.g., to preserve its integrity and transparency).

Documentation Styles

Not only do you have to show rigor in selecting your references; you also have to be diligent about observing the preferred documentation style in your discipline. Your instructor will ask you to observe the most common documentation style in your field. If you're writing an undergraduate or graduate thesis, your college or university will have strict rules about documenting and formatting. If you're submitting anything for publication, you have to observe the editorial guidelines to the letter as well—guidelines that may vary from journal to journal, even in the same discipline (publishers may reject a manuscript without reading it if it is not in the correct style). You must be flexible, in other words: focus on learning the style most common in your discipline, but be aware of alternative documentation styles. Interestingly, hewing to style (or completely to style) seems to be one of the most difficult skills for students to master. Using a reference document written to correct style is often useful.

---

10 Tips for Following a Style Guide

1. Pay attention to referencing format (both within text and in the bibliography). Every detail matters: italicizing, page format, spacing, punctuation, capitalization, and order of information. How many authors are included? all of them? three or six of them plus "et al.?" ("Et al." stands for "et alia" = "and others"). Is the journal title abbreviated? Is the doi (digital object identifier) included in the reference if the reference was retrieved online? Ensure you're following the formatting consistently.

2. Pay attention to how scientific notation is used in text. While one could argue that scientific notation is scientific notation no matter where it's used, it might still differ from style to style or journal to journal—for example, you might see "Tmax" or "T_{max}" or "t_{max}" If you want to be accurate, "t_{max}" is correct, but see how often it turns up that way in press.

3. Pay attention to italicizing. Does your style guide italicize foreign words (*in vitro*) or genus and species (*Staphylococcus aureus*) or Latin notation (*et al, ie, eg*)?

4. Pay attention to punctuation. Is it e.g. or eg? Is it i.e. or ie? Is it S. aureus or S aureus? Or *S. aureus* or *S aureus*?

5. Pay attention to hyphens, compound words, and suffixes. Is it "membrane-bound" organelle or "membrane bound" organelle? What about prefixes? "Pre-hypertensive" or "prehypertensive"? What about compound words? Is it "health care" or "healthcare"? "Website" or "web site"?

6. Pay attention to how numbers are written. Do you spell out "three," or do you just use "3"? Is this done all the time, or are numbers up to 10 spelled out but numbers above ten written numerically?

7. Pay attention to ordinals. Is it "first," "1st," or "1st"? Is it "first, second, third" up to ten, and then 11th (or 11th), 12th (or 12th), and 13th (or 13th) after ten?

8. Know proper capitalization and roman versus (or vs.) arabic numbering (to add to the fun, note that "roman" and "arabic" are not capitalized when referring to numeral style). Is it "Phase II trial" or "phase II" trial or "Phase 2 trial" or "phase II trial" or "phase 2 trial"? ("phase 2 trial" is correct).

9. Pay attention to placement of punctuation. Do commas go inside quotation marks or outside of them? What about semicolons and colons?

10. Pay attention to the Oxford (or sequential) comma (see "Writer's Toolbox"). Is it used or not?

Style manuals are comprehensive guides to writing, formatting, and documenting research-based (but not only) works. They are a relatively recent development in the history of academic publishing (by recent, we mean less than a century old), driven by the need for stylistic and documentation uniformity and consistency amidst the explosion of scholarly publishing in the 20th century. These guides are usually sponsored by large, national, well-respected professional organizations that set the standard for a particular field. You may have used MLA (Modern Language Association) style in your high school English classes, a style widely used in the humanities; and you may have even had to use APA (American Psychological Association) in your communication and social sciences classes. Were you to publish an article in these fields, you would be asked to observe either of these two styles, or perhaps a third, the Chicago Style Manual (popular in academic publishing).

In most of the sciences (particularly the biomedical fields), no clear and universally accepted system of references existed until the late 1970s, and when it happened, it was due in part to the efforts of an administrative secretary at the University of Washington School of Medicine. In 1968, Augusta Litwer, tired of retyping (yes, on a typewriter) her boss's papers whenever he had to submit or resubmit an article to a different journal, wrote those journals' editors asking why they couldn't agree on a uniform reference style. It took 10 years, but by 1978 a meeting of the International Committee of Medical Journal Editors convened in Vancouver where the editors agreed on a uniform reference style. The Uniform Requirements (or Vancouver Style), whose standards were published in 1979, form the basis for many of the style manuals used in the sciences today, including American Medical Association (AMA), American Chemical Association (ACS), and Institute of Electrical and Electronics Engineers (IEEE).

Style manuals are quite comprehensive (they tend to run hundreds of pages) and provide stylistic, punctuation, orthography, design, and formatting guidelines adapted for the areas of interest they address primarily (e.g., APA for social sciences, AMA for medical fields, etc.). They cover rules regarding preparing manuscripts for publishing, acceptable ways to present data and visual information, ethical and legal issues related to authorship, grammar and vocabulary issues, as well as technical matters such as typography, editing, and proofreading. They are a handy reference tools for writers, and if you ever find yourself in a position in which writing is a crucial component (hint: if you are flirting with a career in the science and technical fields!), you'd be well advised to have access to a copy of the most recent edition of the style manual adopted in your field—or, even better, an electronic subscription that gives you access to all the inevitable updates and corrections.

A Heuristic Approach to Documentation Styles

Style manuals are generally very comprehensive but one of the crucial areas in which they provide guidance is writing references. What follows is a primer to understanding the basics of referencing in just about any given style.

For in-text citations, your style manual will generally follow one of these two methods:

(a) Harvard system, in which the author's name and year of publication are cited in the text, usually within parentheses (references are listed alphabetically at the end of the manuscript):

For example: The most common indication for Neuromuscular Electrical Stimulation (NMES) use is to minimize or prevent disuse atrophy and to provide muscle re-education **(Peckham et al., 1973).**

(b) Vancouver style, or numbering system, in which the reference is denoted by a number that will help you located the reference in the list at the end of the manuscript (references will be listed in the order of appearance in the text):

> For example: The most common indication for Neuromuscular Electrical Stimulation (NMES) is to minimize or prevent disuse atrophy and to provide muscle re-education.[9]

Either way, your source must appear in the list of references at the end of the paper, and it has to be formatted according to the style manual you follow (alphabetical in the Harvard system, order of appearance in the Vancouver system):

9. Peckham PH, Mortimer JT, VanDer Meulen JP. Physiologic and metabolic changes in white muscle of cat following induced exercise. *Brain Res* 1973;50(2):424–429.

Most, though not all, scientific citation styles demand abbreviations for journal titles. These abbreviations are standard (do not attempt to "guess"!) The International Standard Serial Number website has a good List of Title Word Abbreviations accessible online; so does the National Library of Medicine.

Your bibliographic information should include accurate and *complete* data that will enable readers to find the source easily and also to assess its significance depending on author, date, journal, and other factors. In general, no matter what style you need to adopt, you need to indicate:

- Name(s) of author(s)
- Year of publication
- If an article: title, name of journal or publication, volume and issue of journal, page numbers; URL (Uniform Resource Locator, in other words, the link) or DOI if the article was retrieved online. The DOI has become standard in academic publications with online versions and allows for a more reliable identification of the article. The URL may change or be removed from a site, while the DOI is a permanent identifier, which you can find easily via a search engine.
- If a book: author, year, book name, edition, publishing house, place of publication (chapter, pages, and editor are also frequently included in book references)
- If a book chapter in an edited collection (chapters with distinct authors): author, chapter title, book name, edition, editor(s) name(s), publishing house, place of publication, page numbers

These elements ensure that the reader can safely identify the exact source to which you are referring. Remember to secure all these details *while you do your research*. Otherwise, you'll be faced with the quasi-Herculean task of working backwards to find out all this information, which is no fun at all.

The formatting of the list of references will differ from style to style, discipline to discipline, and frequently, from journal to journal. Fortunately, we now live in a digital age where referencing can be automated. If you invest in a good reference management program, your life will be made considerably easier (well, your *academic* life). Software programs such as EndNote and RefWorks are available sometimes for free as part of your educational package at your institution or for purchase at considerable educational discounts while you are a student. Free reference management programs are also available, such as Zotero or CiteULike. These programs will allow you to input as much bibliographic information as possible into predetermined fields, and then, with the click of an icon, will enable you to produce bibliographic lists formatted in the style of your choice (most, but not all, styles are usually available). Most word processing programs have referencing tools built into them as well.

Always make sure, though, that you use the most up-to-date version of these programs, so that you have access to the latest editions of the style manuals they support.

Some of these programs will also be directly connected to some of the major databases and allow you to import references directly from there. Altogether, your career in academia and beyond will greatly benefit from investing in and learning how to use one of these programs, especially if you have to write longer pieces such as a thesis or a dissertation. In general, we advise you to keep track of your sources in a database generated in the reference management software of your choice. If you are not using such software, at the very least, you should keep track of your sources in a separate document that you can revisit, expand, revise, and eventually append to your manuscript.

Common Styles Used in the Scientific and Technical Fields

The following is a précis of a few of the most common types of citations you are likely to encounter when using a particular style in a particular discipline. Since space does not permit us to cover all the types of references you may need to document, you should always seek out the original source, which will cover all the reference needs for your particular research project.

American Psychological Association (APA)

Currently in its seventh edition, the APA manual is primarily used in the social and behavioral sciences and offers comprehensive stylistic and reference guidelines for authors. It generally follows the Harvard system for in-text citations and references, with the addition of the page number(s) whenever a direct quote or a direct reference to a particular idea or paragraph is made. The list of references is arranged alphabetically with a hanging indent, authors' first names are abbreviated to an initial followed by a period, and the year of publication, in parentheses, precedes the title of the article or book. Articles or book chapters are listed in *text case* (rather than title case), and journal titles (in unabbreviated form) and volume numbers are italicized, as are titles of longer works (books, films, reports, etc.). If the journal is paginated by issue, you should include the issue numbers as well, in parentheses after the volume number.

Sample APA Citations

Article in a journal:
Genskow, K. D., & Wood, D. M. (2011). Improving voluntary environmental management programs: Facilitating learning and adaptation. *Environmental Management 47*, 907–916. doi:10.1007/s00267-011-9650-3

Note: The DOI should be included regardless of whether you retrieved the article online or in print (hard copy). If you found the article in print but the journal does not provide the DOI, omit it. If you found the journal online, add the URL as follows:

Article in a journal retrieved online:
Genskow, K. D., & Wood, D. M. (2011). Improving voluntary environmental management programs: Facilitating learning and adaptation. *Environmental Management 47*, 907–916. Retrieved from http://www.ncbi.nlm.nih.gov/pubmed/21384272

Book:
Munson, B. R., Young, D. F., Okiishi, T. H., & Huebsch, W. W. (2009). *Fundamentals of fluid mechanics*. (6th ed.) Hoboken, NJ: Wiley.

Book chapter in an edited collection:

Miller, G., Dingwall, R., & Murphy, E. (2004). Using qualitative data and analysis: Reflections on organizational research. In D. Silverman (Ed.), *Qualitative research: Theory, method, and practice* (2nd ed.) (pp. 325-346). London, England: Sage.

Website:

University of the Sciences. (n.d.) Mission statement. Retrieved from http://www.usciences.edu/about/mission.aspx

Note: If you know the author of the particular page or article on the website, as well as the date of publication, include them here. Otherwise, cite the organization owning the site and mention "n.d." for "no date."

Council of Science Editors (CSE)

Formerly known as CBE (Council of Biology Editors), CSE style is used widely in the sciences. CSE accepts both citation–sequence and citation–name systems, which differ only in the order of references. In both systems, numbers within the text refer to the end references; in the citation-sequence style, the references are numbered in the order that they appear in the text, and in the citation-name system, they are numbered alphabetically in the reference list and then cited by their respective number in the text. This style requires that you list up to 10 authors, followed by "et al."

Sample CSE Citations

Article in a journal:

Genskow KD, Wood, DM. Improving voluntary environmental management programs: facilitating learning and adaptation. Environ Manage. 2011;47(2):97–116.

Article in a journal retrieved online:

Arnold JU, Smith TY, Ehrens RT, Brightborn FD, Antonovich MK. Endothelial dysfunction in the early postoperative period after major colon cancer surgery. Br J Anaesth. 2017 [accessed 2017 May 13; 118(2):200–206. doi: 10.1093/bja/aew410].

Book:

Munson BR, Young DF, Okiishi TH, Huebsch WW. Fundamentals of fluid mechanics. 6th ed. Hoboken (NJ): Wiley, 2009.

Book chapter in an edited collection:
Miller G, Dingwall R, Murphy E. Using qualitative data and analysis: reflections on organizational research. In: Silverman, D, editor. Qualitative research: theory, method, and practice. 2nd ed. London (England): SAGE;2004.

Website:
University of the Sciences. Philadelphia (PA): Mission statement; © 2001–2017 [accessed 2016 Dec 13] http://www.usciences.edu/about/mission.aspx

American Medical Association (AMA)

AMA style manual, currently in its 11th edition, is based on the Vancouver style, which means references are numbered in the order they are cited in the text, and in-text citations consist of superscript numbers. The year of the publication appears after the title, and journal names are abbreviated. There are no periods after the authors' first name initials, and authors' names are separated by commas.

Sample AMA Citations

Article in a journal:
Genskow KD, Wood DM. Improving voluntary environmental management programs: facilitating learning and adaptation. *Environ Manage*. 2011;47(9):907-916.

Article in a journal retrieved online:

- With DOI (preferred):
    1. Genskow KD, Wood DM. Improving voluntary environmental management programs: Facilitating learning and adaptation. *Environ Manage*. 2011;47(9):907-916. doi:10.1007/s00267-011-9650-3.
- Without DOI:
    1. Genskow KD, Wood DM. Improving voluntary environmental management programs: Facilitating learning and adaptation. *Environ Manage*. 2011; 47(9): 907-916. http://www.ncbi.nlm.nih.gov/pubmed/21384272. Accessed December 12, 2012.

Book:

1. Munson BR., Young DF, Okiishi TH, Huebsch WW. *Fundamentals of Fluid Mechanics*. 6th ed. Wiley; 2009.

Book chapter in an edited collection:

1. Miller G, Dingwall R, Murphy E. Using qualitative data and analysis: reflections on organizational research. In: Silverman D, ed. *Qualitative Research: Theory, Method, and Practice*. 3rd ed. SAGE; 2004:325–346.

Website:

1. State-based prevalence data of ADHD diagnosis. Centers for Disease Control and Prevention. http://www.cdc.gov/ncbddd/adhd/prevalence.html. Last reviewed: August 27, 2019. Accessed December 12, 2019.

American Chemical Society (ACS)

ACS style is preferred in chemistry and related fields (such as pharmaceutical sciences). It is quite similar to AMA, but it allows more flexibility in handling in-text citations (which can be numerical, either in the form of a superscript or of a bracketed number, or Harvard-style—including name and year). For journal articles, the major differences are: authors' names are separated by semicolon, the title of the article is in title case, the year of publication is formatted in bold, and the volume of the publication is in italics. Standard journal abbreviations can be found in ACS's CASSI database.

Sample ACS Citations

Article in a journal:

Genskow, K. D.; Wood, D. M. Improving Voluntary Environmental Management Programs: Facilitating Learning and Adaptation. *Environ Manage* **2011**, *47*, 907–916.

Article in a journal retrieved online:

Genskow, K. D.; Wood, D. M. [Online] Improving Voluntary Environmental Management Programs: Facilitating Learning and Adaptation. *Environ Manage* **2011**, *47*, 907–916. http://www.ncbi.nlm.nih.gov/pubmed/21384272 (accessed Dec 12, 2012).

Book:
Munson, B. R.; Young, D. F.; Okiishi, T. H.; Huebsch, W. W. (2009). *Fundamentals of Fluid Mechanics*, 6th ed.; Wiley: Hoboken, NJ, 2009.

Book chapter in an edited collection:
Miller, G.; Dingwall, R.; Murphy, E. Using Qualitative Data and Analysis: Reflections on Organizational Research. In *Qualitative Research: Theory, Method, and Practice*; 2nd ed. Silverman, D., Ed. SAGE: London, England, 2004; pp 325–346.

Website:
University of the Sciences Home Page. http://www.usciences.edu/ (accessed Dec 12, 2012).

    Note: If you know the author of the particular page or article on the website, as well as the date of publication, include them here. Otherwise, cite the organization owning the site and mention the date you accessed the site.

MSDS (Material Safety Data Sheets)
Formaldehyde; MSDS No. F4223; Baker & Taylor: Baltimore, MD, Dec 12, 1999.

MSDS online:
Formaldehyde; MSDS No. F4223 [Online]; Baker & Taylor: Baltimore, MD, Dec 12, 1999. http://www.bnt.com/formaldehyde.htm (accessed Mar 23, 2012).

Placing References

Journal Articles or Research Papers

Everything that has a source other than yourself has to be cited as soon as it is mentioned in your text (with a few exceptions we have already covered), and you must follow the guidelines of the style manual dictated by your field, university, professor, or the publication to which you are submitting a manuscript. The only part of the paper where you should not place any references is in the abstract, which must stand alone for the sake of database indexing. This is a practical consideration: since the abstract is the most widely read portion of any article, and sometimes the only portion available in a database, it should stand alone and not be disrupted by in-text citations that might confuse readers. The list of references at the end is usually placed on a separate page. Footnotes and endnotes are generally avoided in the sciences, although some publications might allow them (always check first!).

If you happen to write a whole paragraph that relies on paraphrasing ideas from a single source, you should probably include citations after each sentence based on that source. This is the "better safe than sorry!" approach; but if you feel uncomfortable citing the same oeuvre seven times in a row, you should try to minimize that occurrence as much as possible by rethinking your paragraph—for example, by summarizing the ideas from that source in a more concise manner, mixing up sources, and/or adding in more original material.

Meeting Posters and Presentation Slides

There is virtually no difference between references in an article and references in a scientific poster in terms of formatting. However, you are well advised to keep your list of references to a minimum on a poster, since your readers won't have time to read and remember a full list. In that case, reduce the list of references to the absolutely essential elements so that viewers do not mistake other people's work for yours.

If you are giving a live PowerPoint presentation or creating some other form of enduring electronic presentation (e.g., a CD-ROM, archived web presentation, blog, etc.), consider the audience, venue, and conventions of your field when providing references on the slides. At the very least, if data are presented on a slide, enough information needs to be provided in a reference so that the viewers of the presentation can jot it down and find it ("Taylor PR, *JAMA* 2007"; not "Taylor, 2007"). In most cases, more information can be provided in abbreviated form. If you are creating a PowerPoint presentation that is going to be used for educational, promotional, or industry purposes, or in any case in which the slides are going to be reviewed by a regulatory authority or professional organization, near-complete information should be provided as a reference *on* the slide, not on a *final* slide in the program. General points, like "diabetes is a serious disease" don't have to be referenced, but information like "in a randomized controlled trial of 3244 adults between the ages of 18 and 64, HbA1c and fasting glucose were assessed over a period of 6 months" requires a reference. Does it require absolutely the full reference, the way you would write it in your bibliography? That depends on the situation in which you're presenting it, but usually no. If you have an hour to prepare the presentation for your fellow classmates and you all happen to know that reference well and that it's Taylor from *JAMA* in 2007, you can probably get away with "Taylor, 2007." If, however, you're presenting at a meeting because you got a Young Scientist award from the American Diabetes Association, the reference on that slide is going to read "Taylor PR, *JAMA* 2007;298(4):401–402." It's not necessary to include a string of nine authors or the title of the paper, just the basic information. Original research is original research. Not citing it on a slide is tantamount to saying it's your own data. It is not uncommon at large meetings for the final slide to contain the email address of the presenter, whom attendees can contact for more information about references (or anything else), if necessary.

Style guides rarely provide guidance for this; rather it is a dictate of your audience, venue, professor, boss, discipline, or any combination thereof.

Lay Targets (Newspaper, Patient Information Brochures, etc.)

Because most lay audiences have little or no experience with scientific documentation styles, you must exercise caution when placing references in materials intended for general consumption, or you risk befuddling the reader. Follow the standards of the publication or organization for which you are writing the piece—for example, newspapers and magazine will insist on their own way of referencing sources. In technical documents intended for lay targets, such as patient education materials, you may occasionally include references to crucial studies that you think your target audience needs; when you do, try to avoid journal abbreviations by adopting a reference style such as MLA or APA (as long as you are consistent, you are fine). *The New York Times* is likely to say "In an article published this week in the *Journal of the American Medical Society,* researchers determined that diabetes can be assessed by . . ." A patient education brochure, however, targeted to a much broader range in educational background, might get away with "scientists monitor your diabetes with . . ."

Books

References are placed in books just as they are in articles, generally speaking; some style manuals recommend a bibliography to be included after each chapter, others allow one massive bibliography at the end. Because editors have a little more latitude with books, footnotes and, more commonly, endnotes are usually permitted.

Take-home

You should start researching your field *way* before you embark on a particular research project. References are the currency of science. You cite to enhance your authority (ethos), to ensure reproducibility of research, to avoid plagiarism, and to place your research in context. Make sure you cite the most legitimate, appropriate sources possible—by applying criteria in the analysis of your possible sources: Authority, Audience, Scope, Timeliness, Objectivity, and Documentation. Understand the type of sources you can rely on, and become familiar with the databases, journals, and organizations most likely to be useful in your particular field or topic. Learn to ask your librarians for help: they are the original and most reliable search engines of all.

With a few exceptions for general or common knowledge, you should document *all* your sources at *all* times. Always keep track of your sources either in reference management software database or in a separate document, writing down all the pertinent details that will enable a reader to retrieve the source easily. Become familiar with and have access to the style manual most often used in your field—particularly if you do a lot of writing. You are even more strongly encouraged to acquire and become proficient in a reference management software program which will allow you to quickly convert the bibliographies from one style to the other. Finally, learn some basic style rules and conventions regarding the writing and placement of the most common types of references in the documents you are most likely to produce as a student or professional.

Exercises

1. Identify the most relevant databases, the top 10 journals, and the top 3 organizations for a specific topic. Briefly describe each, providing justification for inclusion in your list. Compare notes with your peers.
2. Become familiar with the search process for a database of relevance in your field. Prepare a mini presentation to be delivered in class, individually or as a group, demonstrating search techniques and tips.
3. Follow a journal or organization of relevance to you on Twitter or some other type of social media. Write a mini-report for class on what you can learn from such mini-updates over the course of a week. Can this endeavor potentially be useful for your career or research?
4. A student emails her professor and asks: "I am revising my paper from our meeting today and I have a question. I tried to find what the actual procedure was in an open surgical hip dislocation and it isn't clearly

explained in any of my sources so I Googled it. I came across a website run by an MD where he describes what is done (http://www.clohisyhipsurgeon.com/treatment-options/open-hip-dislocation). I was wondering if I am able to use this as a source being that it isn't from a journal." What would you reply to her, and how would you justify your decision?

Examine the following statements and decide when you need to cite and when you don't need to cite a source for them. Justify your decisions:

5. a. William Taft was the only US President to also serve on the Supreme Court.

b. Invasive MRSA infections that began in hospitals declined 28% from 2005 through 2008.

c. Avogadro's number is 6.023×10^{23}

d. Molecular knots remain difficult to produce using the current synthetic methods of chemistry because of their topological complexity.

e. The National Hurricane Center's official forecasting of where a hurricane will be 1 to 5 days ahead has become more precise in the past decade.

f. Coagulation factor activation may trigger systemic inflammation.

g. The human hand has 27 bones, 14 of which are the phalanges (proximal, medial, and distal) of the finger.

h. Dolphins communicate via clicks, pulsed calls, and whistles.

6. Summarize and/or paraphrase ethically the following passages:

a. Common dolphin nighttime vocalization data had numerous call periods with patterns similar to daytime foraging vocalization patterns: discrete click bouts and few whistles or pulsed calls, with whistles frequently occurring at the start and end of click bouts. Further analysis of these nocturnal call patterns is needed, but the qualitative pattern supports the idea that this population of common dolphins is feeding at night on the deep scattering layer (DSL). (From: Henderson EE, Hildebrand JA, Smith MH, Falcone EA. The behavioral context of common dolphin (Delphinus sp.) vocalizations. *Marine Mammal Sci* 2012, 28(3): 439–460; page 453).

b. Learning, adaptation and improvement rely on suitable data relevant to the issue at hand. For voluntary programs, this requires generating good information about factors influencing actions and decisions of "target audiences" that most influence the resource of concern. Involving those with local and experiential knowledge is important for understanding and placing the data in context and for reviewing assumptions about actions, intermediate outcomes, and environmental impacts. (From: Genskow, K.D., Wood, D. M. Improving voluntary environmental management programs: facilitating learning and adaptation. *Environ Manage* 2011, 47: 907–916; page 914).

c. In exploring how pregnancy-related health problems influenced level of exercise, it was found that women experiencing a multiple pregnancy and women experiencing pelvic girdle pain, nausea (week 17), musculo-skeletal pain (week 30), uterine contractions (week 30), and sick-leave were less likely to exercise regularly. (From: K. M. Owe, KM, Nystad W, K. Bø K. Correlates of regular exercise during pregnancy: the Norwegian Mother and Child Cohort Study. *Scand J Med Sci Sports* 2009; 19: 637–645. doi: 10.1111/j.1600-0838.2008.00840.x; page 640)

7. In groups of four to five students, pretend you are the editors representing each of the styles we described in this chapter (APA, AMA, ACS, CSE). Because you're nothing short of ambitious, you are holding a preliminary meeting to discuss adopting a standard citation style across all journals in the scientific and technical fields. What are the common rules you would agree on when handling citations and references? Which features would you keep, and which would you modify from each of your style manuals? Why? For this exercise, it would help to have copies of the style manuals or at least short guides, many of which are available online, free of charge, or via your library website.

PART II
GENRES OF SCIENTIFIC COMMUNICATION

8
Writing Original Research Papers

© 2020 Max Beck

You've been fortunate enough to work in a lab peripherally involved with a clinical trial evaluating a sleeping pill. You were the only undergraduate on the team, eyes shining, passionate about neurochemistry, biochemistry, and putting people to sleep. "Sleep," you murmur to yourself covetously, "I remember that. Maybe I can do that again someday." You get to work with some of the Big Names, enter things in the computer, compile data, do literature searches, compile bibliographies. One day, to your delight and amazement, one of the principal investigators (PI) stops by your desk. "Hey Gary," he says in a friendly tone about which you should be very seriously and immediately suspicious, "why don't you take a stab at it and write the first draft?" He tosses a pile of data on your desk. You're unfortunately too much in the initial throes of rapture to realize what has just happened, and too late you realize what you have to do. The PI is gone. Your raptures have evaporated, and you call out feebly "Um, I'm a *junior*."

You stare at the piles of materials on your desk in abject despair as your lab director comes over and says it's no biggie, Gary, just write up an IMRAD draft and we'll tweak it from there. You might even get to be an author!" You give her a confident smile and take off for the day, wondering what "IMRAD" is.

What Is IMRAD, Anyway?

IMRAD is not some mystical government term referring to a secret type of defense system; it merely means—you are going to be disappointed here, probably—"introduction, methods [and materials], results, and discussion." Given that these papers have "methods and results," you can probably guess that these are papers written primarily for the purposes of reporting results from an original experiment or study. They are. It is the most common format used to report original research.

Scientific journals include many types of articles: short communications, announcements, letters, or case studies. By far the most common types of papers that you will read in journals or submit in undergraduate, graduate, or postgraduate studies will be review articles (or papers) and IMRAD (or related) articles. "IMRAD" and "review" are concepts that apply to many different forms of communication—not just scientific papers. IMRAD is a conceptual format: a logical progression of information that makes for an effective means of disseminating information—a protocol of sorts. It doesn't necessarily mean it has to be a paper—you will use "*conceptual* IMRAD" in most any kind of writing, say, grant proposals, lab reports, papers in general, experimental design, and even in thinking. Cooking follows a *conceptual* IMRAD process:

1. (Introduction): I am going to bake a pound cake, which will require specific ingredients and will have to be baked at a certain temperature, and I am going to use my aunt's recipe. Pound cake is a popular culinary treat, dating back to the pharaohs (note: don't make things up).
2. (Methods and materials): I am going follow this exact recipe and use flour, eggs, sugar, butter, lemon extract (etc.), mix it in this kind of bowl, using these kind of beaters, for 3 minutes, pour it into this Bundt pan, bake it in this oven, at this temperature, for 45 minutes, and turn it out to cool on this rack.

3. (Results): This recipe turned out one large pound cake. The cake was somewhat moister and denser toward the top (once everted from the cake tin).

4. (Discussion): The wet and dry ingredients should have been mixed together more thoroughly. The time spent baking perhaps could have been longer. However, this process resulted in an acceptable enough product such that my roommates devoured the results within six minutes. This process yielded a product superior to other recipes, although further studies on this activity should be performed, perhaps tomorrow, as advocated by the assessors of the final product. Auntie would be proud.

See? Introduction, methods and materials, results, and discussion.

Chem lab follows an IMRAD thought process, even if you don't write up the experiment. An experimental protocol is an IMRAD-y kind of thing:

1. (Introduction): Use of pH indicators is an important step in many chemical experiments, so that the appropriate pH environment exists for optimal performance of procedures such as elution of somatotropin.

2. (Methods and materials): I am going to use this lyophilized somatotropin lyophilized in this manner, suspended in this pH of this solution, through this kind of column, filled with this kind of beads, at this rate, following this validated protocol by Rachicot et al. [reference].

3. (Results): Column exploded.

4. (Discussion): Explosion of the column was unanticipated, given that none of the components of the experiment was remotely volatile [reference MSDSs[1] for each component]. While the unanticipated result was pretty cool, the lab instructor apparently did not share this opinion. Sadly, this resulted in a semester grade of "D" for chem lab.

Even if you were invited to not write this lab report, and even if it didn't turn out the way you planned, you were following a protocol (albeit probably incorrectly) and practicing conceptual IMRAD.

Try it: Think of any random topic—shopping for cranberry juice, taste-testing yogurt, picking out a paint color, going to a movie—anything you like. Once you have your random topic, design a 60-second IMRAD approach to doing it. Keep it up with different topics until you can pull it off in 60 seconds.

---

[1] Material Safety Data Sheet.

What Do You Do Before You Start Writing?

Visualize your work in an IMRAD format and create an outline. Outlines are certainly easy. I mean, you have an "I" section, an "M" section, an "R" section, and a "D" section. It will have little extras, of course: title, abstract (and keywords), references, and, if it's for a journal, acknowledgements are nice; give a nod to your peers or colleagues who helped and to those nice people at the National Science Foundation who bought you all of your rats and reagents. Disclosures (of any conflicts of interests) may be required for a journal article. Really, the acronym for this should be TAKIMRADRAD (it doesn't have the same ring to it, though).

Some parts are easier than others to write; you're writing an IMRAD paper because you did some kind of experiment or study, and if you did an experiment or a study, then you were following a protocol and using specific instruments or tools to collect and analyze your data, and you were performing specific statistics in your analysis, so there's your "M" section. You know the background information on the disease or atmospheric phenomenon or ecology of your topic, and you know why you did your experiment or study, so there's your "I" section. You got results of some sort (remember, even "nothing happened" is a result): a baked cake came out of the oven, glassware exploded, patients got better, pK profiles were unchanged. So there's your "R." Then you're off into your "D," rhapsodizing about and interpreting your results and where they fit into the context of all things hypnotic, ecologic, atmospheric, pharmacokinetic, or culinary. Add an abstract and references, make sure the title of the paper is compelling and goes with the body of the paper, and you're good to go.

Is it really that easy? No, but the building blocks are somewhat intuitive, and you have all the parts. Some documents are more formal and have more parts (e.g., a journal article) and some are less formal and have fewer parts (e.g., a lab write-up).

The first critical step is to *review the directions*. This is not an optional thing! No skimming! Professors have their own specifications, and each discipline and journal will differ a little bit in a variety of ways: word count, structure of abstract, typeface, line spacing, and margins, documenting references, handling of footnotes and endnotes, captioning figures and a variety of other such captivating details are usually spelled out for you—so take a little while to learn them. It's all right there, in the "Instructions for Authors" section on the journal website.

Suppose you're submitting something to the *Journal of the American Chemical Society*. Can you craft your document in MS Word using your own style and just send it off? Not a chance. Those nice folks at *JACS* will tell you exactly what to do if you go look at their "authors" section—and they don't stop at page number and formatting. "Series or part numbers may not be used, nor may the words "Novel" or "First" appear in the title. Acronyms and abbreviations are not permitted in manuscript titles," they say. Well, there you go. You have to change your paper starting with the title. They'll tell you *exactly* how to write axes, tensors, temperature, planes, vectors, matrices, and a bunch of other things you do not want to have to go back and reformat.

You take a deep breath. All that scary math has to be just so. Your professor smiles at you and gives you an encouraging double thumbs up on her way back to her office. And then she closes her door. And then you spend the rest of the day ascertaining how *JACS* wants their manuscripts. Find out the instructions first. It is easier to write your paper—whether for a course submission or for a journal submission—if you follow the directions as you write, rather than going back to fix it when you're done.

132   Do they want every single line in your manuscript numbered? Do they want
133   it double-spaced? Do they want your text to be center justified? Do they want your
134   font size to be 13-point in Arial typeface? Do they want your figures embedded in
135   the text or added at the end of the article with only placeholders in the text? Do
136   they want your references formatted or unformatted? Do they want you to use this
137   style [Smith, 2008] or this style[12] or this style (2008)? Do they italicize Latin terms
138   like "*in vitro*" or do they set them Roman style, "in vitro?" Do you know what "set
139   rom" means? Maybe you had better look up proofreader's marks to understand the
140   Instructions for Authors. Do they want AMA style or APA style or ACS style or
141   some other kind of in-house style? Is it "i.e." or "ie?" When is it due? If you get it in
142   past the first of the month, will it not even be considered until the following month?
143   What kind of disclosures do they require? What kind of analysis? Do they want a
144   Word doc or a PDF? Have you adhered to page, word, or character (including
145   spaces or not?) limits?

Your professor or journal to which you submit might send your paper right back to you—without reading it—if you have not followed directions. If you have been given a document template, *use it*.

10 Tips for Planning Your IMRAD Paper

1. Think about the research first, and what you would like to say about it. Determine the bottom line of the experiment and the grand (or small but still newsworthy) idea that your research uncovered.
2. Sketch a very high-level, basic outline, essentially putting to paper your high-level view. It does not matter if your paper ultimately does not follow the outline, nor does it matter in which order your write each section when you're outlining.
3. Step away from your outline for a few hours or a day or so.
4. Describe your top-level view to a peer. Does it make sense to them? No? Then it likely will not make sense to your reader.
5. *Read the submission guidelines or the assignment formatting guidelines very carefully*. Read *all* of them. Make sure you understand them and follow them from the get-go. This includes any directives from your professor. Learning this discipline as a student will serve you well as a professional scientist. *Keep submission/assignment guidelines in mind at all times*. Do not write a paper first and then go back and try to format it. You do *not* want to find out that the 750-word abstract you wrote is, in fact, supposed to be limited to 250 words.
6. Start anywhere—it doesn't have to be with the Introduction. That's the advantage of this modular format. In fact, you already have plenty to say for the Methods section, so it is not uncommon to start there, then move right through the Results and Discussion before circling back to the Introduction. There is no right or wrong approach—do what suits you best.
7. It is okay to leave placeholders for information you need to track down later. Don't let missing shards of content interrupt your thinking: flesh out your paper with more details as you go (remember that writing is a recursive process).
8. Write your paper in earnest—finish your first draft.
9. *Read it out loud. Then give it to at least one peer to review.*
10. Revise, revise, revise.

What Goes in Your IMRAD Paper?

Title

The title in an IMRAD paper typically contains the entire gist of the experiment, study, or research being presented in the paper—often with results:

Pickering, G. J. (2009). Optimizing the sensory characteristics and acceptance of canned cat food: Use of a human taste panel. *Journal of Animal Physiology and Animal Nutrition, 93*(1), 52–60. [Note: this is APA style.]

In this paper title (which is not fictitious; someone has to do this kind of research), Pickering is essentially telling us that he performed a cat food taste-test using a human panel, to see which kind tasted best. You'll have to read the paper to find out which one won.

The point of a journal article title (any article, not just an IMRAD article) is to get the reader who is tooling around in a scientific literature database (such as PubMed, Scirus, Ingenta, British Library Direct, PubChem, ChemRefer, Compendex, Google Scholar, Web of Science, etc.) to want to read the abstract of the article You can often tell from the title of a paper that it is germane to your research—but then you click on the abstract and find out that it really isn't ("oh man, this is only a chicken-flavored canned cat food taste test!"). So make the title of an IMRAD paper specific to the research being presented: not just "The cat food taste test," not just "Human preference for cat food taste," but add in the fact that it's canned, and not only the bit about "sensory characteristics" but *optimizing* sensory characteristics. Your research wasn't just on how humans think cat food tastes but how to make cat food more appealing in general (ostensibly to cats). When you read the abstract, you find out that the research in fact also entailed assessment of texture. A review paper on the same subject would have a more general title: "Cat food taste perception in humans," and this paper would be broader, covering the entire vast research field in human cat food taste perception.

Incidentally, there are only a few dozen papers on PubMed that have "cat food" in the title, so if you're looking for a research project, this field is wide open.

Abstract

Almost *all* journal articles have abstracts, and other documents may have them as well (including course assignments). An abstract is the short summary of the entire article that precedes the article. The abstract usually comes up when you click on the title in your scientific literature database (PubMed, PubChem, etc.). Do not eschew reading the paper: it is a mistake to think that all the vital information is in the abstract and that all of the information in the abstract matches that which is in the body of the paper. Reading about results out of context might get you into trouble, and you might interpret the results in the abstract as the ones you need when in fact they are not; you might need subresults or secondary endpoint results. Sometimes, errors are made and numbers in the abstract don't match those in the body of the paper (often the paper and its abstract are written separately).

Different journals have different requirements for abstracts (and different professors will have different requirements for abstracts as well). An abstract can be a very unstructured, general narrative, or it can be *structured*, that is, itself a mini-IMRAD paper (think of it as an IMRAD paper that's 250 words long), with corresponding headings (more on abstracts later).

The structured abstract, then, for an IMRAD paper, will have actual subheadings for the "I," "M," "R," and "D," or some variation thereof reflecting the particular idiosyncrasies of the journal. *The Journal of the American Medical Association (JAMA)* which, of course, caters to biomedical research, has tweaked the IMRAD structured abstract style to include, in order, context, objectives, design/setting/patients, interventions, main outcome measures, results, and conclusion. The alignment to IMRAD is as follows: "context" would be the introduction, "objectives," "design," "interventions," and "main outcome measures" are all part of methods and materials, "results" is, well, results, and "conclusion" would refer to the discussion section. *The Journal of the American Chemical Society*, however, does not have a structured abstract per se (i.e., no specific headings), but it does contain the IMRAD elements in its narrative abstracts.

Our friend in the cat food gustatory science field didn't have a structured abstract with headings and numerical values for anything but did provide a detailed summary, including his materials, if not much of his "methods."

Summary

A methodology based on descriptive analysis techniques used in the evaluation of human food has been successfully refined to allow for a human taste panel to profile the flavour and texture of a range of cat food products (CFP) and their component parts. Included in this method is the development of evaluation protocols for homogeneous products and for binary samples containing both meat chunk (MC) and gravy/gel (GG) constituents. Using these techniques, 18 flavour attributes (sweet, sour/acid, tuna, herbal, spicy, soy, salty, cereal, caramel, chicken, methionine, vegetable, offaly, meaty, burnt flavour, prawn, rancid and bitter) and four texture dimensions (hardness, chewiness, grittiness and viscosity) were generated to describe the sensations elicited by 13 commercial pet food samples. These samples differed in intensity for 16 of the 18 flavour attributes, which allows for individual CFP flavour profiles to be developed. Principal components analysis (PCA) could successfully discriminate between samples within the PCA space and also reveal some groupings amongst them. While many flavour attributes were weakly correlated, a large number (describing both taste and retro-nasal aroma qualities) were required to adequately differentiate between samples, suggesting considerable complexity in the products assessed. For both MC and GG, differences between samples for each of the texture dimensions were also found.

For MC, grittiness appears to be the most discriminating textural attribute, while for GG viscosity discriminates well between samples. Meat chunks and gravy/gels differed significantly from each other in both flavour and texture. Cat food products differed in their liking ratings, although no differences were found between homogeneous, MC and GG samples, and eight flavour attributes were correlated with overall liking scores. It is now necessary to determine the usefulness and limits of sensory data gathered from human panels in describing and predicting food acceptance and preference behaviours in cats. For instance, while the sense of taste in cats appears generally similar to that of other mammals, they lack a sweet taste receptor (Li et al., 2006), which may limit the applicability of sweetness ratings obtained from humans. Modification of existing techniques used with human food research, such as external preference mapping (Naes and Risvik, 1996) may be useful. Ultimately, this may facilitate more economical and efficient methods for optimizing cat food flavour and texture and predicting the effects of composition and processing changes on cat feeding behaviour. This will require collaboration between pet food manufacturers and nutritionists, animal behaviourists and human sensory scientists. The results of this preliminary study should assist in this process.

This abstract teaches you an important lesson: no matter how frustrating your research project might be, at least you're not sampling cat food to check for "grittiness."

Note that in the Pickering abstract, references are included. References are not typically included in article abstracts, but again, this will depend upon the journal style you're following or the dictates of your professor.

When do you write the abstract? Well, since the abstract is essentially a summary of the entire paper, common sense would dictate that you write it when you're done. This varies widely by author preference (there is no "right" way to do it). Some authors write the abstract when they're finished; others write it first, to serve as an outline or thought process for the construction of the rest of the paper; and others write it sketchily at first, as an outline, and then "fill in the holes" once the paper is complete.

Introduction

Assuming the IMRAD paper is reporting on an experiment or other type of research, when writing one of these papers you don't just launch right into what it is and how it's being done. You start with a bit of background history, previous research, and why it's important. You're providing context and a rationale for the experiment. What is the history of cat (or animal) food preference? Why is it important? Why use a human taste-test panel? What groundbreaking information is missing in this exciting field?

Pickering's introduction runs almost a full page, describing the cat food industry, behavioral studies in cats, and why humans can be and are used in taste tests. He also provides the objectives for the study—setting the stage for the wealth of information to come:

Introduction

Pet food production is a highly competitive, multimillion dollar industry, and represents a significant share of the internationally prepared food industry. In urban environments, the essentially carnivorous domestic cat (*Felis catus*) and dog (*Canis familiaris*) can exercise relatively little choice as to what and how they are fed. There is therefore a considerable responsibility on pet food manufacturers to develop products that are both nutritious and palatable. In addition, they should be convenient to use, economical, and acceptable to the owner (Booth, 1976). Optimizing the sensory characteristics and acceptance of cat food is, however, challenging. Cats are sensitive to flavour differences in diet, very discriminative in food selection, and clearly unable to verbalize their likes and dislikes. These issues have dogged the industry for decades.

Behavioural studies with cats are carried out by pet food producers. Simple preference and acceptance tests can be conducted to determine the effects of changes in processing, raw materials, flavourings and presentation. These tests are, however, expensive to maintain, time consuming (Booth, 1976), and yield limited and often equivocal data. In addition, individual animal variation, previous diet or experience, and lateral bias complicate the protocols (Rofe and Anderson, 1970). In conventional flavour research using animals, considerable time and effort is also involved in isolating the required quantities of aroma and fractions, and large amounts of material are required because of concentration-dependent quality changes (Booth, 1976). In light of these limitations and the need for more rapid test procedures, the concept of using human taste panels has been advanced.

While there are differences in the physiological and perceptual systems involved in taste between *Felis catus* and *Homo sapiens*, there are also some broad similarities and common sensitivities to stimuli (Boudreau and White, 1978), and evidence suggests that human sensory data can be useful in assisting cat food formulation. Cats use both taste and smell in the detection and selection of food (Bradshaw, 1991). The third chemosensory system, the vomeronasal organ appears to be involved only in the perception of social odours (Hart and Leedy, 1987). Good reviews are available on the sensory capability of cats (Boudreau and White, 1978; Bradshaw, 1991) and on experiential factors that affect feeding behaviour (Bradshaw, 1991). In-house tasting trials using a human taster are commonly conducted by the pet food industry, although there is a paucity of relevant information in the scientific literature. Lin et al. (1998) used a human sensory panel to evaluate the effects of extrusion parameters (fat type, fat content and initial moisture content) on the sensory characteristics of extruded dry pet food during storage. The panel rated the perceived intensities of a set of predetermined aroma, hue and (manual) texture attributes relative to control samples.

The objectives of this current study were to develop a methodology for using human taste panels to assess canned cat food and to develop base-line flavour profiles for a range of commercial canned cat food products (CFP). The relationship between flavour profiles so-developed and data obtained from cat acceptance/preference trials may enable a more rapid, quantitative and predictive indication of the effects of ingredient and processing changes on the performance of products. Evidence from pet food manufacturers indicate that food flavour, rather than colour or ortho-nasal aroma is dominant in influencing acceptance/preference behaviour in cats. Therefore, this study concentrated on retro-nasal aroma, taste and textural attributes.

"Dogged," ha ha! The Introduction contains references: it is a place where you want to invoke the wisdom of previous researchers (as well as their shortcomings) in order to render *ethos* (authority) to your own research. You need context, and sometimes historical context provides much needed perspective. Cat/human comparative gustatory sensation studies apparently date back to those fabulous 1970s, when the boundaries of science were truly being pushed.

Methods and Materials

This section is usually less exciting than the rest of the paper—it may be in a smaller font size, to accommodate all the information that has to be included.

It contains extensive detail on how the experiment was set up—down to the minutiae. This is in the paper to show how *every* possible variable but the test variable(s) has been controlled for so that if there is a change, the researchers can assume the change is due to manipulation of the variable(s) under study. Ideally, another scientist should be able to take the Methods and Materials (M&M) section out of a paper and recreate the entire experiment exactly. It is not entirely uncommon, incidentally, that scientists *do* occasionally take an M&M section and try to recreate results—and when the results cannot be recreated, that's when things get lively, and sometimes frauds or fabricated data are exposed. ("Re-creation" experiments are also frequently performed by scientists skeptical of results in other papers).

It is becoming more and more common for the M&M sections to be housed electronically on the journal website and not included in the paper at all. This can be annoying if you're reading a paper and want all of the information in one place, but on the other hand, it can allow for inclusion of more information than usual since there are typically no space constraints. In some papers, Methods and Materials have been moved to the end of the paper (we'll talk about IRDAM in a bit) so as not to interrupt the "flow" of the narrative.

Methods cover the exact protocol or study parameters used. *Materials* are, well, *things*; all the physical substances and apparatuses that you've used in your study: placebos, chemical substances administered to rats, spectrophotometers, scales, software packages, computers, dyes, microscopy lenses. Precision is crucial: not just "cat food," but what kind?

A:	Homogeneous product marketed as minced-beef based
B:	Homogeneous product marketed as jelly-meat based
C1:	Product C—binary system: meat chunks portion
C2:	Product C—binary system: gravy portion
D:	Homogeneous product
E:	Homogeneous product
F1:	Product F—binary system: meat chunks portion
F2:	Product F—binary system: gravy portion
G:	Homogeneous product marketed as fish based
H1:	Product H—binary system: meat chunks portion
H2:	Product H—binary system: gel portion
I1:	Product I—binary system: meat chunks portion
I2:	Product I—binary system: gravy portion

Pickering avoided use of brand names for blinding and "brand sensitivity." The panel was chosen from university "staff" and—probably not too surprisingly—"students." I imagine compensation was provided, although even as an impoverished student I'm not sure there would be enough compensation for taste-testing "jelly-meat based" cat food. Surprisingly, some 30% percent of prospective panelists declined to participate, leaving an N of nine, consisting of seven males and two females. All of that information is in the Methods section. Institutional Review Board (or its Canadian analogue) approval was probably obtained prior to this experiment ("You're going to do *what*?"), although that was not disclosed.

The experimental protocol was described in exhaustive detail, including subsections on stimuli, panel selection, training and protocol development (with three subsections), and testing and data analysis. Should other scientists halfway around the world want to delve into the science of cat food taste-testing by humans, they should be able to do so.

Often, the M&M section includes a description of the "Statistical Analysis," for example, the type of statistical calculations to be performed on the data obtained by following the protocol; sometimes, details on stats go into a separate section (usually right after M&M) section. The statistics section is equally as *critical* as the M&M section. Details are provided on the kinds of tests being performed and why (ANOVA, ANCOVA, Fisher's exact, Student's *t*, log transformations, regressions, least squares, etc.); any of dozens and dozens of tests will be performed on the (objective or subjective) data to make sense of the information. As with the non-statistics part of the M&M, the stats section might be set in a smaller font, or it might be absent entirely (or mostly) and housed on the journal website.

Results

This is the meat (get it?) of the IMRAD paper (or any "IMRAD project"). *What happened?* You took all those variables, used all those materials and followed all those methods and generated a boatload of data, which were analyzed just so. Results are not *discussed*—not in this section. This isn't where you say "this study makes a significant contribution to the study of. . ."; that's discussion/conclusion/ summary stuff. Here, you *only* report the exact results, all the numbers, ranges, *P*s, SDs (standard deviations). Sometimes these sections (depending upon how long the paper is) will have a great many tables and graphs in it. If possible, show data graphically, rather than in tabular form, keeping in mind that some data *should* just be tabular (subjects, variables assessed, etc.). Having all of this information in there in tabular and graphed format is typically determined by the length of the paper or requirements of your professor or dictates of the journal to which you want to submit your paper (you may be limited to *x* number of tables and figures, for example, and only *y* number of words). When writing a paper, however, your readers are using it for research, and they probably want as much data as they can get. To have a statement in the results section saying that "all comparisons vs. placebo were statistically significant" doesn't help other researchers if they need those numbers for subsequent comparison in their own studies. Overkill is more useful, but if those cranky editors over at your destination journal limit you to 3,000 words and you have at least 8,000 words of information, then there will be content casualties. Sometimes the data that are missing in Results sections are bewildering. It usually takes no extra space to include confidence intervals on a graph or in a table. Be as comprehensive as you can within the limits imposed upon you. Again, if you have a massive amount of data, be aware that some of it might be relegated to an online supplement.

The Results sections *usually* follows the format of reporting results *by end-point* (or *outcome*) by variables assessed as reported in the M&M section. Your research question (overtly or not) is crafted around a null hypothesis: Is there a difference in the variable we studied before and after the intervention we propose? If you were doing a nonexperimental observational study, what results did you get—what were your observations—in order, based on the observations you were planning to assess in your protocol?

Assessing cat food impressions certainly falls into the domain of subjective data—participants in this study are reporting their perceptions. Perceptive reporting is rigorously structured in the protocol, but it's still perception, that is, subjective. You might, for example, use a Likert-like scale to ask your participants "just how revolting was that?" Your participants could score their perceptions:

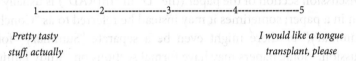

1-----------------2-----------------3-----------------4-----------------5

Pretty tasty   *I would like a tongue*
stuff, actually   *transplant, please*

Pickering, no slouch he, provided a thorough report, including a very large table and no fewer than six graphs. He did, in fact, use a coded Likert-like style for "Like Extremely" and "Dislike Extremely." Conclusion: college students will eat anything. Grittiness and viscosity were assessed as well, as was hardness and chewiness. Pickering did not have a section in Results (or Discussion) titled "Eeyew, Gross."

You can generate a lot of data when you have nine volunteers munching on chewy, gritty cat food, depending upon how you set up your study protocol. In his Results section, Pickering provided detailed graphing of his data in a variety of forms (see Figure 8.1).

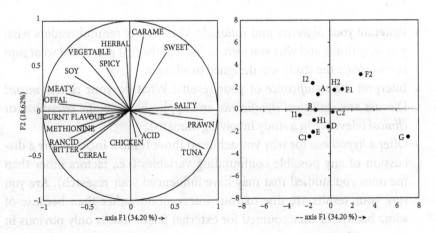

Figure 8.1. Loadings for factors 1 and 2 and scores from principal component analysis of 13 canned cat food products. Each data point (right figure) represents the factor score from the ratings of 11 judges and duplicate assessments. © 2007 Gary Pickering. Used by permission.

And—spoiler alert—it would seem that the highest ratings on the hardness and grittiness intensity scale were the meat chunk portions of products C1 (hardness and chewiness) and I1 (grittiness), respectively. If you use Pickering's protocol and have volunteers on the same subjective wavelength, you should be able to confirm this yourself.

Discussion

The Discussion section of the paper (the "D" in "IMRAD") is usually the last section in a paper; sometimes it may instead be referred to as "Conclusions" or "Summary," *or* there might even be a separate "Summary" following "Discussion." Some papers may have formal sections on "Study Limitations" or "Future Directions" tacked onto the end of the paper, or, if these elements are present, they may be rolled into the "Discussion" section.

This is where you interpret and discuss the results of your experimental or observational study; usually within the *context* of pre-existing research (i.e., first the tree, then the forest). Discussion of a study without placing it into context is not particularly helpful; therefore, every effort should be made to discuss the study within the landscape of the field (this is a type of mini-review; see Chapter 9 for more on reviews).

Here are some possible rhetorical goals for you to consider in the discussion section (some or all of these may be present, with some overlap depending on the type of research and the discipline you're in):

- Reiterate your *objective* and *rationale*: Very briefly remind readers what you were doing and why you were doing it in the first place, and what *gaps in knowledge* the study was designed to address.
- Interpret the *significance* of your results: What do your results *mean*? Discuss any *statistical significance* and applicable *scientific relevance* (or *clinical* relevance in a study involving humans).
- Offer a *hypothesis* for why you achieved those results: Include here a discussion of any possible confounding variables (i.e., factors other than the ones you studied that may have influenced your research). Are you *sure* your results are due to your intervention rather than because of some bizarre or unaccounted for external factor that is only obvious in retrospect?
- *Compare* your results with similar studies (mini-review).
- Explain *similarities and differences* (or concordances): Why do you think your results are similar to or different from other studies'?

- Explain the *limitations* of your study.
- Suggest implications and/or applications of the results.
- Provide recommendations for future research.
- Reiterate the *importance* of this type of research.
- Future directions or recommendations might suggest experiments that have not been done but should, specifically, experiments that would complement the one in the paper. Perhaps the "jelly-meat based" segment of the market is underserved, being as how it isn't gritty or chewy enough, or perhaps its viscosity ranked low on "hedonic response" (which was really the term they used for taste [what does *hedonic* mean?]) because it made volunteers gag.

For example, at the end of his study, Pickering makes sure to

recommend that these components continue to be assessed separately in future sensory research with CFP [cat food products]. A more detailed study examining CFP texture may be appropriate, should this domain prove particularly important for cat feeding behaviour. This may include further consideration of adhesiveness and particle size, shape, and orientation.

Acknowledgments

If you're writing for a class assignment, including acknowledgments might not be considered to be a formal expectation of your paper, although if you had help, you should do this. Acknowledgments are common in capstone projects or theses (again, follow the directions for the project). If you are working off of a grant, you must acknowledge the grantor, complete with grant number (if there is one). If the paper you are writing is for a journal that has other contributors, consider what constitutes an "author" (see Chapter 3 for clarification on that point); keep in mind that acknowledgments are typically made to individuals who are not authors. Pickering, you'll notice, is an author; the volunteers assessing the chewiness and grittiness of the cat food are not, heroic as that contribution might be. Pickering acknowledges an industry supporter and individuals who provided technical assistance. He did not acknowledge his tasters, although perhaps it was in their better interests to remain anonymous.

References

What do you reference in an IMRAD paper? What are you *allowed* to reference? Are you permitted to use review papers or meeting abstracts? Do you have to use *primary* references only? If you're stating basic science, how many references do you need for it, and do you have to cite the original reference? If you're talking about cats and what they eat, do you dig back through 9,000 years of human history to the first Egyptian hieroglyphs depicting cats being fed by humans? Unlikely. But you need to know what should be included before you start.

> "The more references you include, the more scholarly your reader will assume you are. Thus, if you write a sentence like, 'Much work has been done in this field,' you should plan to spend the next nine hours tracking down papers so that your article ultimately reads, 'Much work has been done in this field.[1,3,6–27,29–50,58,61,62–65,78–315,952–Avogadro's Number] If you ever write a review article, EndNote might explode."
>
> —Adam Ruben, "How to Write Like a Scientist"
> Mar. 23, 2012, *Science Magazine*.

Absolutely, positively, without question, ascertain the reference style you need to use *before you start writing*. If you use superscripted reference numbers and then later find out you are supposed to be using the "author, year" format, it will take you an awfully long time to go back and change it all (unless you have magical referencing software that will do that for you), and when you do that, you also open up a window for error. If you are using a manuscript template, the style might be in there. If your professor is giving you guidelines, he might have sample references in the instructions (e.g., for articles, book chapters, websites, etc.). If you are submitting to a journal, the first rule is check their "instructions to authors" for their required reference style. The second rule is "make sure they're really *using* that style." Always check. If you know AMA or ACS or APA style like the back of your hand, you don't want to find out when you're done that the journal that said "we use AMA style" really isn't using it or is using a modified version of it. Most instructions to authors also contain *examples* of how their references are written. Use those. Many journals also do not want references formatted, which means if you use Endnote or some other referencing software (which you positively should do—writing is a fluid, dynamic, nonlinear process; you want your references to travel around with the text they're referencing), you will need to unformat your references before submitting the manuscript (obviously, keep a formatted copy). Your professor may also *require* the use of specific referencing software. *Following directions is critically important in referencing.* Referencing is not difficult, but it is one of the most frequently violated instructions. If you are not provided adequate instruction up-front about referencing, ask. Your librarian can also help you with this.

---

10 Tips for Writing an IMRAD Paper

1. Determine authors in advance and ensure they meet authorship criteria. Provide author disclosures if necessary.
2. Make sure your title has enough information and is engaging enough to get readers to read your abstract when they're using a search engine.
3. Follow the rules for abstract formatting with regard to structured versus unstructured, correct sections, word or character limit, and inclusion (or not) of keywords (not exceeding limit).
4. Ensure that your introduction provides enough background and *context* for your paper, including history, rationale, objectives, and educational gaps the research has been designed to fill, and ensure that it is not overlong and compromising the amount of space you have for the actual science.
5. Make certain that your methods are complete and that they provide sufficient information for replication. If submitting to a journal, determine where the

Methods section goes (e.g., after the Introduction, at the end of the paper, or in an online supplement)

6. Make sure that your results are complete and that you have presented your data in tabular (following assignment or journal dictates for style, including statistics) or graphic format as much as possible (i.e., less prose, more visuals), and that you have ascertained in advance if you are limited to a specific number of tables and graphs by the journal or your assignment instructions. It's important to keep mechanics in mind; make sure you're following the guidelines with regard to table and graph placement (e.g., embedded within the paper, at the end of the paper, or in separate files), color (or not), and file format.

7. Your discussion should be complete and thorough, get your point across, answer your research question, address the objectives and gaps in knowledge set forth in your introduction and place your research into the larger context of the field.

8. Include any "extra" sections mandated by your assignment or journal (e.g., "conclusion," "summary," "future directions," etc.).

9. Acknowledge contributors and individuals who provided assistance, as well as your grantor if you have one, and provide a grant number, if one exists.

10. Write your references using the correct style—*pay careful attention* to this; apart from reference style (e.g., AMA, ACS, etc.) itself, check to see if references should contain elements such as DOI (digital object identifier) numbers, live links to the paper URL, website, or search engines, and ascertain if references are supposed to be formatted (e.g., in Endnote) or not.

What Is IRDAM?

IRDAM (Introduction, Results, Discussion, and Methods) is a variation on IMRAD that is becoming more common, particularly in some basic science journals (e.g., *Proceedings of the National Academy of Sciences*—which actually uses a "Significance," "Abstract," "Results," "Discussion," "Methods" format). The same information is essentially there, just presented in a different order. The rationale for this is that going from an introduction into a discussion of results is a natural, uninterrupted flow of information. This isn't particularly surprising. When people do research and find themselves reading a staggering number of papers, they're often reading papers out of order anyway. In one paper you might find yourself tarrying in the introduction, in another you might care about nothing but the discussion or conclusion. If you're early into research, you might read papers more linearly but might skip the methods until later anyway.

Banishing the Methods section to the end of the paper, however, necessarily requires some adjustments to the now-preceding section. While the Methods section might be cumbersome, it still provides structure and context to the experiment or study—structure and context that is necessary for assessment and discussion of results. In an IRDAM format, then, the tail end of the introduction might continue on for another short paragraph after context, rationale, and objectives are provided. This short paragraph can give a high-level overview of the experimental approach to the study (leaving the details for the Methods section at the end), and a one-sentence overview of results. This sets you up for the Results section.

In this case, rather than taking an umbrella view of the entire study, break the Results section down into each endpoint or objective, and for each of these sections, provide a mini-IMRAD structure: objective-specific question, high-level methods overview, results, and brief discussion of those results, putting them in context. These subsections are very brief and help break the study down into conceptual chunks, where all of the information relevant to a specific aspect of the study is discussed in its own silo. The Discussion section then serves to tie them all together, and then the trailing Methods section provides the minutiae of the details.

IMRAD style has been around for a long time (formally established in the 1970s) and more seasoned scientists are probably more used to it. But IRDAM has its prosaic and pedagogic merits and is likely to become more popular, particularly in the open-access arena. In any event, if you are submitting a paper to a journal, you will need to follow the dictates of the journal—this is why it's critical to read the instructions to authors *before* you start writing. Converting an IMRAD-style paper into an IRDAM-formatted paper to meet journal specs will require some significant retooling (and you might have to do this anyway if you get rejected from one journal and submit to a different journal). As always, follow instructions (from your professor or the journal) and use a recent IRDAM paper as a model.

Take-home

IMRAD is a recipe: an exhaustingly *thorough* recipe, and the coin of the realm for reporting original research. You don't have to write it in order; you can piece it together as it makes sense to you—just ensure that you are containing all of the necessary information you need for each section. Always remember to ascertain style and format before you start writing. If you are thorough, and if you follow directions, your volunteers will not have eaten cat food in vain.

Exercises

1. Write an abstract.
   a. Select an IMRAD paper germane to your area of research (or use one provided to you by your professor).
   b. Do not read the abstract; only read the body of the paper.
   c. Write an abstract for the paper.
   d. Compare the abstract you wrote with the abstract on the original paper. How did you do? Was the abstract you wrote similar? Did you follow directions?

2. Make some graphics and/or write about them.
   a. In a large, complete IMRAD paper *without* graphed data, select a narrative discussion of results. Turn data in a narrative or a table into a graph or graphs if possible, *or* indicate the type of visuals you think best fit the description. How well do the graphs reflect the results? Are they effective?
   b. Conversely, select the graphs/visual portion of the Results section of an IMRAD paper in your field on a topic of interest to you, *without* the narrative portion. Write the description yourself to the best of your abilities, and then compare to the original. What was similar? different?

3. Make a peanut butter and jelly sandwich.
   a. (Don't cheat and do this if you already did it in high school; pick a different common procedure).
   b. What would be the M&M for making a PBJ? (Statistical analysis not required unless you want to add that.)
   c. Compile M&Ms from the group.
   d. Assemble lab supplies: loaf of bread, peanut butter, knife, and "lab bench."
   e. Assemble sandwiches following M&Ms *to the letter;* be *very* exacting.
   f. How thorough were those M&Ms? Did they have enough information?

4. Pick a major representative journal in your field. Go to the "Instructions for Authors" section and *read the entire thing.* Read it carefully.
   a. Summarize the key requirements.
   b. List five things that would completely not have occurred to you.
   c. List three things that surprised you.
   d. List three things you do not understand or have not heard of before.

Discuss your findings in groups and share with the class.

5. Find a short IMRAD paper in your field (or use one provided to you by your professor) and convert it into IRDAM format. Alternatively, do the opposite: convert a short IRDAM paper into an IMRAD paper.

Acknowledgments

Very kind thanks and appreciation to Dr. Gary Pickering, Professor of Biological Sciences and Psychology, at the Cool Climate Oenology and Viticulture Institute at Brock University in St. Catherines, Ontario, Canada for permission to use materials from his cat food taste-testing paper.

9
Writing Review Papers

You are finishing reviewing the notes left by your lab partner when Dr. Wilkins calls you into her office. Dr. Adler is already there.

"So, how busy are you, exactly?" Wilkins inquires.

You panic slightly. You've been working really hard, but have you been up to her standards? Is it possible that you haven't been busy enough?

"Well, I . . . I'm really busy," you say, nervously. "We just finished the assays on—"

"Very good, very good," Wilkins interrupts, charitably. "Listen, we've had a request for a review from *Biomedical Engineering*." She waves a recent shiny *BME* issue at you as she continues talking. "They know about the work we do here and they've asked if we could do an article on ROS and angiogenesis. Since your group has presented on this in the past, we thought we'd ask you if you want to work on it. We don't want you to get bored!"

You stare at her in disbelief. Bored? "Um. Yeah. Sure. I've got *plenty* of time."

"Great!" she chirps, "because we already told them you'd do it. They need it by the beginning of April."

"It's the middle of February," you point out, as a cold sweat starts to trickle down the middle of your back.

"The three of us will be coauthors, but you will be the *lead* author," Wilkins continues. "You've earned it!"

"It's the middle of *February*," you say again. Maybe she didn't hear you the first time.

The sordid truth, however, is that getting a publication would be awesome and this was a good way to get into a leading journal in your field.

"We'll meet once a week," Adler lied. She was very efficient. "So, what do you say, next Monday at, let's see, 9 to 9:30. Seems to work for everyone." By "everyone," of course, she meant herself. Adler was already typing her appointment into her smartphone. "Bring a tentative outline next week and we'll talk."

Leaving the meeting, you wonder if there is some scientifically possible way to work 14 hours in a day, but you also feel a twinge of excitement, tempered with simultaneously being convulsed with fear—an emotional mélange that gradually changes from "maybe exciting" at the start of the day to "stone-cold fear" by the end of day. Your group has been working on the role of reactive oxygen species in angiogenesis for over a year. Research was slow and meticulous and everyone kept up with the literature in the field— so that was a bonus. But still, there was so much to write about! Where to begin? How long, exactly, should the article be? How will you write this and do your lab work and all your other work obligations (a poster session was coming up in a month, and a deadline is looming for a new grant). You already understand that the brunt of the work falls squarely on you as lead author (*lead author!*), whereas the other coauthors would be there for advice and support (or "advice" and "support"). As a postdoc at a prestigious research university, the pressure is immense to get results, and get them published—fast. Will working on a review diminish your time spent on original research? Of course it will. You have already written a literature review as a chapter in your dissertation but never a stand-alone article. Can you pull this off?

Luckily, you were well trained, as a scientist, to deal with doubts methodically. "Divide and conquer," your advisor used to say when you were working on your dissertation: divide each seemingly monumental task into manageable parcels, tackle them one by one, and you shall emerge victorious. You are more than qualified—you know a lot about the subject—so now all you need to do is find a strategy to turn what you have into a review article that will establish you as an expert on this subject—even though it's only February and you need a finished paper by the end of April. Confidently, you say, "It's the middle of *February*."

This is a plausible scenario of how review essays may originate: enough "buzz" is generated around a topic and enough research conducted that journal editors take notice and demand an assessment of what's happening so far in that particular niche. This chapter will introduce you to review papers: what they are, why they are necessary, and what skills you need to successfully write them. The workload part and the insane deadlines are kind of a given, but you're on your own for those.

What Are Reviews?

Scientists' productivity and (in academic settings) tenure and promotion considerations are ominously tied to the amount of *original* research they do. Scientists set out to produce new knowledge on issues for which society demands a solution, and then selflessly share what they find by writing papers that they can conveniently list on their CVs: heart disease, alternative fuels, climate change. It is not unusual for multiple teams across the world to work and publish on similar issues—the more vital the issue, the larger the body of research. The accumulation of new discoveries on a single subject can, however, be staggering, to the point that it may not be productive to read all the 15,224 papers published to date on it. Instead, one would need a guide through this research labyrinth.

Most bibliographic databases for the sciences have papers numbering into the millions, and in many cases, some of those papers are free. In PubMed you can even tick a box for only free papers for your search term. Do not forget that through your university library you may have access to a database that can access thousands of journals for free.

Enter the review article—also known as the review essay, review paper, or simply, review. This kind of article may not be pushing back the frontiers of science, but it summarizes, synthesizes, contextualizes, and assesses the research that does.

A review usually reports on *recent* trends and developments (or state-of-the-art, cutting-edge knowledge) on a particular topic; some reviews may have a broader scope (as in, reviewing an idea, concept, or practice that has been around for a while), but they also usually end up reviewing the latest research on it. Reviews may not have the sexy "news value" of original research—but while that is *generally* the case, it is not a rule; in fact, many reviews develop interesting and original claims. As such, they serve crucial functions in the sciences:

- *Reviews synthesize a large body of research for training and continuing education purposes.* Sadly, science marches on after you complete your training; after you graduate and practice your profession, you will have to rely on the literature and particularly on reviews to keep you up to speed with the newest developments in your field. If you graduate with 2022-level cutting-edge knowledge, by 2024 you'll be out of date—maybe *way* out of date—if you haven't been keeping up. In the life sciences, especially in clinical fields, reviews are essential for practitioners. A journal could publish thousands of original research papers a year in a single specialty, such as in biologic-based treatments for cancer—which could theoretically require oncologists to spend three-quarters of their time reading them. Oncologists don't have that kind of time, so they may take the review shortcut. Reviews help summarize and make sense of huge amounts of research in a relatively short space. They are the well-lit paths through the thick forest of research—and, as such, they are invaluable not only to clinicians but to scientists in general. As an added bonus, some reviews may also make forays into the history of a particular concept, which may add satisfying insights into our understanding of science and all of her far-flung subdisciplines.
- *Reviews explain and clarify contradictory research results.* It will come as a shock to no one that not all scientists agree on all topics. You may find that acupuncture works better than pain medication in treating lumbar pain, while your colleagues across the country find no such evidence; other scientists will be feuding about the nature of dark matter or black holes— this is why science is so lively. Reviews analyze, evaluate, and compare a pool of studies and provide informed, authoritative conclusions about the current consensus on a topic in a particular field. Unsurprisingly, an

original idea often emerges out of this synthesis—one that can, in fact, be much more surprising or novel than the original research in the first place; see Box 1 for an example of a review that challenged some cherished assumptions of primary care physicians and cardiologists.

- *Reviews provide the basis of best practices in the field—and of clinical guidelines.* Clinical guidelines determine standard practice—such as how you get treated for flu, or how your grandmother manages her diabetes. The clinicians who manage such cases and usually take care of these problems are too busy to read the staggering amount of original research put out every year by clinical journals. Reviews help clarify the implications of the original research that define the "standard of care." With a solid clinical guideline update review under his belt, your grandmother's diabetologist can confidently tell her it is safe to use sucralose instead of aspartame when she sweetens her coffee.

- *Reviews point out gaps in current research and provide guidance for future research.* Reviews may assess what has been done and what remains to be done, thus pointing researchers in the right direction. For that reason, grant proposals usually require some kind of literature review that highlights the need for new research (it may be part of the "needs assessment" or "background" or "theoretical and empirical justifications" section). Such grant literature reviews show that the research idea has potential and is backed up by evidence but some pieces of the puzzle are missing. A similar idea stands behind literature review sections or chapters required by theses and dissertations, capstone projects that require original research. The literature review portion of such projects provides the justification for the current study. Additionally, some healthcare journals are starting to require reviews before publishing clinical trials as well.

Box 1. Cochrane Review Fills Doctors With Self-Doubt

It was the byline that launched a thousand ice-cold daggers into the hearts of clinicians:

"January 20, 2011 (London, United Kingdom) — A new Cochrane review has provoked controversy by concluding that there is not enough evidence to recommend the widespread use of statins in the primary prevention of heart disease" [1] (Heartwire).

If the Cochrane review's conclusions were true, the number-one type of drug prescribed to prevent heart disease was hardly better than placebo, and most, if not all primary

care doctors and cardiologists around the world had given false hope and misinformation to their patients—albeit unknowingly. Statins had been leaders in world drug sales
for years (at the time of the Cochrane publication, one of the statins topped the charts
at over $13 billion in world sales, roughly $4 billion more than the second-leading drug);
thus, news that they were not as effective as originally believed was bound to upset
drug makers (to put it mildly), doctors, and patients alike. Upset clinicians questioned
the article's conclusions. Academicians balked. Authors of studies that had found exactly the opposite to be true (e.g., statins help prevent heart disease) expressed their
disagreement in no uncertain terms. Both *Newsweek* and *The Atlantic* picked up the
story, with a twist: not only were statins ineffective but they were dangerous as well.
Ultimately, clinical guidelines were called into question. Once the media hype sub-
sided, researchers did what they do best—they called for . . . more research.

But fear not. The *updated* version of the Cochrane review on statins for the prevention of cardiovascular disease, published in 2013, helped assuage and vindicate
practitioners, finding that "[t]here was no evidence of any serious harm caused by
statin prescription," and that "[e]vidence available to date showed that primary prevention with statins is likely to be cost-effective and may improve patient quality of
life." Compare this with their 2011 conclusion which read "Only limited evidence
showed that primary prevention with statins may be cost effective and improve patient quality of life. Caution should be taken in prescribing statins for primary prevention among people at low cardiovascular risk."

Phew. Science truly is a lively, dramatic show full of twists and turns.

Types of Reviews

Reviews come in all shapes and sizes—from brief, one-page clinical updates
to "mini-reviews" that may be only three to four pages long, to longer, more
involved and exceptionally thorough reviews. They are more common in the
biological and clinical fields, where the explosion of information has been
the most dramatic in recent decades, but they have been gaining popularity
in the physical sciences as well. They are also common in the social sciences
and in the humanities. There is nothing like a review when you don't know
where to start.

Narrative Reviews

Narrative reviews are the most common form of review; they are extremely
popular in medical and especially nonmedical fields. "Narrative" in this

case means simply that the prose does not follow the IMRAD structure of the research article but instead is divided into sections and subsections according to an outline reflecting the author's interpretation of what is relevant about a topic. The selection of original articles to be reviewed is also up to the author—upon whom we rely to make the best, most recent, most cutting-edge, most informed choices (this is one reason why many reviews are usually invited: the editors will pick an expert who can make the most careful source choices). The sections and subsections of a narrative review and their arrangement are at the discretion of the author, but they should emerge organically from the available literature. Thus, a strong outline is key to a narrative review. Clear markers of a narrative review's structure (headings, subheadings) also allow readers to scan and zoom in on the parts of the paper most relevant to them. The abstract of a narrative review is typically also descriptive (see Chapter 10) and in paragraph format.

The Oldest Reviews Indexed in PubMed

One of the oldest review articles indexed in PubMed is in fact a lecture given by Dr. Henry Lyman and published in its entirety with the title "A Review of the Present State of Exact Knowledge Regarding the Causation of Epidemic Infectious Diseases" in *Public Health Papers and Reports*, 1878, followed closely by R. Garner's "The Brain and Nervous System: A Summary and a Review" in *Journal of Anatomy and Physiology*, 1881—both available from PubMedCentral (PMC). Narrative reviews have since come a long way since the early prosaic efforts of Drs. Lyman and Garner. These articles, apart from being lengthy, give no clue of the article's organization (no discernible outline); include no headings, subheadings, figures, or tables of any sort; and are certainly meant to be assiduously perused rather than cursorily scanned and mined for information. And back then, apparently, bibliographies were a thing of the future.

Systematic Reviews

Due to the overwhelming number of clinical research studies published in the past decades, the need arose for a systematic way to select and organize studies to address specific questions. Systematic reviews are exactly that: *systematic*. The authors develop very specific criteria for the selection of the papers they review, to ensure comparison of proverbial apples to proverbial apples. Criteria may include such things as databases searched, language(s), year range, specific inclusion keywords or terms germane to the review (e.g., age, sex, disease, geography, epoch, wavelengths, time periods, intensity, salinity, etc). They papers chosen for the review have to meet all of the criteria. The 256 papers may have "met the criteria," but upon review, the authors may only end up with 27 of them for inclusion, because they all didn't *really* meet *all* the criteria *exactly*. The papers are compared (usually in tables) and summarized. Systematic reviews aim to eliminate such potential bias by reintroducing the logic of IMRAD—with the emphasis on "M" for "methods." By using a specific, reliable, and reproducible protocol, the process of article selection is transparent and readers can judge for themselves whether the inclusion or exclusion criteria are relevant for their purposes (by contrast, the "protocol" of a narrative review is usually not reproducible). Systematic reviews, therefore, include:

- a clear set of objectives
- a rigorous Methods section explaining the selection of sources (usually through a systematic search)
- an assessment of the validity of the sources, including an evaluation of any bias
- a Results section describing the findings of the studies reviewed
- a discussion the implications of the findings, including recommendations.

In short, systematic reviews are just like any other IMRAD-style research article, except their "data" are other research articles. The IMRAD structure is also reflected in the structured *abstract* of a systematic review. Because they can account satisfactorily for a variety of variables and can control bias better than narrative reviews, systematic reviews have become the gold standard review source in clinical research. In fact, scientists have started to advocate using systematic reviews in disciplines where they are not at all common, such as engineering.

The most recent standards for conducting systematic reviews and meta-analyses (see next section) for health research are detailed in the PRISMA tool (Preferred Reporting Items for Systematic reviews and Meta-Analyses), available on the EQUATOR network (which contains a number of other reporting guidelines). The goal of the PRISMA statement was to correct "poor reporting" of systematic analyses and to provide specific guidelines for conducting them (see Figure 9.1 for a complete checklist of systematic review components).

Section/Topic	Item #	Checklist Item	Reported on Page #
TITLE			
Title	1	Identify the report as a systematic review, meta-analysis, or both.	
ABSTRACT			
Structured summary	2	Provide a structured summary including, as applicable: background; objectives; data sources; study eligibility criteria, participants, and interventions; study appraisal and synthesis methods; results; limitations; conclusions and implications of key findings; systematic review registration number.	
INTRODUCTION			
Rationale	3	Describe the rationale for the review in the context of what is already known.	
Objectives	4	Provide an explicit statement of questions being addressed with reference to participants, interventions, comparisons, outcomes, and study design (PICOS).	
METHODS			
Protocol and registration	5	Indicate if a review protocol exists, if and where it can be accessed (e.g., Web address), and, if available, provide registration information including registration number.	
Eligibility criteria	6	Specify study characteristics (e.g., PICOS, length of follow-up) and report characteristics (e.g., years considered, language, publication status) used as criteria for eligibility, giving rationale.	
Information sources	7	Describe all information sources (e.g., databases with dates of coverage, contact with study authors to identify additional studies) in the search and date last searched.	
Search	8	Present full electronic search strategy for at least one database, including any limits used, such that it could be repeated.	
Study selection	9	State the process for selecting studies (i.e., screening, eligibility, included in systematic review, and, if applicable, included in the meta-analysis).	
Data collection process	10	Describe method of data extraction from reports (e.g., piloted forms, independently, in duplicate) and any processes for obtaining and confirming data from investigators.	
Data items	11	List and define all variables for which data were sought (e.g., PICOS, funding sources) and any assumptions and simplifications made.	
Risk of bias in individual studies	12	Describe methods used for assessing risk of bias of individual studies (including specification of whether this was done at the study or outcome level), and how this information is to be used in any data synthesis.	
Summary measures	13	State the principal summary measures (e.g., risk ratio, difference in means).	
Synthesis of results	14	Describe the methods of handling data and combining results of studies, if done, including measures of consistency (e.g., I^2) for each meta-analysis.	
Risk of bias across studies	15	Specify any assessment of risk of bias that may affect the cumulative evidence (e.g., publication bias, selective reporting within studies).	
Additional analyses	16	Describe methods of additional analyses (e.g., sensitivity or subgroup analyses, meta-regression), if done, indicating which were pre-specified.	
RESULTS			
Study selection	17	Give numbers of studies screened, assessed for eligibility, and included in the review, with reasons for exclusions at each stage, ideally with a flow diagram.	
Study characteristics	18	For each study, present characteristics for which data were extracted (e.g., study size, PICOS, follow-up period) and provide the citations.	
Risk of bias within studies	19	Present data on risk of bias of each study and, if available, any outcome-level assessment (see Item 12).	
Results of individual studies	20	For all outcomes considered (benefits or harms), present, for each study: (a) simple summary data for each intervention group and (b) effect estimates and confidence intervals, ideally with a forest plot.	
Synthesis of results	21	Present results of each meta-analysis done, including confidence intervals and measures of consistency.	
Risk of bias across studies	22	Present results of any assessment of risk of bias across studies (see Item 15).	
Additional analysis	23	Give results of additional analyses, if done (e.g., sensitivity or subgroup analyses, meta-regression [see Item 16]).	
DISCUSSION			
Summary of evidence	24	Summarize the main findings including the strength of evidence for each main outcome; consider their relevance to key groups (e.g., health care providers, users, and policy makers).	
Limitations	25	Discuss limitations at study and outcome level (e.g., risk of bias), and at review level (e.g., incomplete retrieval of identified research, reporting bias).	
Conclusions	26	Provide a general interpretation of the results in the context of other evidence, and implications for future research.	
FUNDING			
Funding	27	Describe sources of funding for the systematic review and other support (e.g., supply of data); role of funders for the systematic review.	

Figure 9.1. Components of a systematic review, as outlined in the PRISMA statement.

The Cochrane Collaboration

The most famous collection of systematic *clinical* reviews is the Cochrane Library, a result of the Cochrane Collaboration, which provided the impetus for observing a rigorous methodology in such studies. Archie Cochrane (1909–1988), a British epidemiologist, deplored the lack of an organized way to sift through medical evidence. In 1979, he famously and portentously wrote: "It is surely a great criticism of our profession that we have not organised a critical summary, by specialty or subspecialty, adapted periodically, of all relevant randomised controlled trials." Little did he know what little ball of wax *he* started rolling. The Cochrane Collaboration, established in 1993 and named in his honor, is an international organization preparing and disseminating systematic reviews of evidence in healthcare issues (see www.cochrane. org). In contrast to most systematic reviews, Cochrane reviews contain one extra element: a plain language summary (usually right after the structured abstract), which is relatively jargon-free and tailored to a general lay audience as opposed to a scientific audience. This meets the organization's goal of disseminating medical information to the public. Cochrane reviews are a gold standard in health and medical science research.

Meta-analyses

Systematic reviews are often, though not always, accompanied by meta-analyses. The need for meta-analyses became apparent with the proliferation of studies asking the same research question and using a limited number of subjects (e.g., can aspirin control arthritic pain better than acetaminophen?). A search for these terms may yield a number of disparate studies enrolling any number of subjects. Rather than read each of these studies and make inferences based on each particular case, you may want to conduct a meta-analysis: pool all the subjects of similar studies together and analyze the results statistically; the idea is that an analysis of 500+ subjects may yield more reliable or significant results than an analysis of a study with only 12 participants. PRISMA defines meta-analysis as "the use of statistical techniques to integrate and summarize the results of included studies.... By combining information from all relevant studies, meta-analyses can provide more precise estimates of the effects of health care than those derived from the individual studies included within a review."

More robust statistics is a particularly seductive aspect of meta-analyses. However, pooling the results of 20 mediocre studies does not yield stronger evidence—much like combining 20 cups of weak coffee does not yield stronger coffee. The selection of studies is paramount, so hold the foul temptress of statistical robustness at bay and choose your studies carefully. Once you have your candidates, you are then faced with the rather complicated decision of if and how to combine data; this hinges on statistical, clinical, and methodological considerations. As for the clinical and methodological considerations: they depend on your team's (a) research question(s) and (b) evaluation of bias in the studies selected. Whatever your team decides, however, it is important to state your criteria clearly in the Methodology section to ensure transparency, reliability, and replicability. In medicine, meta-analyses are considered to be formal epidemiologic studies. We need to mention a caveat here: some researches love meta-analyses, some look at them askance, and some positively hate them, believing them to be oversimplified. Meta-analytic reports get published and may influence opinions, practice, recommendations, or even guidelines—sometimes for the worse. Meta-analyses, for example, may come to a conclusion not widely accepted or agreed upon by the scientific community. We recommend accepting them for what they are, without necessarily taking them as fact. What they—and systematic reviews—are truly useful for is a list and summary of important studies that can save you a lot of time and clue you in to references you should find and read yourself. In other words, systematic reviews and meta-analyses are good research resources.

You may have noticed some things about systematic reviews and, especially, meta-analyses. They might be assembling and assessing pre-existing information, but they are following strict protocols and generating unique results. So are they really reviews or are they original research? Most journal editors consider systematic reviews to be original research, and meta-analyses are even considered to be a kind of trial in medicine (ie, definitely original research). So they lurk in a grey area, called "reviews" but more properly characterized as an original research. Both of these types of publications are usually quite substantial and of significant utility in research.

Narrative or Systematic Reviews?

Supporters of systematic reviews (SRs) regard them as superior to the traditional, nonsystematic, qualitative narrative reviews (NRs) because SRs are designed to minimize bias. By applying inclusion and exclusion criteria so strictly, however, SRs may end up excluding a vast number of relevant studies. Say you start working on an SR on lupus and begin a broad, cross-database search that yields approximately 400,000 results. Once you start getting nit-picky and strictly applying your inclusion and exclusion criteria, you've whittled your studies down to exactly 4, since you only wanted studies that were (a) large randomized control trials (b) comprised of Asian women who were (c) younger than 30 years of age, (d) suffering from lupus, (e) treated with a combination of drugs and yoga, (f) for a minimum of eight months with follow-up, (g) don't like karaoke . . . you get the picture. Your criteria have to be rational enough for a large enough study. Researchers have also argued that NRs are not only useful but indispensable since they may fulfill other research roles that SRs are not well suited for, such as asking specific clinical questions, comparing and synthesizing conflicting research, explaining mechanisms, and "translating" research between disciplinary traditions. In the end, the goal of ultimate objectivity that SRs aspire to may be elusive, and traditional reviews can also minimize bias and be quite rigorous.

Reviews as Integrated Parts of Larger Works

Narrative and systematic reviews and meta-analyses are self-standing genres. However, as we have hinted, you can find well-developed review sections in larger documents such as grant proposals or theses and dissertations. In all of these documents the literature review section or chapter serves the important purpose of positioning the researcher(s) and the study within the field. Rhetoric researchers call this an attempt to find one place in the larger scientific conversation, or within one's "discourse community" (see Introduction and Chapter 2). A well-written review section enhances your ethos as an expert knowledgeable of the field and your predecessors, painting you as uniquely qualified to propose new research, build on previous research, or fill a gap in knowledge. Careful selection of the best sources that apply particularly to the study being described or proposed will determine the quality of the review section: you must include the "big names" and established authorities as well as those works that are specifically related to your study. You must be able to understand, synthesize, and evaluate previous work in terms of methodological design, data interpretation, conclusions, and impact. Conversely, failure to include well-known or relevant studies or to synthesize and assess properly may result in quick disqualification by a grant review or dissertation committee. Even the typical research article (IMRAD) includes a smaller review section—in the introduction and/or in the discussion of the results. No matter where you have to include them, reviews are vital to any academic endeavor.

---

10 Tips for Reviews

1. Start with an intriguing, well-formulated question.
2. Narrow down your topic and scope to something specific and manageable in size.
3. Search for recent reviews on the topic to mine their references, then extend the scope of your research. If possible, apply inclusion/exclusion criteria

(e.g., exclude case studies). Ensure that you include the most recent references possible in your review—that's one reason reviews are valuable, because they contain the most up-to-date information.

4. Arrange your primary research studies summaries in a table to get a "bird's eye view" of similarities and differences between studies, as well as of the comparative strengths of their arguments or evidence.

5. Create an outline based on the major themes that emerge from your table or idea-mapping.

6. Create your own research space in the introduction by clearly defining the scope and goals for your review.

7. Map for the reader the trends, conflicts, and convergences you find in the research.

8. Try to tell the *story* of research on your topic.

9. Don't be afraid to make evaluative judgments when warranted (e.g., when a study seems too limited and/or you have doubts about the methodology).

10. Point out directions for future research based on the gaps or needs you have perceived by reviewing previous studies. Additionally, you may make concrete recommendations if you asked a specific question.

How to Write Reviews

Because it is likely that you will encounter this genre early and often, most of the rhetorical strategies for review essays we present here apply to narrative reviews rather than systematic reviews. While we focus on reviews as a self-standing genre, most of our advice is also applicable to review sections of grants or theses. Writing a review may seem like a daunting task—especially when it's due at the end of April and it's mid-February now—but remember the advice we discussed in the beginning of the chapter: divide and conquer. Review essays are manageable if you tackle them one piece at a time:

Step-By-Step Guide for Writing a Review Essay

Step One: Formulate a topic and a goal

Step Two: Search the literature and select and manage your sources

Step Three: Summarize the research

Step Four: Create an outline based on the themes/topics emerging from your sources

Step Five: Write your paper following the outline

Step Six: Revise and polish

Step One: Formulate a Topic and a Goal

You should look to your own areas of expertise or passion when choosing a topic. Don't know where to begin? Try Thomson-Reuters' Sciencewatch (www.sciencewatch.com), a site that tracks the "research front maps" as well as the "fast-moving fronts" in a variety of scientific fields—that is, clusters of highly cited papers on topics that dominate the national and international research agenda. Alternately, look at recent issues of the premier journal in your field (ask a professor if you are not sure which one it is) and get a sense of what topics are "hot" or "trending" right now (yes, that happens in academia as well). Try browsing through your favorite database; type in a keyword or two and (a) see how many hits you get, (b) how many and what kind of variations there are on the topic, and (c) how many recent papers there are on the topic. Start with a broad topic (e.g., grazing species) and then gradually ask questions about it until you are able to narrow it down (e.g., "A biogeographical review of Tsessebe antelopes (*Damaliscus lunatus*) in south-central Africa").

Example: In our scientific writing class, Ben said that he was interested in air pollution. A great topic—but hardly a focused one. After group discussions, Ben decided to pursue available evidence regarding the effects of air pollution on the vegetation. Narrowing this further, he produced the following research question: "What impact does air pollution have on North American trees?" If you feel ambitious, or if you get a sense that there is enough research out there, you may further narrow this down to deciduous trees, for example. Not enough? Oak trees. Not enough? Pin oaks. Not enough? Pin oaks in the eastern United States. Not enough? Pin oaks in Virginia. No, *western* Virginia. Okay, western Virginia in the Blue Ridge Mountains. Wait—on east-facing slopes.

As you start your research, you may find that your focus will shift, based on the availability of the information you review. You may have to tweak. Zoom in. Zoom out. Your title should reflect your focus as closely as possible:

Sample Review Titles (from PubMed and ScienceDirect)
Biology (general)
- Behavioral decisions made under the risk of predation
- Antioxidants, oxidative damage and oxygen deprivation stress

- Biotic pollination mechanisms in the Australian flora
- The silence of genes in transgenic plants
- Renal filtration, transport, and metabolism of low-molecular-weight proteins
- Overcoming substrate inhibition during biological treatment of mono-aromatics: recent advances in bioprocess design
- The biology of leptin

Chemistry

- Doehlert matrix: a chemometric tool for analytical chemistry
- Multivalency in supramolecular chemistry and nanofabrication
- Ligand K-edge X-ray absorption spectroscopy: covalency of ligand-metal bonds
- Magnetic anisotropies in paramagnetic polynuclear metal complexes
- Novel proteins: from form to function

Physics

- Inhibited spontaneous emission in solid-state physics and electronics
- Review of physics and applications of relativistic plasmas driven by ultra-intense lasers
- High-temperature superconductivity in iron-based materials
- Physics of liquid jets

Medicine

- The neuropsychiatric burden of neurological diseases
- The impact of stretching on sports injury risk
- Fibromyalgia and obesity: the hidden link
- Obesity, metabolic syndrome and esophageal adenocarcinoma: Epidemiology, etiology and new targets.
- Femoroacetabular impingement: a review of diagnosis and management.

Engineering

- Integrating environmental impact minimization into conceptual chemical process design
- New materials for micro-scale sensors and actuators
- The ensemble Kalman filter in reservoir engineering

Notice that there is substantial variety in review paper scope. Some focus on explanations of mechanisms, some on the most recent advances, some on specific populations, and some on connections between two or more discrete concepts. What other possibilities are there? How would you characterize your own topic?

If you are dealing with clinical issues, you may take what the British call "the PICO" approach, an acronym that stands for Patients, Interventions, Comparisons, Outcomes:

P: Which *patient* population do you have in mind (think of relevant attributes to your study: adults? children? the elderly? male? female? athletes? mine workers?). What diseases or conditions are you investigating?

I: What *intervention* do you want to investigate? Is it observation, diagnostic, or trial? If treatment, is it surgery, a drug, a diet, a series of physical exercises, a combination thereof?

C: Are there *comparisons* to be made, differences to be established? What is the intervention compared to (placebo? nonintervention? surgery?). For example, do you want to compare a nontraditional intervention (e.g., meditation) to a traditional one (e.g., antihypertensives) to see how it affects a certain population's blood pressure?

O: What *outcomes* are relevant? What is measured and how (e.g., morbidity, mortality, quality of life, pain, recovery time, ability to walk/stand, swelling, range of motion, blood pressure, weight, adverse events, etc.)?

As you decide on a strong, focused topic, consider your goal in relation to your target audience. Ben's goal, for example, was to educate environmental researchers about pollution-related damage to trees but also to motivate them to search for ways to reverse that effect. Yours might be to help physical therapists determine which rehabilitation path they should recommend for patients who have undergone knee surgery. This goal will drive your review and will help you achieve consistency and coherence in your paper. Remember that you are not writing an encyclopedia entry or a general introduction to the topic (although you will necessary include definitions and the like), but rather are trying to answer a specific question about the topic by relating two or more aspects: an intervention to a population; a species to a geographic area or to a human intervention; a chemical compound, physics concept, or biological entity to an application or circumstance of some kind.

Most helpful at this stage is to formulate your topic as a question: How can hippotherapy help children with cerebral palsy? What is the role of surfactants in creating new asthma drugs? Which principles of quantum physics can we use to achieve teleportation? Is there a connection between anxiety and autism? This will help set up the structure of your paper and it will certainly help with finding sources.

Step Two: Search the Literature and Select and Manage Your Sources

We covered research skills in Chapter 7, so now you're a pro. However, being able to navigate databases is only the first step. Recall that you must include recent and relevant research articles and that the references of a recent article on your topic can provide you with a rich mine of sources as well. Pay attention in particular to recent review papers that are on similar topics, for two reasons: (a) you don't want to replicate work that has already been done, or you want to know how to put your own spin on the topic and avoid plagiarism; (b) you can mine their references section. Look for authors who consistently research your issue (but who may publish in a variety of journals, some interdisciplinary) and for works produced by labs, institutes, and departments that are famous for their research on a particular topic. Also, although you are not writing a systematic review, it may behoove you to set up some minimal inclusion/exclusion criteria, in collaboration with your instructor; for example, you may dismiss case studies from your initial pool as being too qualitative and idiosyncratic, or you may want to include only papers from journals with a certain impact factor. If you decide on such criteria, you will *have* to disclose them in your paper.

A clear sign of novice writers and researchers is that their bibliography is incomplete, is poorly chosen, is dated, or strays too far from the purported goal of the paper. Don't let that happen to you. Some sources are nice to have and others are *need* to have. Focus on the latter. When in doubt, consult your professor or mentor for suggestions.

Step Three: Summarize the Research

"Wait a minute!" you say, brightly. "That's easy! The articles are already summarized for me—look at these abstracts!"

Not so fast.

We have already discussed the obvious appeal of the review essay—why read a hundred articles when you can get roughly the same information from a single paper of reasonable length? Do not think for a minute, though, that review essays are the science version of CliffsNotes (formerly Cliff's Notes). Summarizing essential information is merely the first step. The real work— and your original contribution—comes from the selection, analysis, synthesis, and evaluation of that information. You need to connect the dots among carefully curated research studies and draw the correct conclusions. A review essay is not a data dump—that would be both tedious and counterproductive—but an attempt at finding order, logic, trends, and future directions for research on a particular topic. In other words, the goal of the review essay is to increase understanding rather than to parrot back "data."

> The goal of the review essay is to enhance understanding, not rereport data.

Otherwise, why would anyone bother to read your review, when she can spend many enjoyable hours reading abstract after tantalizing abstract in a database?

Reading and summarizing for a review essay is an active and involved process, so take notes along the way. If you have been judicious in your article selection, all your sources will have a common denominator—a research question, a type of intervention, etc. You may take notes directly in a summary table (see Table 9.1).

Table 9.1. Sample summary table template

Study	Objective	Methods	Results	Conclusions/ Recommendations	Strengths	Limitations
Author, year	Purpose of the study and/or research question	Materials, participant characteristics, study design, duration, controls	Brief summary of main findings (include key numerical values)	Authors' conclusions and recommendations	Any design strengths? Is the paper cited by other researchers?	Any design flaws? Statistical errors? Understated or overstated conclusions?

By no means should you take this summary table format to be set in stone. Rather, use it as a basis for your own table: what are the categories most important to you? You can always add or subtract columns. (For the Strengths and Limitations columns, revisit the tips in Chapter 8.)

Don't write in complete sentences in these tables for two reasons: (a) tables should not be cluttered but rather offer a snapshot of the essential info that will help your synthesis; (b) it will be much harder to plagiarize (however inadvertently) if you don't copy sentences verbatim from the articles when you take notes. Table 9.2 is an example of a summary table created by a student reviewing pharmacologic interventions for menstrual migraines.

Again, your table headings may look very different, depending on your field of study. The table may even evolve as you read, and you may add categories or concepts that will apply to some of the studies and not to others—and that is okay too. You may also add a column that scores the strength and reliability of

Table 9.2. Diagnosis and treatment of menstrual migraines. Summary table by Tanner McCalley

Trial/year	# of patients	Medication	Dose	Treatment cycles, timing	Results
Szekely et al., 1989	118	naproxen sodium	550 mg bid	Baseline untreated cycle followed by 3 treated cycles on days -7 to +7	Naproxen sodium is more effective than placebo
Allais et al., 2007	20	naproxen sodium	550 mg qd	Baseline untreated cycle followed by 3 treated cycles on days -7 to +7, then 3 treated cycles on days -5 to +5	Baseline no. of attacks: 1.7 – 0.11 Month 3: 1.2 – 0.10 ($p < .001$) Month 6: 1.1 – 0.07 ($p < .0001$)
Silberstein, et al., 2009	179	frovatriptan	2.5 mg bid 2.5 mg qd	3 cycles 2 days before anticipated menstrual migraine Duration 6 days	Migraine incidence 37.7% during bid treatment, 51.3% during qd treatment ($p = .002$), 67.1% with placebo
Brandes et al., 2009	410	frovatriptan	2.5 mg bid 2.5 mg qd	3 cycles 2 days before anticipated menstrual migraine Duration 6 days	0.92 headache-free treatment periods with bid treatment, 0.69 with qd treatment, 0.42 with placebo ($p < .001$ and $p < .02$ vs placebo)
Mannix et al., 2007	Study 1: 287 Study 2: 346	naratriptan	1mg bid	4 cycles 3 days before anticipated menstrual migraine Duration 6 days	Mean 37% of treatment periods w/o migraine per patient with naratriptan vs 24% with placebo ($p < .05$)
Ferrante et al., 2004	11	phyto-oestrogens	Genisteine 56mg diadzeine 20mg PO qd	3 untreated cycles followed by 3 treated cycles	Month 3: 2.2 days per month ($p < .005$)
Calhoun 2004	11	conjugated equine estrogens	0.9 mg qd on days 22 to 28	2 consecutive 28-day cycles	77.9% reduction in number of headache days per cycle

the study according to evidence and methodology (e.g., a large-scale study may rank higher than a case study). Or you may choose to create a matrix: a more detailed table listing the studies in which you "check off" features or answers to specific questions that you're asking, leaving cells blanks for those studies that don't address those features or questions (e.g., evidence for a link between high cholesterol and polycystic ovary syndrome may be strong in some studies and weak or nonexistent in others—check the appropriate boxes). Once you complete the table, you'll have a comprehensive picture of what your sources are about, and you will start noticing similarities and differences, as well as common trends and concerns that the authors address consistently across the board. These issues should guide your outline and your synthesis.

Step Four: Create an Outline Based on the Themes/Topics Emerging From Your Sources

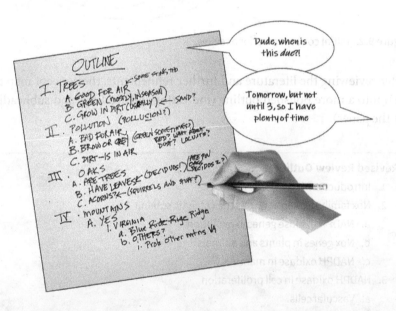

If you were already an expert on your topic, you could have probably proceeded with the outline long ago—before you had to read and summarize your sources. If you are not an expert on your topic, you need to do your best to focus your topic and select the most relevant sources in order to generate a logical, useful outline. You may also organize your ideas visually in concept maps or idea trees—a fancy way of saying that you can jot down your main points as loose clusters around a page and draw connections between them (lines, arrows, circles, etc.). The resulting diagram may help you identify the main subtopics associated with your area of research. Having a good, audience-specific goal will help you focus only on those topics and subtopics that are relevant to the goal of your review. A sample initial concept map is shown in Figure 9.2.

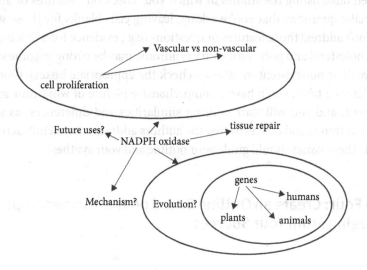

Figure 9.2. Initial concept map for a review.

After reviewing the literature and further refinements, the concept map may turn into a more polished outline (indicating the headings and subheadings for the paper).

Revised Review Outline

1. Introduction
2. Nox family protein acting as the catalytic subunit
   a. NADPH oxidase genes in evolution
   b. Nox genes in plants and animals
   c. NADPH oxidase in mammals
3. NADPH oxidase in cell proliferation
   a. Vascular cells
   b. Non-vascular cells
4. Mechanisms of NADPH oxidase regulation of cell proliferation.
5. Role of NADPH oxidase in tissue repair processes
   a. Wound healing
   b. Angiogenic responses
6. Perspectives for tissue engineering

When you compose your outline, rely on your review of the literature and jot down the common trends or themes you observe and you consider important to include in your paper. When you revise your outline, think of the readers and how much will they need and want to know, so that you accomplish your stated goal for the review.

Step Five: Write Your Paper Following the Outline

Some of the following suggestions are inspired by Laurence Greene's (2010) wonderfully detailed *Writing in the Life Sciences*. Essentially, Greene suggests taking a rhetorical approach to writing the scientific paper, identifying several major rhetorical goals for each section of the paper and several rhetorical strategies for each goal. Here, we simplify and streamline this approach into a set of rhetorical strategies.

Writing the Introduction of the Review

Let's look at this example of an introduction for a review article on ovarian cancer and nanotechnology:

Sample Review Introduction	
Rhetorical strategies	**Introduction**
Sentence 1: Directly introduces the problem and establishes relevance. Note the clearly marked abbreviation and citations.	(1) Ovarian carcinoma is the leading cause of death from gynecologic malignancy, which is due to its late initial diagnosis in addition to recurrence of ovarian cancer (OVCA) associated with resistance to therapy [1–4]. (2) Nanotechnology has great promise in addressing existing problems to improve diagnosis and therapy of OVCA. (3) Although the nanomedical field is relatively young, we recently witnessed an expansion of research in this field especially with applications in cancer. (4) Nevertheless, the literature and current reviews on nanotechnology are mainly oriented around nanomaterials or nanosystems with emphasis on their characteristics and their general applications in cancers or other diseases. (5) The focus of this review is on OVCA with challenges of its diagnosis and therapy and how recent nanotechnological developments are contributing to overcoming these challenges. (6) This review first presents OVCA and the challenges in its detection and therapy, introduces the readers to nanotechnology, then presents the potential of nanotechnology in enhancing diagnosis and therapy of OVCA and ends by presenting conclusions with future directions.
Sentence 2: Introduces a potential solution to the problem identified in sentence 1.	
Sentence 3: Provides background on the topic, directly referring to recent developments (reinforcing relevance and timeliness).	
Sentence 4: Identifies a gap in the literature.	
Sentence 5: Points how the current paper will fill that gap and explains the goals of the paper.	
Sentence 6: Previews the organization of the paper.	

As you can see, the introduction contains typical rhetorical moves meant to give readers a clear idea why they should read your paper—and what they should expect from it. The introduction is where you "create a research space" (John Swales) and invite your readers to come on in and follow along. These are the most common steps in the introduction (and no, by no means do you have to use all of them—but we guarantee that you must use at least some of them):

1. *Identify and define your research issue.* As in most cases in scientific writing, a direct approach works wonders (just as in the previous example).

2. *Explain the relevance, importance, and timeliness of the issue.* This is where you remind readers that your issue has a significant impact and is relevant to your readers as well as our society in the present moment.

3. *Provide essential background information* that readers will need to understand your research issue. The amount of background information you provide varies, of course. Two pieces of advice: (a) look for review articles in your field and determine how much essential background, in general, goes directly in the introduction and (b) when in doubt, keep it short. (Actually, follow this second piece of advice at all times.)

4. *Reveal the knowledge gaps in your research topic.* Here you remind readers that you are adding to the overall knowledge of the field by filling a vacant niche. You may point out that guidelines or recommendations on your issue are "fuzzy," missing, or antiquated; that little is known about the effectiveness of a procedure; that no one has compared x and y before; that recent discoveries warrant a reexamination of the issue; etc.

5. *Summarize contrasting concepts and theories* that underlie your research issue. In some cases, you might want to do a review because there are two or more conflicting schools of thought on an issue. If so, briefly review them here.

6. *Present the purposes of your review.* This introduces the specific focus and goal(s) of your paper.

7. *Preview the organization of your review.* This will allow readers to understand, at a glance, what is ahead. It's a good idea, just like it's a good idea to have a roadmap when going somewhere you've never been before.

Writing the Body of the Review

The coherence and cohesion of your paper will depend on the strength of your outline and notes. Since writing is a process (and a messy one at that, as you might have figured out by now), you may feel the need to make adjustments to the outline along the way. That is okay, as long as you keep true to your core research question and your goals. Here are some of the most common steps for crafting the body of the review paper—we note these are not "chronologic steps" but rather features to be included in some shape or form in the review:

1. *Provide essential background knowledge* about the studies, critical evaluations, and arguments that are central to your review. Unlike the introduction, where you provide a general, cursory background about the topic, here you can go into the history or some important details or background knowledge related to your topic and explain in more depth the key concepts or terms the readers need to understand.

2. *Summarize* the studies relevant to each section of the review, based on the outline you have created. When you summarize, make sure that you describe the research issues, objectives, methods, results, and conclusions of your studies, as well as evaluate their relevant strengths or limitations. Do not be afraid to include numbers and figures: they are both essential to your audience.

3. *Synthesize.* Synthesis should occur throughout each section. You engage in synthesis every time you:
   - provide a brief overview of the subtopic at the beginning of each section
   - characterize common traits of all the studies included in the section (such as methodology, goals, underlying mechanisms)
   - generate claims based on more than one study
   - acknowledge and respond to viable counterarguments or limitations to your central claim

4. *Evaluate.* Not everyone agrees that this is a necessary component of all reviews, and, indeed, in many cases you may want to eschew such judgments, or you may find them unnecessary. However, if you are reviewing a claim of some sort (e.g., that high field temperatures negatively affect laser performance), you may want to assess how strong the evidence is supporting this claim.

5. *Include visual aids as necessary.* We strongly recommend that you polish your summary table into a usable visual aid for your paper. This practice is already standard in some journals, and, if the table is well done, it can be a tremendous boon to the reader. We also strongly encourage you to use visual aids whenever possible and appropriate—and wherever they can be beneficial to your reader. Think of relevant illustrations, tables, or graphs that make your data come alive.

Dealing with References and Citations

Our scientific writing students are always a bit shocked when we tell them that they should not use direct quotes in their reviews. It seems a little counterintuitive—after all, you are using the data in those studies, and you've learned in basic composition that you have to cite, quote, and reference diligently. But, as we have explained in Chapter 7, in the "hard sciences," you do not quote verbatim from other scientific papers. Direct quotations are almost never used in the disciplines of physics, biology, engineering, and medicine, whereas they are often used in the social sciences and the humanities. Instead, you will have to rely on summaries and paraphrases—which you will still have to cite diligently in the citation style preferred by your discipline. Furthermore, especially in the biomedical fields, nonintegral citations are preferred over integral ones—that means that often information about the author and year of the study is eliminated in favor of a numerical citation:

> *Integral citation:* Liu et al. (2009) found that mice dendritic cells presenting with *Schistosoma japonicum* antigen to T cells showed a marked decrease in inflammatory response when compared to a control group.
> *Nonintegral citation:* Mice dendritic cells presenting with *Schistosoma japonicum* antigen to T cells showed a decreased inflammatory response.[11]

Relegating information about the authors to the reference list serves two functions: (a) it keeps the prose "clean" and succinct, and (b) it highlights the data in the subject position, and not the researcher.

Of course, every once in a while you will see a scientific paper that deviates from this norm—but only if the quote is memorable enough or in some way vital to the paper.

You are probably going to use a lot of reporting verbs in your paper. Table 9.3 lists the most frequent reporting verbs used in the disciplines.

As you can see, there are differences among disciplines; researchers also found out that the "softer sciences" tend to use more reporting verbs than the "harder sciences."

Writing the Conclusion of the Review

Three central rhetorical strategies are specific to the conclusion:

1. *Briefly recapitulate your key points.* Remind readers of the major claims and information you presented in the paper. Do *not* repeat random sentences from the introduction or body of your paper—this is poor

Table 9.3. Ranking of top six reporting verbs by discipline (1 = most frequent; 6 = sixth most frequent)

Discipline	Verbs and frequency					
Rank	1	2	3	4	5	6
Biology	describe	find	report	show	suggest	observe
Physics	develop	report	study	find	expand	
Electrical Engineering	propose	use	describe	show	publish	develop
Mechanical engineering	describe	show	report	discuss	give	develop
Epidemiology	find	describe	suggest	report	examine	show
Nursing	find	suggest	report	identify	indicate	show
Medicine	show	report	demonstrate	observe	find	suggest

Source: Christine B. Feak and John M. Swales, *Telling a Research Story*. University of Michigan Press, 2009.

form. The conclusion should elevate the main points of the paper and reinforce them in a few well-constructed sentences.

2. *Discuss implications for current practice and recommendations future practice.* Your paper may actually make a difference in the way people in your field operate. This is where you explain how.

3. *Suggest future directions for research.* Point to the gaps in the research so far and create a research agenda for the scientific community you are addressing. This is, arguably, one of the most important contributions of the review essay to science. By highlighting what has been done thus far, the reviewer is well-placed to make suggestions about what still needs to be done.

These strategies are not so different from those used in the conclusion of original research articles. The conclusion of the review, however, is usually much more elaborate—which makes sense, since the review covers a lot more data than a single study.

Writing the Abstract of the Review

As usual, you will write this last. We have noticed a recent tendency to use structured abstracts for the narrative review. This is, however, not necessary—or at least, not the norm for now. A brief descriptive abstract stating the topic and research question, mentioning the main issues discussed in the paper, and previewing your conclusion(s) will suffice. The following is an abstract for the NADPH paper whose outline we discussed earlier. Note the rhetorical moves made in the abstract to entice readers to read the whole study!

Sample Review Abstract

Sentence 1: introduces the topic and establishes relevance and timeliness.	[1] The superoxide generating enzyme NADPH oxidase has received much attention as a major cause of oxidative stress underlying vascular disease. [2] However, there is increasing evidence that oxidant signaling involving NADPH oxidase has other important roles in cell biology. [3] Nox family proteins are the catalytic, electron-transporting subunits of the NADPH oxidase enzyme complex. [4] It is now clear that reactive oxygen species (ROS) generated by NADPH oxidase participate in intracellular signaling processes that regulate cell differentiation and proliferation. [5] These mechanisms are important in tissue repair and tumorigenesis, diverse conditions where cell proliferation is required, but when poorly controlled the generation of ROS is obviously detrimental. [6] Indeed, NADPH oxidase-mediated cell proliferation has been observed in a wide range of cell types including those found in blood vessels, kidney, liver, skeletal muscle precursors, neonatal cardiac myocytes, lung epithelial cells, gastric mucosa, brain microglia, and a variety of cancer cells. [7] NADPH oxidases act not as isolated elements downstream of a particular pathway, but rather may amplify multiple receptor tyrosine kinase-mediated processes by inhibiting protein tyrosine phosphatases. [8] Therefore, NADPH oxidase-mediated redox signaling may represent a unique intracellular amplifier of diverse signaling pathways involved in tissue repair processes such as cell proliferation, wound healing, angiogenesis and fibrosis. [9] Recent studies also suggest that NADPH oxidase is involved in differentiation of stem cells. [10] As occurs in unresolved inflammation, however, hyperactivity of this enzyme system leads to tissue injury. [11] Thus modulating NADPH oxidase may have significant impacts on regenerative medicine and tissue engineering, such as growing heart muscle.
Sentence 2: establishes the central claim of the paper.	
Sentences 3 and 4 explain the mechanisms underlying the claim.	
Sentences 5–10 further summarize and explain the relevance and implications of recent findings.	
Sentence 11 explains the larger societal significance of the findings.	

Keywords: Cell proliferation, NADPH oxidase, Protein tyrosine phosphatase, Stem cells, Tissue engineering, Tissue repair

For more detailed instructions on abstracts, see Chapter 10.

Step Six: Revise and Polish

Have you followed all the steps we have outlined so far? Good. Now it's time to rewrite everything.

We're only half-joking. Follow the revision practices we have outlined in Chapter 2; work with a peer reviewer. Leave the manuscript alone for one or two days; come back to it refreshed. We also suggest that you take the time to learn what *not* to do in a review paper this time around.

What to Avoid When Writing a Review Paper

- The "String of Beads"

That's what we call the mindless listing of summary after summary, sandwiched in-between a lackluster introduction and an non-illuminating conclusion. Avoid this at all costs. Always aim to synthesize, guided by the overarching goal for the paper.

- The Little Orphan Annie (missing citation)

You read this really important/interesting/amazing fact about your topic that you want to stick in, but you forgot whence it came. Or you are a zipping through your sections and at some point it becomes unclear who did what experiment. You check your notes (you're nothing but conscientious!) but in vain. Remember that *not* citing a reference is a form of plagiarism. We sincerely hope that you have kept a working bibliography in place as you were working on this paper—as we recommend throughout this book.

- The Headless Horseman

You have so much to say about your studies! Who cares about placing them in context? *Of course* everyone already knows why pegylated nanoparticles are important! Needless to say, a good introduction can make the difference between a confused and a satisfied reader. Explain why your readers should care in the first place. Your due diligence here will pay off in the end.

- The Colossus with Feet of Clay

You've said all you had to say and you're running out of steam. You are at a loss when it comes to a conclusion, so you slyly pass off some of your introductory sentences as concluding remarks. We have seen this in our students' papers, and it's usually an indication that the student is confused about the overall purpose of the paper. It's a good time to revisit your original goal and reinforce it here.

- The Locked Door

A good title is essential, or otherwise you are rendering the rest of your paper rather cryptic. Titles like "Concussions" or "Nanotechnology" have puzzled us again and again in students' drafts. What about concussions? long-term management? prevention? developmental issues? in what population? for what purpose? Avoid the "mystery review" syndrome and choose a precise, descriptive title for your paper.

Once you are done revising—polish. Make sure your list of references is complete and correctly formatted in the citation style preferred in your discipline; make sure that all the works cited in the references are actually cited in the paper. Mark your headings and subheadings clearly. Check for typos and inconsistencies in language or formatting. Get an external proofreader if you can.

Take-home

Review essays are tools for making sense of vast amounts research. They keep scientists and clinicians updated on the latest developments in their fields and provide justification and direction for future research. They can also be crucial components of academic capstone projects or grant proposals.

The main types of review papers are narrative and systematic; systematic reviews and meta-analyses are more structured and akin to original research papers. The organization of narrative reviews should emerge from the careful reading and annotation of the studies reviewed, whereas systematic reviews and meta-analyses follow more of an IMRAD format.

In writing a review essay, you demonstrate mastery of a narrow, specialized, and timely topic in your particular field—with the ultimate goal of summarizing and furthering knowledge of that topic.

Exercises

1. Choose a broader subject (e.g., nanotechnology or diabetes) and formulate as many focused review topics as you can in the space of 5 minutes. Write down your answers and discuss them in groups, with the purpose of choosing the best question for your research project.

2. In groups, identify the premier research journal in your field and search it for a recent review article (note that not all journals publish review articles). Discuss the rhetorical strategies of the article and provide examples. Identify one strategy that you would use in a paper and explain why.

3. Create a summary table template by adjusting the column headings accordingly to suit your field and your current project or topic of interest. Explain to the class (or your group) why you made those choices.

4. In groups, identify the top two to five research journals in your field; look for their "For Authors" section, which usually gives details about the type of manuscripts they receive. What guidelines do they offer for submitting *review* papers, if any?

5. In pairs: search for review paper abstracts in a relevant journal for your field. Provide the abstract to your partner but not the title. Ask your partner to generate a credible review title that would suit the abstract. Compare with the original title and discuss the strengths and weaknesses of your own titles.

10

Abstracts and Summaries

© 2020 Erica Perez

There is nothing like the satisfaction of penning a 40-page manifesto on your analysis of the wastewater treatment facilities in the Chesapeake Bay watershed. You got into the environmental engineering program and now you are now so totally into sewage your friends look at you funny. Forty pages! No problem! You have upstream and downstream water quality analyses, sediment analyses, effluent analyses, analysis of the comparative efficacy of biodegradation and sludge dewatering processes, and you write the most awesome paper ever for Eng291. Now your prof casually suggests you should get it published, perhaps in *Water Science and Technology*. No problem! You crank it out, a work of art, with tables and graphics, hewing to word count. Suddenly, the lab fills with ominous music. There it is: journal instructions for authors.

Abstract: no more than 200 words briefly specifying the aims of the work, the main results obtained, and the conclusions drawn.

Keywords: 3–6 keywords (in alphabetical order) which will enable subsequent abstracting or information retrieval systems to locate the paper.

Abstracts

Okay, no problem. Sure, you can think of 19 different keywords, but once you start winnowing out "pressure swing adsorption oxygen generator" and "polyacrylamide flocculating agent" you can deal with it. But when you attempt the "200 word" thing, you write a draft, and the first draft is two and half pages long. You try to condense it into 200 words and come up with a near-aphasic rendition of "coliform bad. Phosphate up, bad. Ozone if necessary. Water— wet." And by the time you're done it reads like *A Child's Incomprehensible Primer of Wastewater Management*, and *Water Sci Tech* will look upon it with disfavor.

How do you do it? How do you stuff a year's worth of data, a 40-page bloated-with-information paper into 200 words? Well, for starters, you don't. Not if you have a 200 word limit, which is about two-thirds of a page, double-spaced, in Times New Roman, 12-point font, with one-inch margins all around. Ensure that you are following the journal (or professor's) typeface, font and line formatting requirements. In most word processing programs you can select a chunk of text and click on "word count" and it will give you an exact count (in MS Word, your totally running word count is in the lower left corner of your screen). If the journal specifies 200 words, do not write 204. If the journal specifies three keywords, do not provide six.

Why Do I Need to Write an Abstract, Anyway?

When you thumb through a catalog, do you see a picture of some hiking boots and say "those look cool, I'm going to buy them," or do you read the blurb about it and then find out they're solid leather, full-shank, three-pound backpacking boots that require resoling every 200 miles and waterproofing every 10 miles and are suitable only for walking on dirt and not rocks—and cost $375?

A menu will not just say "fish;" it'll say that it's flounder rolled in egg whites and poached in scotch and covered with capers, raisins, pomegranate seeds, candied pineapple, orange Jello, wilted kale, whipped cream (fresh!), and a variety of other things that will make you not order it.

Your abstract needs to include enough information to *sell* the paper's idea to the reader and make them read the rest. When you're searching the Environmental Science and Pollution Management database for wastewater treatment literature, do you see an article *title* and say "that looks perfect, I'm going to read that article," or do you read the abstract first to find out if it's truly germane to your research and a useful resource for the IMRAD paper you're trying to write?

The title of your paper has to be interesting and comprehensive enough to get a "foot in the door," that is, people interested enough in that title to click on it and read the abstract. *Then* you want your abstract to be interesting and thorough enough to get people to read the paper. An abstract that says "water quality analyses were conducted near wastewater treatment plants around the Chesapeake Bay" isn't going to arouse the same kind of interest as "upstream and downstream effluent water samples were assessed for phosphate, nitrate, pH, chloride, and coliform bacteria for 14 wastewater treatment plants discharging into four different estuaries in the northern end of the Chesapeake Bay." That kind of abstract is going to pique the interest of the water quality analysis specialists, and of those specializing specifically in phosphates, nitrates, pH, chloride, or coliform bacteria—or the ecologic issues associated with them; it will tell the prospective audience that both upstream and downstream water was assessed, including effluent, and it will provide a more precise location of the treatment plants, attracting the attention of the environmental engineers interested in that particular area.

In a few sentences, you can provide enough of a gist to interest a potential audience—which is what you want to do. Give your audience enough information so that they can read your abstract and decide if the paper is pertinent to their research or other needs. Word limit notwithstanding, you need to provide enough of a descriptive overview—not a vague "we kind of looked at this stuff."

It is easy to write a short abstract. In the 2011 paper by M. V. Berry et al. titled "Can apparent superluminal neutrino speeds be explained as quantum weak measurement?" in the *Journal of Physics A: Mathematical and Theoretical*, the abstract read "Probably not." The paper, of course, went on to wax eloquent (with no small amount of seriously alarming math) about why "not."

You can do it. You really can. You'd be astonished at what you can put into a single sentence and still have it be coherent. Prior to writing your abstract, though, you need to have a solid understanding of the format your abstract has to take. This goes for journal articles, school assignments, and other related documents that may contain abstracts or abstract-like elements, such as meeting posters and grant proposals (and catalog entries and menus). You may decide to bag engineering entirely and go to culinary school—in which case you'd have to become a master of the *one-sentence* abstract.

The "solid understanding" comes from the instructions you're following—very carefully—when you write your abstract. Apart from word or character limits, you may be directed to write *descriptive* abstracts, *informative* abstracts, *indicative* abstracts, or *structured* abstracts (or versions thereof). *Promissory abstracts* are different—they are not written for papers but rather for presentations at scientific meetings. Most publication abstracts fall into one of these categories, but some publications (or professors) may have different requirements—maybe even an abstract of a few sentences. Abstracts for other types of documents (such as meeting posters, meeting abstract books, and grant proposals) might take on even different formats.

Descriptive Abstracts

While any abstract could be, arguably, described as "descriptive," the *descriptive abstract* is typically a paragraph-long narrative. These tend to be shorter than informative or structured abstracts (200 words or less). The narrative usually does not contain specific information or details, just a general overview of the article content ("water quality analyses were performed"). Note that the "general overview" *still contains enough information* such that a prospective reader can decide whether the full article is worth reading.

Informative Abstracts

Again, it could be argued that *any* abstract is "informative." However, the *informative abstract* actually contains more details than the descriptive abstract ("upstream and downstream water samples from 14 wastewater treatment facilities in four estuaries in northern Chesapeake Bay were assessed for phosphate, nitrate, chloride, hydrogen sulfide, and coliform bacteria"). Some of these abstracts, depending on the article content, might actually contain data.

Indicative Abstracts

More common in review papers and book chapters (if the chapter contains an abstract at all), the indicative abstract contains less information (primarily since review papers do not report original research) and serves more as an outline or table of contents, alerting the reader to what's in the paper rather than providing conclusions about something. Rationales, methods, and results are absent—identifying what will be discussed in the paper is used instead. A general background may be supplied, as well as perhaps why the content is important to understand, what is known, what is unknown, and what needs further study.

Structured Abstracts

These are abstracts that have discrete sections—usually determined by the journal. These sections often have their own headings and spacing and typically are seen in IMRAD papers and systematic reviews, following the IMRAD structure (i.e., introduction, methods and materials, results, and discussion). Many journal abstracts are super-structured; *JAMA: Journal of the American Medical Association* (and origin of AMA style—although abstracts have nothing to do with style), since it deals with topics that are strictly medicine-related, requires its abstracts for articles reporting clinical trial results to contain several sections: Context, Objective, Design, Setting, Patients or Other Participants, Intervention(s), Main Outcome Measure(s), Results, and Conclusions.

Abstracts used in grant proposals also tend to be highly structured (and are typically strictly determined by the grantor you are petitioning). Grant proposals, of course, also contain information about why the money is needed, what it is needed for, and what is expected to be achieved in the study. In a nutshell, the grant proposal also contains a persuasive rationale accompanied by eloquent begging (active, rather than passive, voice should be used).

How Do I Know What Kind of Abstract to Write?

While *JAMA* may provide detailed specifications on the kind of abstract that must be submitted with papers, many journals give you terse directions such as "200 words." This is not particularly helpful when you're trying to craft your abstract, and sadly, it may also happen in the courses you are taking. What do you do, then, if the directions are insufficient or vague? It is always a good idea to go directly to your target journal and start reading a bunch of abstracts.

Here's a hypothetical one from *Water Science and Technology*, the very journal to which you want to submit your paper.

Abstract

Bacterial content of dewatered activated sludge was studied in four wastewater treatment plants of similar size and BOD_5. All plants had high-sulfate water intake without open lagoons. Digestion tank capacities were based on a solids concentration of 2% with supernatant separation performed in a separate tank. Aerobic sludge digestion was assessed from the standpoint of digestion, supernatant separation, sludge concentration, and sludge storage. Sludge samples were centrifugally dewatered, and bacterial content of the supernatant was measured. Bacterial contents of the supernatant were analyzed by 16S rDNA 454 pyrosequencing. Two pathogenic bacteria predominated, *Mycobacterium* and *Vibrio*. Microbial extracellular polymer (ECP) played a significant role in sludge dewaterability, with concentrations of ECP higher in anaerobically digested sludge as compared to raw sludge. Digested sludge contained 25% less ECP than raw sludge, and maximum dewaterability of sludge was calculated as 23 mg ECP g^{-1} SS for raw sludge and 11 mg ECP g^{-1} SS for digested sludge. ECP from digested sludge had higher protein with respect to carbohydrate than raw sludge.

Keywords: 454 pyrosequencing; aerobic sludge digestion, bacterial populations; dewaterability; microbial extracellular polymer (ECP); water content

What kind of abstract is this? It has more information in it than your standard descriptive abstract, yet it is clearly also not structured. This is an *informative abstract*. It falls tidily within the 200-word limit (167 words), and by reading it, you can tell exactly what the paper is going to be about. Note the keywords: most journals request up to six of them. Keywords need to be specific enough to be easily retrievable in a database search (many are phrases for that reason). Do not use very general, and therefore nearly useless, keywords (e.g., "activity" or "brain" are too general; "brain activity" is better, though it could still be more specific).

If you read several abstracts in *Sci Water Tech*, you'll get a sense of what you're supposed to be writing: how detailed (or not), how descriptive (or not), how conclusive (or not), or how technical (or not). Don't just read one abstract, read several. See what kind of variation there is. When you do this, you will become more comfortable and confident (honest, you will) about writing your own and less overwhelmed and daunted by the prospect of stuffing 40 pages of information into a little bitty paragraph. Comfort and confidence, however, is only going to come if you read *several* abstracts. Do not forget that if this paper gets rejected by *Sci Water Tech* and you want to submit it somewhere else, you may have to restructure the abstract entirely—you will have to start your reading-other-abstracts-to-get-comfortable process all over again.

How Do I Actually *Write* the Abstract?

Indeed, how *do* you write it, put enough information in there to make it informative but not sound like the aforementioned aphasic primer? One way is conceptualizing your abstract in terms of *rhetorical moves* or *goals*. Table 10.1 lists a few that you might try (based on John Swales's Create A Research Space [CARS] model). Not all of these components are present at all times in all abstracts, and the level of detail you offer depends very much on the discipline in or journal for which you are writing.

Table 10.1. Possible Rhetorical Moves for Abstracts

Move #	Typical Labels	Implied Questions	"Cue" phrases
Move 1	Background/ introduction/ situation	What do we know about the topic? Why is the topic important?	"...typically..." "...generally..." [use of statistics] "...not well understood" "understudied/poorly studied"
Move 2	Present research/ purpose	What is this study about?	"To determine whether...," "We asked whether..." "To answer this question..." etc.
Move 3	Methods/ materials/ subjects/ procedures	How did we conduct this study? (level of detail dependent on, discipline, journal, type of paper)	"We conducted [an experiment/ double-blind study etc.]" "We interviewed/ measured/ collected, etc...." "X was studied by..."
Move 4	Results/ findings	What was discovered?	"We found that..." "Our results show..." "Here we report..."
Move 5	Discussion/ conclusion/ implications/ recommendations	What do the findings mean?	"We conclude that..." "Thus..." "These results indicate that..." "We recommend..." "important," "novel," "useful/of use," "promising"

Good scientific writing is characterized by both economy of word and economy of style. The more *precise* and *concise* your writing, the better. Use only as many words—*relevant* words—as necessary. We elaborate more on these skills in the Writer's Toolbox, where we present our two cardinal rules for scientific writing: (a) write what you mean (the M rule) and (b) make every word count (the C rule). The abstract will test your ability to get to the point, be clear, precise, and concise—in other words, it is the perfect application of the M and C rules. Let's revisit the hypothetical *Sci Water Tech* abstract:

Digested sludge contained 25% less ECP than raw sludge, and maximum dewaterability of sludge was calculated as 23 mg ECP g^{-1} SS for raw sludge and 11 mg ECP g^{-1} SS for digested sludge.

In a single sentence, the authors have communicated (a) *what* was analyzed (digested sludge), (b) *how* dewaterability was affected by ECP, and (c) *what* were the results (ECP g^{-1} SS for raw sludge and 11 mg ECP g^{-1} SS for digested sludge). This sentence has 27 words in it (exclusive of abbreviations). Note that you must explain abbreviations the first time they occur, even in abstracts (usually). This is a very precise, concise sentence. Less concise, it might read:

Raw sludge had a higher concentration of ECP than digested sludge; this had different effects on maximum dewaterability, depending upon ECP concentration; digested sludge had 25% less ECP than raw sludge.

In this 31-word version, information from the preceding sentence is unnecessarily repeated (without specific results), compromising economy—without necessarily being clearer.

Another way to reduce word count—if appropriate—is to use the *active voice*, rather than the *passive voice*. The following sentence uses the passive voice:

Sludge samples were centrifugally dewatered, and bacterial content of the supernatant was measured.

This sentence could be tightened by using the active voice instead: "We centrifugally dewatered sludge samples and measured bacterial content of the supernatant." Additionally, apart from the less economic, passive "by the authors" (i.e., as compared with "we"), there is a certain amount of redundancy that would be unnecessary in this particular target audience. If one is evaluating dewaterability, one can assume that water content is being measured. So, tight as this abstract is, it can be even tighter, without loss of information necessary for it to be a useful abstract.

Using the first person in your writing humanizes your work. If possible, therefore, you should avoid using the first person in your writing. Science succeeds in spite of human beings, not because of us, so you want to make it look like your results magically discovered themselves.

—Adam Ruben, "How to Write Like a Scientist,"
Mar. 23, 2012, *Science Magazine*

Scientists often prefer passive voice and may be resistant to calls for inserting themselves in the prose ("We measured . . ."). Check the journal instructions, read papers in your target journal and your discipline to see acceptable style variations, and use your judgment. In general, use the active voice wherever you can, but recognize that in some cases passive voice works better (we discuss the active vs passive voice conundrum in the Writer's Toolbox at the end of this book as well).

Abstracts—no matter what kind—should *not* contain citations or references, although you might see this in some published papers. This helps you cut down on word count but, more importantly, makes the abstract stand on its own and be easily intelligible when it is retrieved from a database.

When Do I Write the Abstract?

There are a few ways you can approach this. While there's no dogma, you might want to write the abstract last; you'll have a better sense of what goes in it and in what order. You can also write an outline-like placeholder for the abstract that you can fill out as you're developing your paper or later when you've finished your paper. Keep in mind that trying to write a complete abstract at the level of detail required by the journal (or your professor) is a challenge if the paper isn't yet written, and even if you try it, it is very likely that it'll sustain some significant editing when you're done, so that it "goes" with the paper.

The abstract and its keywords are what are going to be *abstracted* by *abstracting services*; these are the words (or phrases) that are going to turn up in the searchable databases. In some cases, the abstract might be the only thing that your intended audience will be able to read. It is not uncommon for disparities to occur between the abstract and the body of the paper (usually during the drafting stage), so always revisit the abstract after you've made the final revision to make sure there is content consistency. When you're doing your research and reading papers, it is important not to use information from the abstract instead of the body of the paper. Apart from revisions that will make the abstract not match the body of the paper, taking information out of context (which is what you'd be doing if you're using only the abstract as your source of information) can cause some big problems. In general, *make what you write compelling enough for your audience to want to read the whole paper.*

Are Meeting Abstracts Like Abstracts for Papers?

It depends. If you're including an abstract that is essentially identical to a paper you have in press, containing information on research that has been conducted and concluded—that is, you have results and a conclusion—then yes, it is similar to an abstract for a journal article. However, many meeting abstracts are *promissory* in nature, in that the content they contain is, well, "coming, we promise." This is not unusual at a scientific meeting, since one goes to scientific meetings to learn about the latest science, much of which is in progress (hence, "latest"). A meeting abstract might be theoretical, or more of a proposal, or may be (usually) reporting on "results obtained thus far." You don't really go to a meeting to find out about what is already published. These abstracts are written well in advance of the meeting—deadlines can be six months prior or more, although some organizations offer LBA (late-breaking abstract) options with shorter deadlines. This is another reason they may not have complete data.

Meeting abstracts are to be written—as with everything else—according to the directions provided by the society conducting the meeting and publishing the abstract book (or, these days, an e-version thereof): often *very* strict word counts (or worse, character counts) and frequently very structured. For some meeting abstracts, you may be permitted to include figures or tables (and the instructions for abstract submission will tell you if this is allowed, and if so, how many, and how many words or characters they're "worth" in space). If you were fortunate enough to have an abstract accepted at SETAC (Society of Environmental Toxicology and Chemistry), it might contain a very brief introduction, include significantly abbreviated methods, and be result-heavy, with a few bullet points for a conclusion.

10 Tips for Writing Abstracts

1. Identify the abstract style (descriptive, informative, etc.) for the paper.
2. Use a rough abstract draft, then complete the abstract when you're done.
3. Hew to word or character limit!
4. Reduce word count by combining information into single, lucid sentences; consider using the active voice if appropriate or mandated by your professor or target journal.
5. Choose the most important keywords, and do not exceed limit.
6. Ensure that the abstract reflects paper content.
7. Remember that meeting abstracts rarely look like paper abstracts . . .
8. . . . and neither do abstracts for grant proposals or meeting posters.
9. Make the abstract interesting and informative enough so that your reader will want to read the full article.
10. Never use just the abstract as a source of information for your own research; always read the full paper. Meeting or congress abstracts are legit sources of information.

Summaries

"Summary" simply means an abbreviated version of something. In papers, summaries typically occur at the end (although some journals refer to abstracts as summaries). You can summarize a paper for a lit review, a book for a book review, your résumé into a biographical sketch, your research for a grant proposal or a lecture into lecture notes—these are stand-alone summaries. You can summarize a much larger document into "key points" for an executive summary or summarize multiple presentations at a large scientific meeting into a meeting summary. These are all vastly different things; the only thing they have in common is that they are shorter, well, summarized versions of a whole.

Summaries for Scientific Papers

Basically, the abstract and the summary sandwich the paper into a digestible whole. If a whole paper (IMRAD, review, or any other kind) is a meal, then the abstract is the appetizer, the paper with all its main components is the main dish, and the summary would be dessert.

While an abstract provides information about *what's coming* in the paper (in more or less detail, depending upon the type of abstract), the summary provides information about what *was* in the paper. Very importantly, this last discussion takes the conclusions or findings from the paper and *puts them into context* in the larger scheme of the field and *emphasizes the significance* of the information.

Try it: What was the last thing you worked on in a science class (no matter what—paper, lab experiment, presentation, data collection, research, etc.)? In 60 seconds, state the main point of that activity. Can you do it? You should also be able to answer the question "What's the point?" for everything you purposefully write.

Summaries typically reiterate the background or rationale as well as the results or conclusions of the paper—very succinctly. Typically (although not always) no new information is added; this section should not contain references, since it does not have any new information in it, although references are certainly provided for other studies that place your research in context. Research, or studies, or reviews, however, are of limited utility to your readers—or to the field in general—if they are not placed into context. Science does not occur in a vacuum. Your paper is a piece of a big jigsaw puzzle, and you need to talk about where your piece goes. After you have established context, point out the significance of the research (or the focus of the content if it's a review paper). Remember, if you did an experiment and didn't get statistically significant results, that's still important information. It still needs to be placed in context, and the importance of lack of anticipated statistical significance should also be discussed.

A useful segue (although not necessarily always present in papers) is to make recommendations for future research. Now that you have these results or conclusions, what else needs to be done? In which direction should subsequent research go? Or not go? These are useful recommendations and may help scientists design studies or conduct research on specific topics in the future. Maybe you've already started on that recommended research yourself; maybe you already are getting promising results. Maybe you can toss in a teaser. *"Let's find out:* that's next on our agenda."

Keep in mind that just as there are different kinds of abstracts, there are different kinds of summaries for different kinds of papers. A review paper summary won't look like an IMRAD summary or like the two-sentence summary in a meeting abstract or poster (technically not papers, but these aren't standalone summaries either). The journals to which you submit your papers may dictate your paper sections, as might your professors for your assignments. The final section in a great many scientific papers is called "conclusion" instead of "summary," and they're actually the same kind of thing. Technically, they're both correct by *definition*, in that "conclusion" means "the end of something," as well as "the outcome of something." If, however, you want to get *semantically* picky about things, in science when you say "conclusion," your readers or interlocutors are going to think "outcome," not "ending." To be accurate, an IMRAD paper should have both a conclusion *and* a summary, but a review paper should only have a summary. Is it wrong if it's not done this way? No, and you might be at the mercy of the dictates of the journal or your professor anyway.

Executive Summaries

An executive summary is a special type of summary that has specific meaning and format that is quite different from that of the paper summary. It is almost always a stand-alone document (occasionally appended to a longer report, depending upon the destination and use of the longer report). Stand-alone notwithstanding, it is not an abstract—even though it comes first, if appended to a larger document—and may not refer to a single experiment or study—or maybe not even to science at all, except perhaps in the abstract (e.g., it might be about a pharmaceutical marketing plan). Executive summaries are created such that information in the longer detailed report (such as a marketing plan, or massive grant proposal, or clinical trial study report, or lengthy federal environmental document) can be absorbed rapidly, by individuals with limited time—usually (surprise!) executives.

The summary itself has a modular structure to mirror the sections of the document, with headings and even subheadings and various other document design features to facilitate speed of reading (e.g., bulleted lists, dividers, etc.). Good executive summaries are only a fraction of the length of a full report—usually no more than 10%, though this number is highly variable—but they are definitely longer than the abstract for a scientific paper. Executive summaries are the most and often (sadly) the only part read of longer, complex documents with multiple parts. They need to be concise, highlight the bottom line, and help the readers make decisions or reach rapid conclusions about the topic.

Executive summary Haiku:

Consensus report

Has three hundred six pages

You want five bullets

—Anonymous (fortunately)

Obviously, this kind of summary has to capture the *main points* of the entire report. Any detail from the complete report that can be assumed to be common knowledge to the audience or that is of limited importance is likely to be left out of an executive summary. Often, a section is devoted to "key findings" or main points, usually in list format. An executive summary also differs from an abstract or a paper summary in that it typically *makes recommendations* based on information on the report (e.g., "the upper Chesapeake watershed is a more critically endangered ecosystem than originally estimated, and more aggressive control of phosphate levels is necessary in the immediate future. Given the fragility of the ecosystem, the most effective approach to phosphate control would be . . . ").

Further, an executive target audience might be very different from a scientific target audience. If you write an executive summary targeted to the heads of all of the R&D departments in a large environmental engineering company, you use a different voice, level of sophistication, and vocabulary than if the summary were targeted to the marketing or PR departments. Bear in mind that you might have multiple target audiences; if this is the case, ensure that you write something everyone can read and understand. What does your audience *want* and *need* to know?

If you are tasked with producing a draft of an executive summary, create a document that has a single, short paragraph, describing the rationale of the project and summarizing what was done. Then, try an organized and logically grouped, (or bulleted) short report with subheads, summarizing one key concept or main point per bullet or subheading. Keep in mind that not all executive summaries look the same. Conceptually, they are (key points, then recommendations), but in actuality, they vary by whim of the higher-ups, intended target, use of the report, and content.

Meeting Summaries

A *meeting summary* is quite different from paper summaries and executive summaries. You have done no research, conducted no study. Suppose you were lucky enough to accompany your lab director to the SETAC World Congress this year. "Surely," she says, as you're each checking into your hotel rooms, "you didn't think you were going to spend four days flitting about the talks and poster sessions and snagging free pens off the exhibit floor?" Your guard doesn't go up yet, because your attention is split between the fact that you're going to have an hotel room and that there's a mysterious place called an "exhibit floor" where, apparently, one can snag free pens. "You're on phosphate patrol," "Phosphate patrol?" you say, *now* on your guard. "You're writing the meeting summary for the local chapter," she says, and she heads off to the elevators after telling you not to eat the $45 peanuts in your hotel room snack bar.

You can see what meeting summaries look like online. For your particular field (or close enough), choose an organization that is having a meeting *right now*. Google "conference coverage" with the name of the organization (e.g., "SETAC conference coverage"). Notice that the summaries being uploaded are only a matter of *hours* old—they're being written and submitted *during* the meeting.

After confessing abjectly that you have no idea what a "meeting summary" is, she tells you "go find out about phosphates. Talks, sessions, presentations, posters. Talk to experts, talk to researchers. Take notes. Write it up. Then you can go get some pens."

You go back to your room and study the meeting program. One of the keynote speakers is talking about phosphates, so you assume that's important news. You check for other presentations that are related (searchable conference apps are very useful for this). Sure enough, there's some new research out there. You identify five separate key "phosphate stories" you decide to track. You go to the sessions and take careful notes (don't video or record anything or take photographs without express permission—read the meeting program for their rules about this. You do not want to have your cell phone taken away or get booted out of the meeting). You made sure you attended the sessions relevant to your topics and then perused relevant poster sessions, getting the opportunity to speak to poster presenters, some of whom who had also been session speakers. You listened. You asked questions. You took painstaking notes. You got copies of posters if they were available (again—no photographing without permission; these days more and more posters also provide a QR code you can use to access the poster later), acquired business cards. On the exhibit floor, in the middle of your bewildering swag frenzy, you talked to exhibitors who made phosphate-sampling products or who had chemical analysis companies. Besides acquiring some really cool branded rulers and calculators and water sampling bottles and coffee mugs and USB drives and tote bags, you also acquired some journal article reprints and actually got to speak with people knowledgeable about phosphate analysis.

Back in your hotel room, you gaze longingly at those $45 peanuts and then start drafting your summaries *right away*, while the sessions are fresh in your mind—and you can still decipher the detailed scribbled notes you took as backup. Background, rationale for studies, experimental specifics, findings, main conclusion. You list source articles, if there are any—if the news is new enough there might not be any. You include the names of the presenters and include quotes, if you managed to get any. You write five meeting summaries related to phosphates, all based on "breaking news" from the conference. There were topics that weren't "new," strictly speaking, on phosphate analyses in specific regions, sessions on new methods for phosphate measurement, sessions on the interaction of phosphates with particulate materials, phosphates and eutrophied systems, phosphates and laundry detergents, and a host of other sessions on other well-known phosphate issues. But one of the keynotes was speaking on the heretofore unrecognized danger to the Chesapeake ecosystem as a consequence of steadily and rapidly increasing phosphate levels, and there were four other sessions on it as well. Another three sessions—and half a dozen posters—were devoted to the topic of phosphate detection in higher latitudes as a presumed consequence of climate change. Another on a series of bizarre crustacean mutations from multiple locations linked to phosphate levels in the water. Some other sessions related to the retraction of one of the biggest landmark phosphate studies ever published back in the 1980s—and the subsequent havoc the conclusions from that study wreaked on the field ever since.

In short, you identified the truly *new* information and focused on that (phosphates in detergent is not big news). Your summaries were narratives but *concise*—almost a combination of a journal abstract and an executive summary in style but with no personal input, interpretation, editorializing, or recommendations. This included the "voice" you used to write the summaries; no overtones of shock, disgust, excitement, or bewilderment—just the facts. The "voice" you use can be editorializing in of itself; even if your words are as pure and free of bias or influence as you can possibly make them, how you combine them must also be of steady journalistic quality.

As with executive summaries, not all meeting summaries are the same. Some contain graphics—some even contain post-session (permitted) video interviews with the speakers themselves. Meeting summary format will be dictated by the vehicle through which the information is disseminated (e.g., on an association website, a news blog, Twitter, YouTube, an online journal or association update, etc.), the scientific field in which the reportage occurs (some sciences are necessarily more visual than others), the target audience (lay audience? students? general scientists?), and, of course, the content itself.

---

10 Tips for Writing Summaries

1. Write the appropriate kind of summary for the document you are preparing.
2. Do not introduce new information in a summary; summaries should not contain reference numbers (except for context).
3. Briefly reiterate background and rationale.
4. Discuss the information in the paper in the larger context of the field.
5. Emphasize the significance of the results or information in the paper.
6. Suggest directions for future research.
7. When writing executive summaries, identify your target(s) carefully before you start.
8. Keep executive summaries short.
9. When writing meeting summaries, take thorough notes and acquire as much extra information as is practical (e.g., from posters, presenters, exhibit floor, etc.).
10. Write a meeting summary immediately after (during is even better) the meeting.

Take-home

Abstracts and summaries are both short and factual, with abstracts being more of a here's-what's-coming chunk of text and summaries being a here's-what-you-just-saw chunk of text that places things into context and emphasizes significance. Abstracts take different formats depending upon what they're abstracting: IMRAD versus review papers, book chapters, meeting posters, grant proposals. If you're trying to write one in a particular style, read a bunch of similar ones first. Executive summaries serve as the CliffNotes version of the 84-page report everyone needs to know about but that no one has time to read. And meeting summaries, well, they represent the combined skills of a detective who wants "just the facts" and a newspaper reporter who knows how to report them. If you are a long-winded writer, writing abstracts and summaries will cure you of that habit.

Exercises

1. Write an abstract.
   a. In a group, each individual selects one journal article (relevant to your field of study). Remove the abstract and hand the abstractless paper off to one of your peers. You'll be getting one too. Read the paper, then write an abstract for it. What kind of abstract goes with your paper?
   b. As a solo exercise, do the same with an abstractless paper provided you by your professor.
2. Compare an abstract with the body of its accompanying paper. Use a paper from your field. In your comparison, decide if:
   a. The abstract is descriptive, informative, indicative, or structured.
   b. The abstract was correctly written for its style.
   c. The information in the abstract was correct.
   d. There was enough information in the abstract.
   e. The abstract was well written (particularly from the standpoint of economy of word—can you tighten it up?).
   f. The abstract was compelling enough to make you want to read the paper.
3. Write a meeting abstract.
   a. Identify a major meeting relevant to your field (e.g., American Physical Society, American Chemical Society, American Society of Cell Biology, etc.) for which you can access a meeting abstract book (a great many are accessible online free of charge).
   b. Review abstract formats carefully, and read instructions for abstracts if you can find them on the association web page.
   c. Choose an IMRAD paper from your field from which the abstract has been removed by your peers or your professor.
   d. Abstract the IMRAD paper for the meeting abstract book.
4. Critique a summary. With a paper provided you, assess the summary for:
   a. Accuracy and completeness of summarizing the paper.
   b. Context (Is it there? Is it compelling?).
   c. Significance of research/study/experiment (Is significance discussed? Did they "get it right?" What other "significance" might there be?).
   d. Do you think the summary is missing something? What? Is anything confusing?

5. Write an executive summary. For your specific field, find a paper or the most recent *detailed* guidelines or recommendations for a practice, process, or condition (or something similar). Government websites are a good choice here (HIPAA is 145 pages long, for example). Find a massive document, and craft an executive summary. This is a lengthy exercise and might be better accomplished as a team.

11

Writing Proposals

You haven't seen your advisor in days. Or is it weeks, maybe? Rumors are that she's curled up over her laptop in her ~~lair~~ office, piles of papers on her desk, office plant dying a slow, miserable, death. Upon further inquiries, you discover that no one knows if she has even left the office in the last two weeks. Word is that copies of the document she's working on are periodically circulated among grad students and faculty, sent back, only to reemerge for further revisions. You might have glimpsed at a copy titled "Draft 87" and shuddered involuntarily: who writes 87 drafts of *anything*? What *is* that mysterious document? You've heard it referred to, in hushed voices, as "The Grant Proposal," a document that ensures the lab your advisor runs is funded, including paying the hordes of graduate assistants—as well as a few lucky undergrads, like yourself—to assist with the research the lab conducts. You are also vaguely aware that your assistantship in the lab next year hinges on that funding. You really love your lab job, plus you need the money! You need your advisor to be successful once again in acquiring support but are also thankful that you are not the one who actually has to make it happen—the whole endeavor seems a little intense. You tremble, thinking that if you stay in research you might one day have to face this peril yourself.

Who knows what career in science you will eventually end up pursuing—could be teaching, research, administration, or sales. Two things are certain, however: (a) no matter what your chosen post-collegiate career, it *will* involve writing; and (b) there is about a 90% chance that you will have to write a proposal asking *someone* for money, or your employment will be contingent upon someone asking for money. Science is expensive. Be prepared to make a case that your project is worth it. It's eloquent, structured begging for a good cause—simple as that.

Why Learn About Proposals?

In case you're not entirely convinced that you should learn about grant proposals, here are some more reasons. As science majors, some of you might be working on research projects funded by grants. In that capacity, it would behoove you to understand what a grant involves—that includes the work and considerations leading to the award, as well as the work and responsibilities that come with the award. Furthermore, you may be called on to assist in some capacity in the writing of a grant—an invaluable experience to have under your belt, by the way (don't shun it if one comes your way). The advice that follows may come in handy for you in other ways in the future—if you intend to pursue a career in the sciences beyond an undergraduate degree, you will be soon be encouraged and in some cases even expected to apply for grants of your own; in some cases you will have to have a research agenda before you are even accepted into such a program. Not all available grants will require massive narratives, but the rhetorical principles will stay the same. Finally, you might be required to write proposals or prospectuses for undergraduate or graduate theses, in which case you should pay special attention to the last section of this chapter.

The Rhetorical Goals of Proposals

You may be more familiar to other usages of the word "proposal"—for example, marriage proposal. Don't laugh—marriage proposals actually share some common characteristics with the grant proposals your research livelihood is dependent on. In essence, they both target a very specific audience (e.g., the significant other) and perform a request ("Will you marry me?") backed up by reasons ("I love you!") and buttressed with promises ("I will spend the rest of my life trying to make you happy!") and evidence (a diamond ring!), in hopes of getting a positive answer ("Yes! I will!"). Grant proposals for educational or scientific purposes (usually awarded by government or large nonprofit organizations), as well as business proposals or technical proposals (very common in the industry), work on the same basic principles—except with a lot more science and budget numbers built in and fewer diamond rings. (The rate of rejection is also a lot higher than for marriage proposals, but let us not be deterred by this for now.) All proposals share a few common characteristics:

1. They are highly *persuasive* documents. Every proposal's goal is to get funding for a particular project, thus keeping the people in charge of the project happily employed and provided with the opportunity of contributing significant information to their particular field of science.

2. They are *promissory* documents. If accepted and funded, they promise to deliver results that will ultimately further the goals of science (and humanity as a whole). Well, that's the plan, anyway.

3. They target a very specific and narrow *audience*. Proposals need to be exquisitely tailored to the needs and requirements of the granting agency. The funders usually have very specific goals, needs, and areas of concern: how are you addressing those? Neglecting your audience's requirements will throw you out of the contenders' circle faster than you can say "proposal."

4. They usually have very *rigid formats* dictated by the granting agency. This is the part with which most writers usually obsess, but that needn't be the case. Yes, you do need to conform to the letter to the technical requirements of the proposals (must the font be Calibri 11.5 kerned at .6 with 1.25-inch margins and a gutter of .2? Totally doable, but it's one of the last things you should fret about).

5. Their outcome is always *uncertain*. Unlike marriage proposals (where there's a solid chance your beloved will say "yes"), grant proposals

have a better chance of being rejected than accepted (this, of course, depends on where you're submitting the grant). Once upon a time, research proposals enjoyed a respectable 40% acceptance rate from federal granting agencies, but by 2010 that acceptance rate had sunk to about 10%, only to rise again to about 21% in 2014. That means that for a proposal to get funding, it has to be outstanding—not just good, not even *very* good. Grant proposal submissions are at the mercy of forces you cannot control—namely, acceptance rates—so even flawless, perfectly targeted writing may not yield the outcomes you want.

6. They are often resubmitted before being accepted. You'd have serious doubts popping the question *again* after an initial refusal, right? Not so with grants: you can always resubmit. In fact, many of the accepted grants are resubmissions that carefully incorporated reviewers' suggestions. In this respect, grant proposals are much like journal articles that have to go through the barrage of peer reviewers and editors. In some cases, reviewers invite you and encourage you to resubmit; in others, they may suggest much more extensive revisions before they would consider your proposal again. Either way, you will learn a lot from the process. The grant process is a fairly circular and recursive one: one submission is likely to lead to a resubmission after review; and getting one grant increases your chance of getting another. So it goes!

Antoine Lavoisier (1745–1789) was a French aristocrat and chemist co-credited with discovering oxygen, arguably one of the most important scientific discoveries of all times (he was building on Joseph Priestley's research, but it was Lavoisier who came up with the name of the gas). He was also incredibly wealthy, which meant he could pursue his extravagant experiments in chemistry to his heart's content without pleading for money from rich patrons. Unfortunately, his riches were his downfall: he was guillotined during the French Revolution.

Identifying Sponsors

There was a time when the ability to pursue science depended entirely on whether you were a gentleman (and occasionally a lady) of means—or at the very least you had a rich patron who could fund your extravagant pursuits in chemistry, natural science, lethal radioactivity, alchemy, heretical ideas like heliocentrism, or other noble scientific endeavors. Whether you're an alchemist or a garden-variety vaccine researcher, it is incumbent upon us to advise you that you are far more likely to be funded by government or private organizations rather than by a single wealthy benefactor.

So let's say you have a great research idea, but it would take equipment, materials, space, travel, and a team of dedicated people to pursue it, all of whom need wages. How do you get someone to pay for that? Obviously, you need to identify sponsors. If your idea has to do with basic research, a governmental organization is probably your first and best bet, but you shouldn't limit yourself to that. There are plenty of private organizations that may be willing to fund you as well, as long as your objectives align with theirs. How do you find them? Going through your organization's office of sponsored research (or whichever office at the institution oversees grants) is a great place to start. They will likely coach you on how to navigate sites like www.grants. gov, how to identify current and open calls for proposals, and how to identity private sources of funding. Community of Science is another wonderful database listing all open calls for proposals from a variety of federal, state, and private institutions and professional organizations—and your institution is likely to subscribe to it. National professional organizations in your field or some of their local chapters also offer research opportunities—visit their sites regularly to check new postings.

Don't get hung up on the idea that you must *first* have a million-dollar idea and only *then* seek funding for it. Although that's often how it works, browsing open calls for proposals on some of these publicly available sites will sometimes offer unexpected avenues for creativity or will most likely enable to see your research to date cast in a new light or from an angle you hadn't thought of before. In other words, you may be able to get inspired by some of the calls you see and seek opportunities based on what is currently available (case in point: the 2009 TARP legislation unlocked millions of dollars of US funding to educational research, spurring a flurry of proposals on that front—proposals that might not have otherwise been funded under different economic circumstances).

Be realistic. If you're just starting out in research, do not go to NOAA (National Oceanic and Atmospheric Administration) seeking a quarter of a million dollar grant so that you can set up a mini fishery in your lab. Go to Sigma Xi (a scientific honor society, if you belong to it) and get $250 to buy the fish tanks, pumps, and other supplies you need.

The Key to Writing a Successful Proposal: Understand and Anticipate Your Audience's Needs

To have a real shot at being awarded a grant, try putting yourself in the funding agency's shoes. What if *you* wanted to provide a bunch of money to research—and you had to select from among the projects presented to you? What would you like to know? Here are some of the burning questions you might have:

- Why should I fund *this* particular project? (Or: Does this project align with my agency's mission and interests?)
- How is this project better/more interesting/more promising than others? (Or: Is there anything out there that would deserve my money more than this?)
- Why is this the best *person* to run this project? (Or: Who are you and why should I trust you with my money?)
- Why is this *institution* the best place to run this project? (Or: Does your lab have the capacity to complete the project? And can you handle a budget?)
- Why is this important *now*? (Or: Why is this a *priority*?)

These are reasonable questions, and you can expect the funding agency to scrutinize your proposal for answers to them. Therefore, one of your major tasks in writing a proposal is to understand the needs and requirements of the sponsor and make sure you address their concerns in a clear, linear, logical, and persuasive manner that will make your project rise above all others. This is the rhetorical game, and your success at it is measured in clear yes/no terms (you get or you don't get the funding—unfortunately, there is no second prize). Yes, your main interest is to get your project funded and off the ground, but to get there you have to learn to present it in the most favorable light, which means you have to step outside your project and learn to look at it through the lens of an outsider. You are a salesman, essentially, learning your customer's preferences so you can show them how your product meets their needs.

Half the battle in getting funding is won just by writing a proposal the funding agency *wants* to support. That is why it is so critical *before* you start writing to understand the goals, interest, and background of the funding agency, to scrutinize the call for proposal for keywords and clues, and to communicate from the very beginning with key contact personnel. You should also be well informed of the issues and hot topics and trends in your field, because you want your project to be cutting edge, pushing the boundaries of science, rather than rehashing old territory.

It would also help to make sure that your writing style and goals fit with the agency's goals. Chat with the grant officers to find out how things really work. Email and sometimes even phone contacts are usually provided in the call for proposals—take advantage of that to steer your proposal on the right path from the beginning. Ask the advice of your peers and colleagues, particularly those who have submitted grants before. Remember to never underestimate the power of peer review. You'll be surprised by the things you learn. For example, when you formulate your objectives, how specific do you have to be? Seek the wisdom of experienced grant writers and principal investigators to become as well prepared as you can for the task.

If available, enroll in grant writing courses that may be offered by your institution's office of sponsored research or by the granting institution itself (many of them are online and free, and you can do them at your own pace). The more educated you are about this process, the better.

Request for Proposals

Part of your prewriting grant preparation process is to become intimately acquainted with the Request for Proposals or Applications (RFP or RFA) document put out by the funding agencies. If the funding agency is a smaller or private organization, often the RFP will be relatively short and more or less straightforward. If you are dealing with a government organization, expect the RFP to be tens of pages in length (single-spaced and in tiny print), explaining in exhaustive detail all the requirements for a successful proposal, from eligibility to formatting. Pay close attention to the eligibility criteria: if you don't meet them, the project is a non-starter. Additionally, the grantors often invite applicants to address them directly with questions, and you should take full advantage of this opportunity.

Letter of Intent or Inquiry

Quite often, sponsoring agencies will strongly encourage you or even require you to submit a letter of intent or of inquiry (LOI) in advance of the actual proposal. This is, in fact, a mini-proposal, no more than two to three pages in length, containing the kernel of your project (background, need, objective, specific aims, strategy and goals, leadership/organization, budget, significance). It is a sensible plan: the letter will offer you a first crack at making your case in writing in a condensed form, and the responding agency can quickly steer you in the right direction. If the research has merit and is aligned with the interests of the agency, they can encourage you to submit a full proposal (which they will be prepared to review); and if it's not something they think they can fund, they will let you know all the same. Either way, it's a win-win scenario that can spare both the requestor and the grantor a lot of time and pain in the future. The agency will usually offer you very specific guidelines in terms of content and form for the letter—and you want to make sure you follow them.

Because it is streamlined, abbreviated, and high-stakes, the LOI can be a difficult document to write *well*, in other words, properly targeting the funding agency, in particular the perceived interests and level of expertise of the reviewers. If the LOI is a "mini-proposal" of sorts, it should mirror all the required parts of a proposal.

Suggested Outline for the Letter of Intent

1. Opening Summary
   - Who are you and what do you want to do?
   - How much is being requested? Are you requesting partial or full funding for your project?
   - What is your timeline?
2. Statement of Need
   - What issue/problem will you address? What gap in knowledge will it fill?
   - Why you have chosen this particular methodology?
   - What impact will this project will have, and who will it serve?
3. Project Activity
   - Give an overview of the activities involved.
   - Will you collaborate with other department/institutions/agencies?
4. Outcomes/Evaluation
   - What specific outcomes you hope to achieve, and how will you measure them?
5. Organization Information
   - Why is your institution the most qualified to carry out this project?
6. Budget
   - What will the total project cost, and how much will you request from the foundation?
   - What kinds of activities will need funding?
   - Include other sources of funding, both cash and in-kind. Especially indicate what your institution will contribute.
7. Conclusion
   - Offer to answer questions or provide additional information.
   - Include a contact name and contact information.

Assuming your LOI has a favorable response—in other words, if you get a letter encouraging you to submit a full grant—you should get to work in earnest!

Components of a Grant Proposal

The following is a list of the general components of a grant proposal; however, keep in mind that this list can vary depending on the agency, type of grant, the RFP, etc.:

1. Abstract or Summary
2. Narrative
   a. Background
   b. Statement of the problem or need
   c. Statement of research question(s) and/or hypotheses
   d. Objectives and specific aims
   e. Methods and procedures
   f. Limitations
   g. Significance
   h. Assessment
3. References
4. Description of facilities, equipment, etc.
5. Biographical sketches of the investigators
6. Budget (and justification for the budget)

Depending on the project, you may also be required to provide proof of Institutional Review Board (IRB) approval if you are working with human subjects, or the approval of an Institutional Animal Care and Use Committee (IACUC) if animals are involved. (You would have to go first through your institution's IRB or IACUC. Most research institutions have such boards or committees who must maintain their certifications in good standing to be accepted by the grantors.) Numerous forms specific to the granting organization complete the picture. The grant proposal will also likely include attachments and appendices, which may contain tables, figures, maps, lists, and other documents that might be of interest to the reviewers or help them understand the proposal fully.

Budgets are specific to the project and are arcane affairs that must be worked out in conjunction with your institution. Usually a narrative budget justification must also be provided. Vitae of the main investigators and other key personnel are standardized (a model is helpfully provided, usually, by the sponsor) and are often accompanied by biographical sketches that emphasize the stakeholder's experience with the type of research proposed—as well, as importantly, any other current and pending support. Finally, a description of the facilities and the needed equipment is also an essential element in your persuasive arsenal: you are demonstrating that you have the physical space and resources to conduct the research, and you know exactly what you need to complete the research.

The hard, persuasive work of the grant occurs, however, in its two key sections: the abstract or summary and the narrative. We look at those in detail next.

Project Summary

All proposals have a "Summary" page that explains, in no more than 500 words or one full page, the whole proposal. Conciseness and persuasiveness are key goals in this section, which is probably the most rewritten section of all, and for a good reason: it is often the most *read* section of all. If you can't make your case persuasively here, chances are that few will bother to read past it and actually peruse your assiduously put together methodologies and budgets. Follow the grant instructions to the letter—this is where you are not simply abstracting but truly selling your grand idea to a diverse audience, not all of whom are experts in your field. As such, make sure you cover all or most of these rhetorical moves in a summary as you would in an abstract:

- Clearly state the problem you are hoping to solve or the need for the research you are proposing;
- Describe the promising idea(s) your research will build on;
- Describe your objectives;
- Describe your methodology;
- Describe expected outcomes or how your research will advance knowledge; and
- Describe greater societal benefits (the broader impact).

We repeat: this is probably the most important page of the proposal. Every single sentence has to make a compelling case for why this research should be funded. Nothing should be obscure or confusing.

Let's see how all these principles play out in the executive summary of a successful proposal (source: National Institutes of Health [NIH]):

A Combined Computational and Experimental Approach for Structure Prediction of Foldamers

The research proposed here aims to enhance rational design of foldamers through development of an approach and enrichment of information for structure predictability of an important class of foldamers. Foldamers are oligomers that fold into well-defined secondary structures in solution, providing a potential for a variety of **novel** applications, and thus have been the focus of numerous experimental undertakings in the past two decades. For many classes of foldamers, specifically for those relying on local conformational control, the initial design is based on structural features of certain small molecule patterns. Literature examples, as well as preliminary studies obtained in this group have shown that conformational preferences of arylamide model compounds significantly depend on specific nature and location of aromatic ring substituents, as well as the environment. The main objective of this proposal is to establish information transferability between the foldamer building blocks and the final foldamer structure. The effort will build a databank, with structural and energetics information on foldamer building blocks and oligomers encompassing various structural features. The systematic pool of information obtained includes the effects of varying hydrogen-bonding (H-bonding) ability, size and location of the aromatic substituents, as well as oligomer chain lengths, on foldamer conformation and force field parameters. In addition, this information will be provided for several different environments, such as water, methanol, chloroform and gas-phase, to address a variety of media for possible foldamer applications. The proposed project will also provide the scientific community with design tools that address specific foldamer requirements. The research proposed here is a critical step in a long-term goal to enable a rapid and accurate prediction of foldamer conformations and related molecular interactions.

By investigating electronic and structural features of a number of systematically chosen arylamide foldamer segments, the minimal structural unit that contains the critical information needed for structure prediction will be determined. This effort will result in an increased understanding of non-covalent forces, specifically H-bonding and π-conjugation, which will reach beyond the foldamer scientific community, due to their general nature. The specific issues to be addressed through the proposed research are: obtaining a quantitative ratio of H-bonding strengths for intramolecular H-bond systems within a systematically chosen pool of *ortho*-substituted arylamides; quantitative comparison of strengths of several types of shared H-bonds with respect to the corresponding single H-bonds, i.e. assessment of cooperativity of H-bonds sharing an acceptor or a donor; level of retention of intramolecular H-bonding in protic solvents of varying H-bonding ability; quantitative assessment of influence of π-conjugation on conformational preferences and related molecular rigidity; effect of aromatic ring substituents on π-conjugation between peptide group and aromatic rings; obtaining arylamide-specific torsional parameters and charge treatments. While the focus is on arylamide model compounds, the findings will be general, and thus applicable to a wide variety of chemical and biochemical systems.

Explanatory title is provided

The writers start with the overarching goal of the project.

Definitions are helpfully provided.

Note the expert use of "signal" words and phrases such as "novel," which play a persuasive role. Can you identify others?

Writers prove they've done their homework, hinting at a literature review and their own expertise in the field.
The scope of the project is further explained.

The significance of the project is explained (also "signaled").

More details about the project are provided in a systematic, linear manner.

The wide applicability of the project is again emphasized.

To achieve these goals, a combination of computational chemistry, synthesis, NMR and IR spectroscopy studies is planned. The research will be conducted as a collaborative effort, involving 3 groups of researchers at 2 universities: *ab initio* and molecular dynamics (MD) simulations will be conducted by Dr. Pophristic's group at the University of the Sciences in Philadelphia (USP); Dr. Moyna's group will synthesize model compounds and conduct NMR experiments, while Dr. Teslja at the Fairleigh Dickinson University (FDU) will carry out the IR part of the project. The experimental assessment of the computational results will guide the parameter optimization process and lead to more accurate prediction of foldamer conformation through MD simulations. Such research organization will enable close interaction between undergraduate and graduate students, and postdoctoral fellows working on computational and experimental subprojects. Importantly, the nature of the project which calls for frequent interchange of experimental and computational results, will strongly promote FDU/USP collaboration.

The results of the research will be disseminated via peer reviewed publications and presentations at scientific meetings. The data will ultimately be available through a publicly accessible, searchable databank. Based on the general nature of the information to be provided by the proposed research, as well as its applicability to foldamer specific research, it is expected that both the broad scientific community, e.g. chemists, biochemists, and material scientists, as well as researchers working on specific foldamer topics, will benefit from this research very positively.

The methodology is briefly explained.

Further benefits of the project are highlighted, in this case responding to a specific guideline in the RFP regarding training and collaboration.

Assessment and deliverables are explained.

The beneficiaries of the project are named. Broad impact of the project is further emphasized.

Notice that while the language is relatively technical, the writers take care to explain the goal, significance, and scope of the project in language that educated reviewers (scientists who are not necessary specialists in the field) would understand, and the line of reasoning is logical and linear. In other words, the big picture is well defined as to be both clear and persuasive.

The Narrative

Background

The background, like any good introductory move, should set the stage for the project you are proposing. You need to reread the RFP very carefully so you can address all the points the agency wants you to address. The background needs to be very detailed—it is a chance for you to review the literature (make sure you give clear references!) and show that the field is primed for *your* research. Chapter 9 can offer some pointers in that direction. You should not feel the need to review *all* research on your topic; however, you should survey the top 15 to 20 papers or works that inform your research. Select the most research and demonstrate your familiarity with the field. Briefly describe the progress to date researchers have made on this particular problem: basically, explain what is known on the issue and what is not known, so that your audience understands where you come in. Bonus points if you can demonstrate that you've already done preliminary and promising studies on the issue.

Statement of the Problem or Need

You should state in no uncertain terms why your research is necessary. Why, exactly, should we research the vision of the mantis shrimp and how they process color, for example? It is up to you to make reviewers care. A gap in knowledge is your starting point: build on it to show how you are filling a void in the current state of science and, hopefully, paving the way to innovation and the Greater Good.

Statement of Research Question(s) and/or Hypotheses

Clearly state the question driving your research and your preliminary hypotheses. Formulating an interesting research question is often a difficult process; it is tacitly accepted that an interesting question is often worth more than its answer(s). Asking a good question is an art, and it reveals where your interests lie; it is also a good place to put a spin on your research that your granting agency would find intriguing.

Objectives and Specific Aims

These need to be very carefully formulated and should not overlap: in other words, the objectives should be self-standing, so that achieving one objective is not dependent upon the achievement of another. The main difference between objectives and aims is that aims tend to be subordinate to objectives—they are more specific and detailed and you can cluster a few of them under each objective. Your stated objectives and aims should specifically solve the problem you have articulated.

Methods and Procedures

How will you achieve your objectives? This is one of the most crucial sections of your proposal, one that you have to lay out in utmost detail and as logically as possible. You may have a variety of methods for achieving your goals, in which case you should describe them all chronologically. Make sure your experiments or actions are achievable and relevant and articulated correctly and clearly.

Limitations

Is your research going to solve any and all issues related to the problem at hand? Probably not. Even if you think it will, it's a good idea to foresee and acknowledge the limitations. You will need to present a report on your grant results someday so you'd better be prepared to acknowledge, earnestly, what your research can and cannot do. Also be prepared to account for any factors that can limit your success.

Significance

What good will all these efforts do? What is the impact on the larger scientific community—as well as the community at large? This is probably the part of the narrative where you get to showcase your persuasive prowess. Make sure you use "signal" words to specify why your project is *new, critical, necessary, vital, important,* and *beneficial* in any way to specific and broad audiences alike. There is no need to be bombastic, but make sure your audience understands the importance and necessity of the project as well as *you* do.

Assessment

This is an increasingly important area: how will you assess that the grant has fulfilled its objective? What will you be able to deliver that will prove that the goals you have established have been achieved? It is helpful to project into the future: where would you like to be three or five years from now? Again, the RFP should provide you with further clues as to what the funding agency would expect from you. Depending on your project, your team should be able to deliver some of the following: articles, conference abstracts, presentations, posters, patents, databases, reports, etc.

References

The list of references should not be intimidatingly long (20 would be a safe average, although this can vary significantly based on the length and complexity of the grant), and it should consist of the most relevant and recent research informing your project. Format it in the style most commonly used in your field or as dictated by the grantor.

Responding to Sponsors' Decisions

Let's say you and/or your advisor worked hard and the proposal got accepted. Hurray! You can now proceed apace with your research.

Lightfield studios © 123RF.com

But in the relatively more likely case that your proposal gets kicked back, after the requisite period of mourning you should read the reviewers' report carefully and use it to revise and resubmit the grant. If you're lucky, the reviewers will encourage you to resubmit. Usually, the reviewers will score your proposal both holistically (an overall impact score) and according to a variety of criteria which are usually disclosed in the RFP. For the NIH, for example, these criteria (rated on a scale from 1 = exceptional to 9 = poor) are: Significance, Investigator(s), Innovation, Approach, and Environment, with additional criteria added as necessary for the protection of human subjects, etc.

Revisions should address the concerns of the reviewers, which may include requests for clarifying objectives/aims, scope, rationale, methodology, significance, personnel, or budgets. You don't get a chance to revise before a consideration is made—you will need to resubmit in the next cycle. Although you might not get the same reviewers on the second round, you should specifically address the reviews and criticisms you received in the resubmission draft (some authors include a separate section clarifying the revisions to the draft; this may also be required by the grantor).

© 2020 Max Beck

Sometimes the reviews will be less encouraging, but make this a learning opportunity. Read them for clues as to where your project might be changed or strengthened. Make sure your enthusiasm and faith in the project shines through this second (or third, or fourth) draft as well. After all, the proposal is where you build a case for your brainchild—and you hope to inspire the same level of excitement to your reviewers. Again, peer review is critical. Share reviewer comments with your peers, and have them review your revised proposal. As with life, there are no guarantees that all this hard work will eventually get funded—sometimes through no fault of your own. Science just moves on sometimes and by the time you get to review your topic is no longer "hot" or a recent discovery has made it less relevant. Reviewer teams change, and they have different sensibilities and priorities. Or it is simply not that proposal's time now. No worries: move on to the next one.

---

10 Tips for Writing Grant Proposals

1. Follow the funding agency's guidelines to the letter—for both content and form.
2. Ask for as much feedback as possible from peers and/or from grant officers whose job is to address your questions early in the process.
3. Be extremely logical and linear. You know your stuff; give your reviewers the best shot at understanding it as well.
4. Be specific and realistic. Do not overpromise, and explain clearly what it is that you promise.
5. Clearly situate your project in a continuum between past and future studies, showcasing how it builds on previous work and how it will lead to future advancements.
6. Emphasize the importance of your project for the scientific community as well as the populace at large.
7. Revise early and often.
8. Spruce up on your project management skills: have a checklist with deadlines. You must be extremely organized and enlist the help of your team and Office of Sponsored Research at your institution to make sure you have all the parts of the grant ready by the deadline.
9. Do your research: make sure your references are recent, relevant, and, inasmuch as possible, authored by respected/established researchers in your field.
10. Triple and quadruple-check that everything conforms to the RFP. We've said this before, but it's that important. You don't want your proposal to be rejected from the start over a technicality such as a missing cover letter, for example.

Special Cases: Thesis Proposals

In case you need to or want to write an honors thesis or conduct a more elaborate undergraduate research project, you might need to write a thesis or research proposal; the same applies if you want to continue with graduate studies, which usually require you to write a thesis or dissertation. While not quite as elaborate as grant proposals, thesis or dissertation (research) proposals, or prospectuses as they are sometimes called, still need to answer the same fundamental questions as any grant proposals. Again, the primary directive you should operate under is Know Your Audience, as your adviser and the committee members may have their own idiosyncratic requirements. In general, however, they will want to know:

1. What is your topic?
2. What are you trying to do? (goals and scope of research)
3. Why are you doing it? (rationale and background)
4. What has been done so far on this issue? (brief literature review)
5. How are you going to do it? (methodology)
6. What do you expect to find, broadly speaking? (expected outcomes)
7. Have you tested this to make sure it's going to work? (sample/preliminary/partial analysis of data using the methodology you propose)
8. What's the plan? (outline of the remainder of the thesis and tentative calendar)
9. Did you do your research? (list of references)

As long as you offer detailed and rigorous answers to these questions, your research proposal will be on solid ground. The degree of detail required, though, will vary depending on institution, department, field of study, and committee. Again, ask your adviser what is reasonable at this point in your research. If possible, ask to see an example.

Since most theses or capstone projects emerge organically from preliminary studies, papers, or coursework, based on research questions tested during your course of studies, you'll probably have *something* to show or illustrate at this point—in terms of data or results that are intriguing and promising enough to persuade your committee to give you the green light to go ahead with your project. Although you are not asking for money, you are asking for the time, feedback, and guidance of a number of committee member for the foreseeable future (a minimum of two years, usually, although it can be more); moreover, you are also essentially entering a contract between you and your institution that acknowledges that your work, if and when successfully completed, entitles you to a graduate degree, with all the rights and privileges coming from it. Furthermore, your results are representing your institution, which in return commits to backing you up (often financially through assistantships and fellowship) in recognition of your promise. That is why both you and your committee need to be extra careful before proposing and respectively accepting a project upon which your future career hinges (your advisor's too, to a smaller but not negligible extent).

Take-home

Research or grant proposals are essential to the very endeavor of science. Because, quite often, basic research does not result in immediate commercial gains, scientists must rely on external funding to advance their research agendas. For that, they must persuade a panel of peers that, in fact, their research has significant societal benefits, a solid scientific basis, and the requisite resources to solve those basic societal needs. Logical, linear thinking, great organizational and management skills, superior attention to detail, and unwavering enthusiasm are required (not to mention patience).

Exercises

1. Individually or in groups:
   a. Pick several topics for research in your field and formulate three research questions suitable for a grant proposal that highlight different aspects of the topic.
   b. Based on one of your research question, formulate three objectives and aims for a possible grant.
   c. Write a statement of need for your chosen research question (aim for one or two paragraphs). What would you emphasize in such a statement?
2. Analyze a grant: ask for one from a professor (funded or unfunded) or download a sample one from grants.gov, the NIH, or the National Science Foundation. How many of the components we have identified can you find? Are they all clearly marked with headings and subheadings? In some other way? Identify specific "signal" words that play a persuasive role in the proposal.
3. Pick two likely agencies that could possibly fund a grant you might propose—a government agency and a private one. Look at a recent guide to writing proposals and/or RFPs that these two agencies put out and find at least two major differences between them. Discuss why you think those differences exist.
4. Play the Agency Game. In teams, make up an agency that could possibly fund research in your general field of studies. Give it an official title. Then sketch out a RFP in which you provide the following general guidelines:

    a. General areas of research you expect to fund; what projects you would prioritize. Be specific!

    b. Eligibility: who should apply?

    c. Criteria for assessment: how would you evaluate incoming proposals? Share your RFPs with the rest of the class. What are the differences and commonalities? Why do you think the differences have occurred?

5. Individually or in groups: interview a professor/researcher at your institution working on a grant proposal or who worked on a grant proposal in the past. Write a list of questions you want to ask him or her, set up the interview (in real life or online/via email), record the answers, and share with the rest of the class.

12
Presentations

Proof your tats

Anatomy and Physiology (A&P), you thought, would surely be a nice, predictable pre-med class to take—lots of memorization and dissection. It's a tough class, one of the "weeder" classes, and the prof has a peculiar sense of humor. He's one of those profs who automatically puts you on guard when he asks a question and a slow smile starts to spread across his evil face. You come into class, and he has a piñata hanging from the ceiling. You look nervously at your classmates. Surely it's filled with body parts; that's the kind of stunt Dr. Atkins would pull. Still, you obediently take turns and whap at it with a bat, until it breaks open, spilling perfectly innocent-looking candy on the floor. And little folded up pieces of paper. "One each!" he chirps. You unfold yours, and it says "tattoos." Your lab partner looks up and says "I got *Candida*. I mean, on the paper." Everyone looks around nervously, whispering things like "pain," "third-degree burn," "sweat," "tanning." You're a good student and have done your reading: "Oh, this is all integumentary system. He's doing the integumentary system. *Skin*." "No," he says, "*you* are." You blink. He says, "Surely you don't think *I'm* going to give all the lectures this semester."

"I *So* Don't Want to Create a Presentation, Much Less Give One"

Ah, but you want to be a scientist, right? While, granted, in some fields you can hole up in a lab somewhere—perhaps in the bowels of a large research institute or museum where no one can find you—it's unlikely that you're going to make it through school without giving the occasional presentation. It could be a presentation to your class, or the lab in which you're working, or a local club or association or—particularly if you're headed for research—at a meeting. Your advisor will make you give 15 minute presentations at national meetings, or craft posters for poster sessions, or create presentations to be delivered via webinar or through other electronic means. Maybe you'll give a talk to elementary school kids on "Science Fun Day," or maybe you'll be interviewed by a local newspaper. Maybe you'll be using slides. Maybe just some notes. Maybe it will be interactive, and you'll be writing on a white board or flip chart. Maybe you'll be tagged to fill in for your prof on a lecture he can't make for his intro bio class.

"Presentation" just means "to present something." A livelier way to communicate. And, while the end product might not be writing per se, writing will certainly be involved prior to delivering your opus, in whatever form it takes. If the thought of standing up in front of other people (or even making a virtually anonymous online presentation) convulses you with fear, just don't think about that part until you are finished with your writing and preparation. *Then* you can focus on the stage fright. Stay coherent while you're writing.

How Do You Start Thinking About Presentations?

The topic may be chosen by you or for you; in itself, it will dictate some level of complexity. However, the really crucial aspect of preparing a presentation is the following.

Target Your Audience Accurately

To whom are you speaking? Your A&P class? a biology department sem-inar? little kids in an elementary school? a professional scientific meeting? Consider what your target audience knows already, and write to that level. You also need to consider what your target audience *wants* and *needs* to know. If you're talking to your A&P class about tattoos, what you're presenting is clearly part of a unit on the integumentary system. You won't focus on just tattoos per se; you'll discuss the related anatomy of the skin layers, show his-tologic sections, and discuss the procedure in some detail. Your talk "goes" with the other "skin" talks. If you're talking to a fourth-graders, however, our advice would be to *not* show them close-ups of a tattoo needle piercing the skin and depositing ink into the connective tissue of the dermis. At a medical conference, depending upon the specialty, you won't give a general A&P-style talk on tattoos; you'll focus on pathology, infection, "tattoo regret," or other medical tattoo-related topics. If you're presenting to your journal club in your department, you may be inclined to inject (see what we did there?) a little levity in places.

> If you're fortunate enough to be working on a topic like tattoos, you have no dearth of "levity" available. A simple online search for "tattoo fails" can keep you entertained for hours. You might also want to check out Vail Reese's "skinema.com," your go-to source for Hollywood Dermatology (check out "Villains with Problem Skin"). There is also the classic Saturday Night Live documentary on the removal of "really cool lower back tattoos," available from our friends at NBC (NBC.com, "Turlington's Lower Back Tattoo Remover").

Understand Your Topic in Depth Before You Start

We are stating the obvious here, but you also need to *know your stuff.* If you're writing about tattoos, you are not going to prepare by reading "just enough to get by" or enough to fit 10 slides. The thing about presentations is that your au-dience frequently asks questions. You don't just need to know the basics; you need to understand the anatomy of the skin, the composition of the ink, how it may affect skin tissues, how it is administered, the different kinds of ink there are, the long-term effects of tattooing, and tattoo removal, among other things.

Assess Your Delivery Medium

You know to whom you're writing, and you know what you're writing about, but the next thing you have to consider is *how it's going to be presented.* "Presentation" is an umbrella concept. Are you talking for 60 minutes in a seminar for your department? 15 minutes at a scientific conference? Is it going to be live? Is it going to be online? Are you going to be in a classroom, or a ballroom that holds 150 people? Are you going to be sitting at home with your cat in your lap, delivering a webinar?

It's important to think about these things before you start: how much time you have, in which environment you will be presenting, and through which medium you are delivering the information. If you are giving a 15-minute talk, you can't have 30 slides. And if you are creating a presentation that is supposed to run for 60 minutes, including a 15-minute Q&A, do not create a 60-minute presentation; create a 45-minute presentation.

If you're giving a 60-minute talk at a meeting, you're going to have a captive (mostly) audience for 60 minutes. If you're presenting a meeting poster, you might have 10 to 15 seconds at most to get them interested enough to stop at your poster and then maybe only a few minutes to discuss it or answer a few questions. You can give a presentation on the same topic (e.g., tattoos) in all of these scenarios, but the scenario (including target audience) is going to dictate how you write and create it.

A "presentation," then, *isn't just writing.* It has an extra prewriting element of communication that papers don't have, that is, scenario planning. So, before you even start to create your presentation, you know:

1. Your target audience is your peers in an undergraduate A&P class.
2. They have a basic understanding of the integumentary system, but that's it.
3. They're going to be curious about the biologic aspect of tattoos (i.e., not, under these circumstances, cultural or artistic aspects).
4. You know you have 15 minutes, *plus* 5 minutes for a Q&A.
5. You know you are expected to create PowerPoint slides (and provide handouts, if necessary).

You need to reflect on all of this before you start writing.

1. You're writing to peers in an A&P class. This is somewhat easy, since you're in the same group. You'll use language sophisticated enough for you to

understand it yourself, but which is not overly "medical." Your "voice" will be that of a college student: informed, but not at the level of a specialist.

2. You will focus on the aspects of the integumentary system relevant to tattoos. While you won't get into the different layers of the skin (your peers should already know that); you might have them in a diagram. You will focus on the dermis, since that's where tattoo ink is deposited. You will discuss dense irregular connective tissue, lymphatics, blood vessels, and nerves.

3. You will get into detail about how the ink is deposited, where it goes, how it stays there; you will discuss colored ink and the ink that can now be removed with lasers; you will discuss infection and other risks, and tattoo removal. Write this down. Put it in an outline as you do your background reading, as you think of the biologic topics you want to address. You can sort out the order later.

4. You have 15 minutes. How long is that? Are you a dawdler or a rapid-fire speaker? Do you know? If you are a slow speaker, you might have to limit a 15-minute talk to eight slides. If you are a rapid-fire presenter, you might be able to get away with as many as one slide per minute (you can figure out how fast you are when you practice your talk). There is *no rule of thumb*—although there is a lot of advice out there: one minute per slide, two minutes per slide, three minutes per slide, and so on. There *can't* be a rule of thumb. A conceptually complicated slide will take longer to talk about than a slide with three general bullet points on it. When you're making slides, it's usually safe to start with an *estimate* of 1.5 minutes per slide and go from there, increasing or decreasing slide number depending upon the content of the individual slides and how quickly you speak. The purpose, format, target audience, medium, style, context, and length of the talk will also determine slide number and speed. The *actual* rule is: "only as many slides as can be presented comfortably without being rushed, but that contain all of the information needed." Then you need to practice your slides . . . over and over and over and over and *over* again. Do not skip this step. It is critical, not just to know what's in your presentation but to ensure that it's not too long. You'll get a sense of whether you're a fast speaker or a slow speaker. What if you're a slow speaker, and you cannot cover even 10 slides in 15 minutes? What do you do then? Pare your content down: distill it into broader concepts; combine slides, if it makes sense. If you're fast, it might be nerves. Rather than add content, try to relax, or expound more upon the slide content that's already there. Fast or slow, you need "enough of the right content," not too little and not too much.

5. You need to have engaging slides (whether in PowerPoint or some other presentation software). Twelve straight slides with bullets are not going to get you points for "creativity."

Many of the early stages of target audience identification are important in *anything* you create, not just presentations. However, target audience, when giving a presentation, needs to be considered in a slightly different context—that is, how you'd *talk* to them, instead of how you'd *write* to them.

"Scientific writing" isn't only about writing; it's about communication in general. To be successful in any profession, whether you're lurking in the dark recesses of a lab doing benchwork or speaking as a clinical researcher to an audience of 500 doctors at the American Academy of Dermatology (AAD), you're going to have to be able to *communicate effectively*. How something appears in a journal is not going to be expressed (hopefully) the same way by a meeting speaker, nor would either of these two things be communicated the same way on a meeting poster. Live presentations "work off of the slides." They are not scripted (or at least they should not be; standing at a lectern and reading off notes is deadly and staggeringly off-putting). Meeting posters have slightly more detail than a live presentation but less than a journal article and should be stand-alone presentations, accompanied (or not) by the poster author(s) for informal discussion.

How it appears in a journal article	The way a speaker says it and how it looks on a slide	What it looks like on a meeting poster
"The discovery of selective photothermolysis has enabled the targeted destruction of tattoo pigments with only minimal damage to the surrounding tissue and limited risk of adverse effects, which contrasts previously used nonspecific methods."	"When Fred Constantine's lab first came up with the photothermolytic approach, the consequences of tattoo removal became pretty much avoidable." [IMAGE ON SLIDE OF TATTOO PROCESS OR OF SOME TATTOO-REMOVAL ADVERSE EFFECTS]	Photothermolysis (Constantine et al., 2010) decreased incidence and severity of AEs.[1]

Wenzel SM. Current concepts in laser tattoo removal. *Skin Therapy Lett.* 2010;15(3):3–5.

[1] Reference at bottom of poster

Do you notice any kind of trend in the way this information about photothermolysis is presented? Can you guess why? Consider two things: delivery medium and time. Same information (assuming appropriate context for each), vastly different ways to present it. Paper: complete information. Speaker: informal delivery in a running discussion, working off of slides of varying complexity and content. Poster: abbreviated for economy of text on a poster of limited size. Notice how the delivery medium also contributes to the "voice" for each.

Live Presentations: PowerPoint, Your "Frenemy"

Don't make me use PowerPoint! You must, you must. While there are alternatives, PowerPoint is still coin of the realm when it comes to live scientific presentation. If you're collaborating with other people on the slide deck, everyone has to have the same presentation software, and that will most likely be PowerPoint (although time does march on). If you give a presentation at a meeting and are directed to bring your presentation on a thumb drive to be used in a conference computer, that conference computer presentation software will likely be PowerPoint, and they might even have given you a template to use. Once you get used to it and *really* learn how it works, you will find that it can be extremely powerful and can simplify your life tremendously. Even if you are conversant with PowerPoint, take a tutorial—you'll be astonished at what you can do with it.

In presentations and essays, Edward Tufte—the celebrated designer of visual information—whimsically referred to PowerPoint as "fluff," "clutter," a "mockery by Microsoft," and a "cruel hoax" (he was not a fan). His presentations are conspicuously devoid of PowerPoint. Tufte is a statistician and artist and Professor Emeritus of Political Science, Statistics, and Computer Science at Yale University. No slouch, he. He wrote, designed, and self-published four classic books on data visualization. Visit his website; he's quite a guy (www.edwardtufte.com).

Create Effective Slides

As with any scientific communication effort, you need to capture your audience's attention. You cannot do this with 15 yawn-worthy slides containing nothing but bullet points. But when you're faced with creating a PowerPoint presentation, your first hurdle is going to be that nice, new, blank PowerPoint file you've just opened, and, just like that nice, new, blank word processing file you opened when you tried to start writing a paper: what do you do? How do you start? What is an "effective slide?"

Since you have done all of your up-front due diligence researching your topic, start with an outline, knowing it can be shuffled around later. Don't stare at that blank file and think you're going to be writing a finished presentation. What you're starting to create is your conceptual framework—one idea (or subidea) per slide. A slide presentation is *not* a paper, nor is it a "paper put onto slides" (in fact that would be a terrible idea, and a common misdemeanor in neophyte writers). Your slides should be "idea snapshots," and here are the things you should consider:

1. Make each slide a "snapshot." A slide should be almost stand-alone in the content it delivers (even if it is under a larger topical or conceptual umbrella). You don't have a paragraph of narrative or even full sentences, and your slides are variable in nature: text and graphs and tables and illustrations and photographs and maybe even video or interactive components mixed in together. For some topics, this can be a challenge (e.g., tattoos offers several fabulous graphics opportunities, molecular genetics less so, but you can do it—use your visuospatial skills!). Avoid slides that have "(continued)" or "cont'd" in the titles. Rethink presenting that information in single-slide digestible chunks. Make them *engaging*, including the slide title. That can be challenging, because you do not want excessively long slide titles. Some slide development pedagogues advocate using a question as a title to "grab" the audience. You can have a slide simply titled "Background," or you can ramp it up to "Tattoos are an ancient artform," or "The philosophy of tattooing," or "Historic tattooing practices" or "How old is the practice of tattooing?" or "Why do people get tattoos?" The use-a-question advocates maintain that if you do that, your audience is thinking "yeah . . . why?" and they're waiting to hear the answer. The short-and-sweet advocates ("Background") maintain that you're providing your audience with a one-second snapshot of slide content so that they know what you're talking about and pay attention to you and the slide content. There isn't a "right" way to do it; it's personal preference (or your prof's dictates). Then there is the question of consistency: do you use questions for every slide? Or mix it up: questions, statements, or single words? There are varying philosophies on this topic as well; follow your instructor's advice or, absent that, the needs of your audience to the best of your abilities. Also: you've seen slide presentations. Which ones "worked" best?

2. Bulleted slides have their place. The only overarching recommendation is that they should not be overused. Again, 15 straight slides containing nothing but bullet points will have your audience trying to look attentive,

when they are in fact daydreaming about what they're bingeing next on Netflix. A bullet point is a *point*. A thought, a concept, an idea—a *fragment* of information. *Fragment.* It should not be long; it should not consist of whole sentences; it should not have periods at the end. If it is necessarily long, create subbullets. Do not have single bullets or subbullets (i.e., they're plural for a reason—have a minimum of two bullets on any slide, and a minimum of 2 subbullets if you require subbullets). Do not put 19 bullets on a slide; hew to the PowerPoint default-size text window (you can turn that off, but I recommend that you do not). As with the title, be consistent with your font size from slide to slide. If your bullets each consist of only one or two words, consider making two columns.

The Problems That Can Occur When You Get A Tattoo	**Tattoo complications**
• Tattooing is very painful. • There can be reactions to the different kinds of tattoo ink • There is a risk of infection that can happen with tattoos • Keloids – a type of reaction where a big 3-D scar happen scarring • Granuloma nodules can form around what the body perceives as "foreign," like ink particles • 9% of people have allergic reactions to the ink • If a person has "tattoo regret," it's hard to remove tattoos	• Pain • Granulomas • Infection • Allergic • Keloids reactions • Scarring • Tattoo regret

Okay, this is a "gimme" (hopefully): can you tell which is the better slide? Let us count the myriad ways (with, say, bullet points):

- Selection of typeface: Never, *ever* use *Comic Sans* or *Papyrus*. You will justifiably be the object of eternal ridicule. Choose a simple, easy-to-see sans-serif typeface like *Arial*—no prominent serifs, no weird ascenders or descenders. You do not want to distract your viewers from the slide content.
  - A *typeface* is the name of the type you're using. It is frequently incorrectly referred to as a font.
  - A *font* is how the typeface is expressed, e.g., 8-point, *italics*, **bold**, etc.
  - An *ascender* is the part of a letter that extends up above the main body of a letter, like tops of the letters "t," "f," or "l."
  - A *descender* is the part of a letter that extends down below the main body of a letter, like the bottoms of the letters "p," "q," or y."
  - Letters that contain neither ascenders nor descenders are letters like "o," "e," and "c."
  - A *serif* typeface has little tags at the ends of exposed lines in letters; "Times New Roman" and "**Rockwell**" are *serif* typefaces; "Arial" and "Century Gothic" are *sans-serif* typefaces.

- *Case* refers to how letters are capitalized (or not). There is *UPPER CASE*, or *lower case* or *Title Case* or *Sentence case*. Whichever of these you choose for your titles, you need to be consistent from slide to slide. *Sentence case* is the "least distracting."

- "Prevailing wisdom" is to use a *light* sans-serif typeface on a *dark*-colored background (this enhances visibility). So does contrast, such as black on white. This is not dogma. There is no dogma in PowerPoint (strong opinions, absolutely).

- Title: The title on the Slide Of Cringe-Inducing Badness is too long, too awkwardly written, unnecessarily underlined, right-justified, and written in title case, complete with the incorrectly capitalized article "a." Title justification (left, right, center) and style (i.e., sentence case vs. title case vs. upper case, etc.) is a personal preference, but again, you don't want to distract your viewer from the content on your slide.

- Amount of text: Way too much for a single slide. Contrast the chaotic bullet points on the slide on the left with the neater two-column list on the right. You do not need to put every single piece of information on a slide. That's what the presenter is for. The presenter uses the slide as a springboard for discussion or explanation, expanding and expounding upon its tidily written points. Presenters should *not* read the slide aloud; that's deadly (and, unfortunately, common).

- Bullet point construction: The slide on the left is appallingly written and inconsistent in format, punctuation, capitalization, and amount of information contained. Some entries are full sentences; some are fragments; one of them is a single word. For ease of comprehension, particularly in a moving target such as a slide presentation, bullet points should start the same way: with a noun, a verb, a measurable objective, etc. Never start with a numeral (unless you have subbullets with stats or something like that). Avoid starting a bullet with "The"—it's often unnecessary and cluttering. Be consistent with capital letters. If you have subbullets that are serving as different endings to a parent bullet phrase, those subbullets should start with lower-case letters.

- Bullets themselves: In PowerPoint, as with in most word processing programs, you can change what the bullets look like (squares, open circles, diamonds, or they can be changed to numbers or letters). You can also substitute little pictures. You can change the amount of space between the bullet and the start of the text. Notice in the Slide-o'-Badness that the space between the bullets and the text is inconsistent. In some cases, you don't need bullets at all. The slide on the right would "work" without bullets, as long as "allergic reactions" was staggered enough so that it did not appear as separate entries.

. . . and here we are only talking about the words. There are many other ways to make truly awful slides: you can use neon colors for either the slide background or the lettering itself (or both); you can needlessly embellish the font with drop shadowing or textured block letters; you can mix typefaces and fonts; you can animate the letters; you can create silly and useless (and distracting) content builds—with bullet points dropping in from the top, then sliding in from the right, and then spinning up from the bottom. You can pick a slide design that is gratuitously silly, which also detracts from content. If all of your slides have a ghosted tattooed arm in the background, your audience is going to be wildly distracted. Best of all, you can add sounds, like cash register noises, beeps, chirps, trumpets, or alarm clocks. If your slides come across as having been created by an over-caffeinated honey badger, you'll have the attention of your audience, but for all the wrong reasons.

3. Clip art. Don't. *Ever.*
4. Good and bad graphics. Graphics are a powerful thing. A graph or table can tidily capture all of the data in what would otherwise be an horrifically overloaded bulleted slide. Your audience doesn't have to read anything; they get an instant assessment or an instant comparison. The old adage "a picture is worth a thousand words" is not only true; it is of particular importance in a slide presentation, where your audience (depending upon the speaker and the complexity of the slide) may have only 30 seconds to two minutes to digest content. You don't want them reading paragraphs or long lists when you also need them to be listening to you. A slide with a bulleted listing of nine meta-analyses assessing tattoo adverse events (AEs) is either (a) going to be too boring to read or (b) will be read at the expense of the audience paying attention to what the speaker is saying. A tidy list is better. Better yet, use photographs of some of the AEs. There's nothing like a wall-sized image of a gruesome keloid to capture the attention of one's audience.

There are, of course, good and bad graphics. Any clip art, no matter what the reason (with one exception: if you're giving a presentation ridiculing clip art and are showing examples), is bad, awful, and tells your audience that (a) you're a neo and (b) you don't have enough information for the slide. Confusing illustrations or blurry photographs will distract your audience. An illustration that more properly goes on another slide is also distracting. Using images that are too small or too large or stuffed onto a slide that is already overloaded with text is also a mistake. Another common error is to take a large graphic that you want to use and reduce it in size enough to make it fit on a slide, or to take *several* images and reduce them in size so that you can fit them all onto a *single* slide. If you have a table of tattoo AEs, eight of which lend themselves to photography, don't stuff a table and eight photos onto the slide. *It is important to review your PowerPoint deck in "slide show" mode.* It will give you a sense of what your slides are going to look like to your audience when you present, and your slides in "slide show" mode will look different than they do in "normal" mode.

A good rule of thumb about using graphics is: whenever you can. This, however, refers to *good* graphics (tables, graphs, illustrations, photographs, etc.) that contribute to the *content* of the slide. For example, if you have a slide on the tricarboxylic acid cycle, put an illustration of the cycle on the slide, not a bulleted list of steps. If you're talking about medical imaging, such as x-rays or MRIs, use those images. If you've done some fluorescence microscopy, include the micrographs. When discussing tattoos, and you're talking about ink deposition, instead of bullet points or numbers enumerating the steps of the process, you can have an illustration (or even a video) of the needle penetrating the skin into the dermis and depositing ink. The subject of tattoos lends itself well to graphics: pictures of tattoos, pictures and illustrations of the process, micrographs of ink in situ, and photographs of bad AEs. Just about everything in the "skin" category lends itself well to graphics, and this is why your professor chose that unit to hand over to the students for a presentation exercise. What you don't want to do (and this is an error beginners frequently make), is to use any of the following:

5. Gratuitous graphics. While it's true that you should use graphics "whenever you can," those graphics should be *relevant to* and *support the content of the slide*. Gratuitous graphics (and we're not even talking about clip art, here) are often used as filler if there is not enough information on slide or if the creator of the slide deck is worried that the presentation doesn't have enough, well, graphics. These are individuals who have a picture of a liver in a slide presentation on the HMG-CoA biochemical pathways in the liver (um, we know it's in the liver) or a presentation on neurons that has a cutaway diagram showing the location of the brain. When you start off your presentation on tattooing, you do not have a front and back illustration of a person with a label titled "skin," complete with a leader line indicating the skin. You wouldn't have an exploded and

labeled diagram of a tattooing needle apparatus, nor pictures of the different types of tattoo art individuals can get. One or two humorous slides (e.g., a misspelled tattoo) might be called for in a peer classroom environment, but this should not comprise half your slides.

If you find yourself faced with a slide that has too little information on it, and you are contemplating adding a graphic—any kind—to fill in the space, *don't*. Rethink it. Can that too-little text be combined with another slide? Can it be replaced completely with a table or graph or illustration? Everything on a slide needs to "go" together, and gratuitous graphics don't "go." They distract.

6. Tables and graphs. Tables and graphs are neat ways to show data and preferable alternatives to a slide stuffed with unwieldy bullet points. If you're slogging through a narrative in a source paper and there is a string of data embedded within the narrative, look at what's in there and see if it would be better represented by a table or a graph. Is there is a list of clinical trials, a narrative including the brand and generic names of drugs along with FDA approval dates? That would go better in a table. a list of different kinds of tattoo inks and the most common AEs associated with each? Put it in a table. Don't, however, get carried away with your tables. While they are arguably a nice neat, clean way to package information, too much information in a table is disastrous in a presentation. When you find yourself decreasing the font size to 9 point, consider how visible this is going to be from the back of the room (hint: not very).

A graph, however, can stuff a ton of data into a snapshot—a snapshot that will be better retained and remembered by the audience after the presentation. Perhaps you're up to the "tattoo regret" part of your presentation, and you have a bunch of cool data on the shifting norms of tattoo removal. You're using a source paper that at least has the data in a table, so you don't have to pull them out of a narrative, but don't use a table (or bullets) if you can turn it into a picture (caveat: these are not real data).

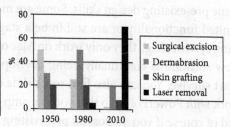

Tattoo Removal Techniques by Year, %			
	Surgical excision	Derma-brasion	Skin grafting	Laser removal
1950	50	30	20	0
1980	25	50	20	5
2010	2	20	8	70

Which of these graphics would be better for a slide? The graph. It's a faster, more visual snapshot of the data. It will also be remembered by more people for a longer period of time than the table with data. Further, this is the kind of slide to which you can gesture, saying. "as you can see, once laser tattoo removal become available, use of the other techniques practically disappeared." Your audience won't be trying to read a table while trying to pay attention your words.

As with gratuitous images, though, take caution not to create gratuitous graphs. A graph is a good tool for *comparing things*, usually discrete bunches of things (in the case of a bar graph) or a progression of things (in the case of a line graph), or a cluster of things (in the case of a scatterplot) or multidimensional things (in the case of a 3D scatterplot or radar graph) or intersection of things (in the case of a Venn diagram). Some data just don't go in a graph or chart or diagram; they should stay in a table or in bullet points. You wouldn't use a pie graph showing the percentage of males and percentage of females who get tattoos; that's a bullet point. You wouldn't create a bar graph of all the AEs that can be associated with tattooing and rank them by percentage. Why? Remember, the graph or diagram has to contribute to the content of the slide. It should *contribute*. A gratuitous two-slice pie chart contributes nothing. Gratuitous graphics are immediately recognized as such by your audience, which is thinking "why did he do that?"

PowerPoint Alternative

Do you *have* to use PowerPoint? You almost never have a choice, since, as we mentioned earlier, everyone is using it everywhere for everything all the time at present. However, there *are* other options. There are some powerfully creative non-PowerPoint slide design options out there, such as Prezi or Keynote, among others. You can create some truly engaging and breathtaking slides with these. There are pros and cons to each of these, largely related to learning curve and expense. Many of these have free trial versions (or are completely free, like Google Slides), but others are expensive. Some require a lot of time, patience, and practice to learn. Some are more or less user-friendly or intuitive, and some require some pre-existing design skills. Some are more reliable than others; some have limited functionality or are still in beta stages, or are limited to specific operating systems (e.g., they only work on Mac or Window machines but not both). New software is continually being developed, and of course some programs don't survive. You can also find some slide design software and templates that work *with* PowerPoint, so that you can amp up your PowerPoint presentations, and of course if you *do* possess pre-existing design skills and software (such as

Adobe Photoshop or Adobe Illustrator), all the better. If you want to venture outside of PowerPoint, do your homework. Go online and research "PowerPoint alternatives," and consider what you can and cannot handle and, importantly, if you'll be able to use the alternative at all within the constraints of your delivery options. Want some creative advice? Visit Presentationzen.com, a positively delightful site for lucid and entertaining presentation philosophies (e.g., Darth Vader would make seriously bad PowerPoint slides; Yoda would go analog and use a flip chart). There are a staggering number of websites providing slide design and presentation design and advice, and some of them are excellent. Keep in mind, though, that if you are collaborating and sharing files, or bringing a thumb drive of your presentation to another location, that not all of your collaborators or presentation destinations will have anything but PowerPoint capability, and your efforts will be for naught.

Know how and where to find images. A Google Image search is all well and good, but there are plenty of sources of free images (often from academic institutions, museums, the government, Wikimedia Commons, or some stock image houses). And don't forget about copyright! You can snag an image of a gruesome tattoo gone wrong from Google for your seminar presentation at school, but you can't do that at a medical meeting for a talk or a poster.

Do I Have to Use Slides at All?

No. Not every presentation calls for the use of slides. Edward Tufte advocates the use of posters and written materials, and waxes eloquent on the large, involved diagrams or maps to which he refers as "supergraphics" (visit Tufte's web page to see examples of these); these would be impossible to display on a slide (although arguably, you could do it on some of the "poster format" PowerPoint alternatives like Prezi). Tufte believes a great deal is sacrificed by pigeonholing a presentation into the limited options afforded by PowerPoint and that this sacrifice is manifested in the decreased quality of the information and its presentation. He is not alone in his opinions.

You do not need slides for pictures. Anatomy and physiology, for example, is a spectacularly visual scientific field. Depending upon the size of your audience, you can pass around casts of body parts, or give an entire presentation standing next to a skeleton, using it to illustrate all of the main points of a lecture on the skeletal system. In presentations that are interactive, it is often better to employ a medium that allows for real-time input, such as a white board, chalk board, or flip chart—or interactive online games (there are lots of sites where you can create puzzles, quizzes, and Jeopardy-like games).

Prior to April 27, 1989, science got along just fine without PowerPoint. Back then, "presentation" meant "a speaker in the front of the room." We have a few thousand years of some fabulous science under our collective scientific belts, disseminated by lecturers who held up skulls, performed experiments at lab benches in the front of the room, dissected cadavers, blew things up, built things, and pointed at plants—no whiteboards, no PowerPoint, no flip charts . . . just a guy in an amphitheater in a crisp white lab coat. Obviously, that worked pretty well. For a really long time.

---

10 Tips for Creating Effective Slides

1. Determine target first; it dictates slide content.
2. Choose typefaces, fonts, and colors that do not distract; eschew "special effects."
3. Craft engaging slide titles (e.g., question, statement, or phrase).
4. Try to stay within template defaults to ensure consistency.
5. Use a combination of slide content (e.g., bullets, tables, graphs, illustrations).
6. Do not use clip art or gratuitous graphics.
7. Do not overcrowd slides; capture a single concept into a digestible "snapshot."
8. Include references on the slides where they are mentioned; do not put them all on a final slide; do not number references sequentially through the deck; start with "1" on each slide.
9. Proofread repeatedly, and use "slide show" mode for final review; ensure consistency in wording, style, numbering, bullets, numerals, ordinals, spelling and statistics. Peer review is critical at this stage!
10. Ensure that the slide content does not interfere with, overlap, or cut off slide template components (e.g., university seal, etc.).

Presenting With a Single Slide: Meeting Posters

Remember, "presenting" just means "to show stuff." It doesn't mean you need to be standing up in the front of the room, using PowerPoint (or something else). A scientific poster is a well-crafted condensation of material into what is essentially a single snapshot. You may have a poster at your university's Research Day. You may be presenting a poster at a local chapter of your professional society, or the PI of the lab in which you're doing some undergrad research may tap you to compose the poster for the Next Big Meeting. Meeting posters are rarely comprised of "review" type material; posters almost always contain new information. It might be the results of a scientific study, or it might just be *interim* results of a scientific study.

Many of the same presentation rules apply for posters, in that they should be engaging, should not consist entirely of text, should contain no gratuitous graphics, and—this is often the tricky part—should not be overcrowded. There are two significant differences, however, between slide presentations and meeting posters:

1. *A poster has to be a stand-alone creation.*
2. *A poster does not have a captive audience.*

This means an individual looking at your poster (assuming the individual is in your target audience) should be able to understand it without a presenter's play-by-play or explanation. True, poster presenters are often on hand to discuss their posters, but they should not be there to *explain* them. And at large scientific meetings, you won't always necessarily be there when your poster is on display. In the context of not having a captive audience with a poster, consider a poster session at a meeting. Depending on how many posters there are, a meeting attendee is going to be cruising along the poster floor at a good clip, trying to get in what she wants to see and still have time for lunch before the afternoon sessions start. A student session with a dozen or two posters might earn a thoughtful stop at each one, but a national meeting with hundreds of posters will not. The big meeting poster-cruiser is going to look for something that (a) is important information she needs or (b) looks really cool. At the AAD meeting, a poster titled "Allergic reactions associated with tattooing" is not going to get any takers. "Dibutyl phthalate and contact dermatitis associated with soot-based black tattoo ink," however, is more of an engaging and informative snapshot. "Ooh!" says the poster-cruiser, "that!" Poster-cruiser happens to be a dibutyl phthalate aficionado, so you've got her eye, but as she winds her way through the crowded poster floor and finds herself nearing four columns of text on your poster, she veers off toward the polycyclic aromatic hydrocarbon tattoo ink poster next to yours, which instead looks like a page out of a graphic novel. The guy next to you has a poster with color on it (but subtle and not distracting), has text organized into very short paragraphs and succinct, informative bullets, has two photographs of tattoos, and one graph. Everything is neatly labeled, none of the graphics have long captions, and he has a max of two references. You stare at yours in contrast, a poor competitor that looks more like the opening chapter to *Moby-Dick*, if Moby were swimming around in dibutyl phthalate.

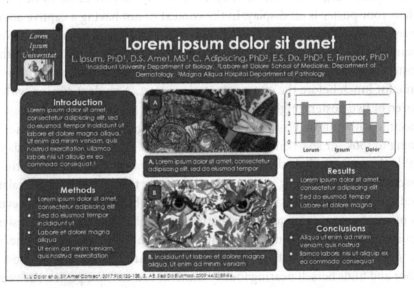

Lorem ipsum dolor sit amet

Lorem ipsum dolor sit amet, consectetur adipiscing elit, sed do eiusmod tempor incididunt ut labore et dolore magna aliqua. Ut enim ad minim veniam, quis nostrud exercitation ullamco laboris nisi ut aliquip ex ea commodo consequat. Duis aute irure dolor in reprehenderit in voluptate velit esse cillum dolore eu fugiat nulla pariatur. Excepteur sint occaecat cupidatat non proident, sunt in culpa qui officia deserunt mollit anim id est laborum. "Lorem ipsum dolor sit amet, consectetur adipiscing elit, sed do eiusmod tempor incididunt ut labore et dolore magna aliqua. Ut enim ad minim

veniam, quis nostrud exercitation ullamco laboris nisi ut aliquip ex ea commodo consequat. Duis aute irure dolor in reprehenderit in voluptate velit esse cillum dolore eu fugiat nulla pariatur. Excepteur sint occaecat cupidatat non proident, sunt in culpa qui officia deserunt mollit anim id est laborum. Lorem ipsum dolor sit amet, consectetur adipiscing elit, sed do eiusmod tempor incididunt ut labore et dolore magna aliqua. Ut enim ad minim veniam, quis nostrud exercitation ullamco laboris nisi ut aliquip ex ea commodo consequat. Duis aute irure dolor in reprehenderit in

voluptate velit esse cillum dolore eu fugiat nulla pariatur. Excepteur sint occaecat cupidatat non proident, sunt in culpa qui officia deserunt mollit anim id est laborum. "Lorem ipsum dolor sit amet, consectetur adipiscing elit, sed do eiusmod tempor incididunt ut labore et dolore magna aliqua. Ut enim ad minim veniam, quis nostrud exercitation ullamco laboris nisi ut aliquip ex ea commodo consequat. Duis aute irure dolor in reprehenderit in voluptate velit esse cillum dolore eu fugiat nulla pariatur. Excepteur sint occaecat cupidatat non proident, sunt in

culpa qui officia deserunt mollit anim id est laborum. Lorem ipsum dolor sit amet, consectetur adipiscing elit, sed do eiusmod tempor incididunt ut labore et dolore magna aliqua. Ut enim ad minim veniam, quis nostrud exercitation ullamco laboris nisi ut aliquip ex ea commodo consequat. Duis aute irure dolor in reprehenderit in voluptate velit esse cillum dolore eu fugiat nulla pariatur. Excepteur sint occaecat cupidatat non proident, sunt in culpa qui officia deserunt mollit anim id est laborum. Lorem ipsum dolor sit amet, consectetur adipiscing elit, sed do eiusmod tempor.

You (lonely).

The dude next to you (thronged by admirers).

But how do you *do* this? How do you *start*? One of the biggest mistakes neophyte poster-creators do is to try to take what would essentially be a stand-alone IMRAD paper, put it onto a single slide, and then blow it up to poster size. You'll see those, we guarantee it. It's okay to think about the poster content in IMRAD terms: introduction, methods, results, and discussion. Obviously, not every study in the world can be characterized this way, but they can all *mostly* be characterized this way if they're experimental. Everything has a background. And if you're doing any work, you have a rationale or an hypothesis, if it's experimental. If you're doing some kind of study—any kind—you have a method. Whatever you're *doing*, that's the method. And of course you have the result of whatever you're doing, whether it's an experimental result or an observation. And anything can have a summary or discussion. How do you fit all of this onto a poster?

- Introduction/background: four to six sentences, no more than one or two references.
- Methods: bare necessities; put it in a (nongratuitous) diagram or table if possible.
- Results: *three-quarters of your poster.* That's why you have a poster. You have something new you want to tell people about. Here's where you make it shine with diagrams, tables, illustrations, photographs, and graphs. If you can get away it, try not to have any text in this section at all apart from legends for graphics. Choose only the important, compelling visuals you need—*samples* of your results, not *all* of them. Do *not* use too many; do *not* sandwich in an excessive number of charts and photos and diagrams. Just enough to make your key points. If you are doing a large study that has multiple endpoints or foci, consider a poster covering only one aspect (e.g., *only* ink reactions, or *only* "tattoo regret" or *only* scarring and keloids, etc., from a larger study looking at all tattoo AEs).
- Discussion/summary/conclusions: as with the introduction, no more than four to six sentences, and if you can bullet your points intelligibly, do it that way instead ("This exploratory study showed 1 …, 2 …, 3 …, 4 … ").
- Save room for: informative poster title, poster authors with affiliations, and references (no more than five, preferably two to three). You may need to add acknowledgments (especially if you're using someone's artwork or photo), a university logo (if required by the school or the department), or acknowledgment of a grantor with a grant number.

Notice that it has all the similar *elements* of an IMRAD paper (or IMRAD-style study or experiment), but you have not, in fact, cut and pasted an entire IMRAD paper and stuffed it onto a poster. You're capturing the main points that get you *first* your 5-second audience (the engaging color and graphics that attract the attention of the person who sees it while wandering around the poster session); *then* your 10-second audience (slows down to read the title to see if she's interested in the topic); *then* your 20-second audience (skims the graphics and reads the summary or conclusions); *then* your 60-second audience (reads the poster); and *then* your minute-plus audience, who finally looks up and makes eye contact with you, inviting the much-desired opening from you: "Hi! Can I answer any questions?" Next thing you know, you're a tenured professor.

10 Tips for Creating Effective Meeting Posters

1. Do not try to stuff an entire IMRAD paper onto a poster.
2. The poster should contain the full title and all authors with affiliations (and university logo, if that's standard for your school).
3. Introduction or background: four to six sentences
4. Methods: enough to get by. Use a diagram instead if possible.
5. Results: most of the poster. Try not to use any words; use tables, graphs, illustrations, or photographs. A mix of these is even better.
6. Discussion/Summary/Conclusion: four to six sentences, bulleted is even better.
7. Keep extra materials limited: references (three or fewer), acknowledgments, grantor.
8. Make each section of the poster easily identifiable (e.g., with larger, bolded headings, or within different boxes or blocks of color).
9. Do not overcrowd poster with text and images. If it takes longer than one to two minutes to read, it's too much.
10. Do not use blinding or clashing colors, patterned backgrounds, gratuitous images or diagrams, or clip art.

©2020 Max Beck.

What If I'm the One Doing the Presenting?

If you're giving a talk in your class or a talk at a meeting, showing a poster at a meeting, or speaking on a webinar or on Zoom—all of this is "presenting." "Stage fright" is out of the purview of this book. Your freshman speech communication classes should come in handy right about now (especially for overcoming stage fright). If, however, you are really prepared, that'll at least attenuate the fear, if only a little bit. Know your stuff, and practice, practice, practice. Even for a poster.

This Again: Know Your Target Audience Cold

By now, you're probably pretty tired of hearing about "target audience." The reason this concept is so tiresomely iterated is because you need to know your target before you attempt any communication at all. With regard to target audience, however, presenting offers both some unique challenges and benefits. If you're off target in a writing piece, you won't necessarily know. If you're off target when you're presenting, you'll know right away. Therein lies the challenge *and* the benefit. Speaking face to face has an unspoken promise: "I'm here, and I'm going to do a good job telling you about this." If you're not doing a good job, you'll be able to tell right away: an audience that is bored (texting or even sleeping), shocked, disgusted, laughing at your typos, or confused comes across pretty quickly, and it can be unnerving. If some invisible person is reading your paper and is bored, you'll never know. If that person is in the front row of your talk, it's kind of easy to tell. That's the challenge. The benefit is much of the same thing; you're getting instant feedback via the body language of your audience (or maybe even interruptions and questions), and it will improve the subsequent presentations you make (i.e., it's experience). The most useful body language feedback you can get is the quizzical, confused look on the faces of some or all of the people in your audience. That helps you alter your presentation on the spot; you can tell where to linger and explain, and where you can trot to catch up. You can even interact with the audience: "Okay, I can see a bunch of confused faces out there . . . want me to go over this again?" or "Whoa—where'd I lose you?" One or two of the less-timid members in your audience will speak up ("Why are you showing a spleen in a talk on tattoos?").

Content level, vocabulary, diction, sophistication, and formality should be dictated by your audience and your venue. Reflect on what they *already know*, what they *don't know*, what they *need to know*, and what they *want to know*. If you're talking to elementary school kids about tattoos, you'd be in jeans and wearing tattoo sleeves, your slides very simple, few words, no tables, no data, no chemicals. No close-up illustrations or photos of a tattoo needle embedded in the skin depositing ink, no close-ups of gruesome AEs, autopsy pictures, or embedded video of skin-grafting surgery. The little tykes will not be interested in Snack Time after that kind of show, and in all likelihood, you will not be invited back. Jeans plus tattoo sleeves plus gruesome photos would probably go over well with your peers (and Dr. Atkins, as long as you covered the necessary biology), but you'll be using more sophisticated vocabulary on your slides, and it will have tables and graphs. In the Student Forum at the AAD, you'll be in a suit with positively *no* tattoo sleeves, and gratuitous gruesomeness, while cool for your peers, would be regarded as gratuitous sensationalism by the AAD attendees, unless your particular topic was about something gruesome. At the AAD meeting, you'll have more staid slides—engaging, informative, interesting, and well constructed—but not "entertaining," not that you won't see those at meetings from time to time. As with other forms of communication, your audience and topic determine what you put on your slides. Once you're confident you understand your audience, then you had better make sure that you really know your stuff.

Know Your Content Cold

Your audience might determine the level of sophistication of your presentation, but regardless of target, you must know your topic inside out and backwards. You cannot possibly explain your slides or spin a fluid narrative if you don't understand your content well. The verbal tics or pauses while you frown at your slides and notes trying to remember what you were going to say are going to be picked up by the audience, and they will correctly interpret them as a lack of confidence and authority on your part. You cannot create a slide presentation by whipping the whole thing up in the morning and presenting it that afternoon. And, much as you think you know and understand your topic, when you practice your presentation (which you should do several times), you'll find you're stumbling when you get to slide 4 because you can't remember if it's the dibutyl phthalate or the polycyclic aromatic hydrocarbon that's associated with the higher incidence of dermatitis, because all you have on the slide is a list of six different tattoo inks. And the *slide* is fine. That's the kind of slide that serves as your background for your speaking; you talk to the slide, expound upon its content. But you *really* have to know that content. When you practice your presentation, and know you know your content, you won't stumble, trip, or fall. Another reason you'll want to know your content cold is because after you breeze triumphantly through your last slide, on time, to the applause of your audience, some upstart members of said audience will have the temerity to ask you questions. You can't practice the questions, and you can't study for them. You have no idea what they're going to be. Some might ask for clarification (or more information) about one of the points you made; some might ask you about other related studies or research; some will ask you things that border on the esoteric; and some "questions" are nothing more than egotistical pontificating. If you know your stuff cold, you will be able to answer most of the questions, or at least most of most of the questions. If they're concrete questions about details and you don't know the answer, be candid about not knowing and offer to find the answer for them. That's completely acceptable. For the esoteric ones (either legitimate or a manifestation of the pompousness of the questioner), counter with "what a fascinating concept," or take her esoteric question, super-size it, and lob it right back at her. If you know your stuff, you will be able to maintain your self-confidence throughout the post-presentation inquisition. If you are fortunate enough to have a volunteer listen to your practice runs, you can get a preview of audience confusion or questions. This is also useful peer review.

So you've nailed your target audience, and you know your content cold, now what do you do? "Always make the audience suffer as much as possible." —Alfred Hitchcock (who probably would have had a *lot* of fun with PowerPoint)

Make Flawless Materials

Ensure that you've followed the guidelines for effective slide construction: economy of word, combination of words and graphics, comprehensible typefaces, fonts and colors that do not distract the audience, good graphics that contribute to slide content, ditching gratuitous content and animations, and eschewing the vile and unprofessional clip art. Ensure consistency from slide to slide, try not to vary font sizes (and don't vary typeface at all), and try not to vary positioning of bullets and subbullets. Have no hanging bullet points ("orphans") or full sentences for bullets. If you are using references, ensure that you start from "1" on each slide (do not start with "1" on the first slide and use them in sequence through to the last slide). If you use a reference on a slide, include the reference at the bottom of the slide. Do not have a final slide containing all of the references. Determine the style you're using: APA? AMA? ACS? Are you being consistent in your referencing style? Ensure that you are.

It's very important, once you've constructed your presentation, to review it *as a slide show*. When you're in slide construction mode, things might look okay. But then when you look at it in slide show mode, you'll see what's wrong—typically an inconsistency (e.g., alignment, font size, typeface, or animation). You'll see where images overlap, where things are falling off the slide or where things simply don't show up. You'll notice you're missing references, or that you copy and pasted something without removing the original, or that you have an extra slide, or that a slide is hidden. Those cool animations you had embedded on your slides? Oops. . . they don't work in slide show mode—oh wait, neither do the audio clips or the interactive software! You're going to have to fix that! If the slide template you're using already has graphics embedded in it (e.g., the crest or seal of your university), ensure that you haven't anything overlapping it or cutting part of it off—that looks sloppy (nudging graphics around or resizing them to avoid this is annoying, but you should do it anyway). You'll catch typos (*use your spell-check*), rogue bullets, extra blank lines or tabbed text, inconsistencies in your titles (are they all sentence case, left-justified?), inconsistencies in your bullets (do they all start with lower case?).

There are also formatting perils. If you're working on a slide deck with several people, that deck could get reformatted somewhere along the line—especially if you're working in more than one country. In PowerPoint, hidden away under the "design" tab, over on the right, you'll see a button for "slide size." You have two options there, "standard" or "widescreen." If you create a PowerPoint presentation in standard size, it'll look peculiarly small if the slide format changes to "widescreen," and if you create slides in "widescreen" they'll be wildly unformatted in "standard" view. Understand in advance which slide size you're going to use, remembering that it might be an external dictate (e.g., by your prof or the meeting where you're going to present it, etc.). If you create it in wide screen and then send it off to a collaborator who is innocently using "standard," your collaborator might "fix it," rendering it inappropriate for wide screen. Ensure that anyone working on or reviewing the deck looks at it using the right size! This is something you will also catch in slide show mode, but it'll also be apparent in normal view. Go through the slide show slowly, and then go through it quickly, focusing on the "big picture" (that's where you'll notice inconsistencies from slide to slide). If you find something wrong (or inconsistent—something perhaps spell-check didn't catch), go back through your deck or use a string search to make sure you've changed *everything* to make things consistent (e.g., were you using "4th" in one place, "4th" in another, and "fourth" in yet another?). All of these finishing touches contribute to the overall perception of the quality of your presentation; a presentation that is well-written to target will come off as an *excellent* presentation if it *looks* excellent. It gets noticed. Get that reputation. Do not forget the invaluable necessity of peer review. If you have no collaborators on your slide deck, ensure that you send it to a peer (or more than one, if you can) to review it. You'll be surprised at what the "fresh eyes" pick up. It is humbling (and very educational) when a peer reviewer lovingly picks up a typo you had in a slide title.

All of these same rules apply for the construction of a meeting poster too, although glitches in consistency are easier to see because everything is in one place. The thing about a poster, though, is that it's static. It sits right there for two or more hours, with way more opportunity for scrutiny by the meeting hordes. While you're standing about waiting politely to talk to one of your poster visitors, that's not when you want to notice that all of your figure legends are in Arial while the rest of the poster is in Times New Roman, nor is it when you want to notice that half of your P values are written correctly (i.e., with no zero to the left of the decimal) and half of them are not, or that you have incorrectly used lower case for "Student's t-test." The casual peruser might notice or not, but *someone* will. That nice buddy of yours who peer-reviewed your slide deck? Let him review your poster too.

Okay, so you've nailed your target, you really know your stuff, and you've made flawless materials. Now what?

Know How to Present and How Not to Present

Here's where the paralyzed-by-public-speaking comes in. Some people are; some people aren't. But you already *have* done public speaking. If you've ever raised your hand in class (or got picked on) and answered a question, there you go. Ever worked in a team? Gone to a lab meeting? Told a joke to your Greek brethren? Start slow, if you think you're nervous. Practice on friends. Practice on a few people in class. Go ahead, volunteer for Story Time at the public library. By preparing carefully for this, you've eliminated many of the things that scare people about speaking:

- You know to whom you're talking. You know you won't be above or below their comprehension.
- You know your stuff. Are you afraid you'll forget your stuff? Turn around and look back at your slide. You have a giant crib note sitting right in front of you.

- You *really* know your stuff. If you're giving a presentation, it's usually on a pretty specific topic, so bear in mind, that no matter how smart the people are in the audience, you know the stuff in your presentation better than anyone. For 30 minutes, you're the expert, and you're ready for them.
- Your slides (or poster) are *gorgeous*. That captures respect immediately.
- You've practiced. *Repeatedly*. And then some more. You have your timing down, you know your slides, you know the whole talk flows. You're ready.

Dress your part, as mentioned; tattoo sleeves for your peers, suit for Student Forum at the AAD. Dress comfortably and in layers; if the room is mercilessly hot, you can take off your jacket, if it's meat-locker cold, you can put it back on. If you're going to be videotaped, avoid any patterned fabrics whatsoever. If you can't wear heels without waddling around like an epileptic duck, wear flats. Are you wearing stockings? Good thing you have that extra pair in your bag when you snag the ones you're wearing nine minutes before you're supposed to go up onto the podium.

You've made engaging slides, now you have to be an engaging speaker. Try never to bring notes; or if you do, use them sparingly. If you've practiced your talk enough you shouldn't need them. Remember, your slides are your crib notes. Do not stand at a lectern and read your notes, not verbatim, not even partially. Don't look down at the lectern; look up at your audience, turning from pointing at something on your slide to back to your audience. *Make eye contact* with multiple people in the audience throughout your talk. Do not stand still. Be animated. Use hand and arm gestures at the very least; walk around if you possibly can. Your mobility is going to be limited by the audio setup, if there is any. If you have a lavaliere mike (the kind of microphone that clips to your shirt), you can pace around like a motivational speaker. Laser pointers have their uses (particularly in anatomy), but they should not be twirled around on a slide gratuitously, detracting from the content and inducing seizures in susceptible audience members.

Take-home

Knowing how to create an effective, engaging presentation is a critical skill, even if your long-term plan is to hide out in the darkest recesses of a lab for the rest of your life. Presentations require the same kinds of skills that other types of writing require (e.g., accurate identification of target audience, knowing your content, and knowing what to talk about), but with the added challenges of requiring a larger variety of "communication" skills and the fact that you're delivering content face to face. While most writing projects follow a think-plan-write path, creating a good presentation requires an earlier step: "pre-communication planning" (taking into account medium and venue) and an extra final step: "communicating in person." It is a highly creative, effective, and satisfying form of scientific communication (that's why people go to scientific conferences!), and it adds an extra dimension to the dissemination of information.

Exercises

1. Using a paper from your field (or provided to you by your professor), choose *one to three informative sentences* from it and turn them into a single slide. If you want to try a tattoo example, use this: "delayed-type inflammatory reactions represent an uncommon adverse event to tattoo pigments. Different reaction patterns, such as eczematous, lichenoid, granulomatous and pseudolymphomatous reactions, have been previously reported, especially in association with metals contained in red tattoo pigments" (Garcovich S, et al. *Eur J Dermatol.* 2012;22(1):93–96.). What should you consider for this slide? How can the information in these two sentences be best represented in a single slide?

2. Look online for some PowerPoint presentations in your field (you can readily find PowerPoint presentations online by typing in a keyword, followed by a space and then ".ppt"—e.g., "calorimetry .ppt" or "lycosid spiders .ppt" or "string theory .ppt," etc.). Find a presentation you think is truly dreadful (this will not be difficult to do) and one you think is really good. Summarize why you think they're bad and good. If working in a group, an alternative is to collectively identify a good one and a bad one, and critique them together.

3. Find a short paper in your field and turn it into a 5- to 10-slide presentation. *You do not have to use PowerPoint* (although it in of itself is a pretty important skill to have) unless directed to do so by your professor. A review paper works best for this exercise. The presentation does not have to be delivered, only critiqued.

4. Find an IMRAD paper in your field—preferably limited to a small study or experiment, or single-endpoint trial—and turn it into a meeting poster. Remember: there is a right way and a wrong way to do this. Critique your posters in a group. An interesting exercise would be to have four or five people take the *same paper* and create posters, and then compare them.

5. Create and deliver a slide presentation from scratch on an assigned topic (it does not have to be in PowerPoint unless required by your professor). In this case, you would be doing your own research, using multiple sources.

13
Writing in the Sciences: Minor Genres

© 2020 Max Beck

It's late at night, in the dark and creepy microbiology lab. Your plates have to be streaked and incubated at specific times, which is why you're in the lab at 1:00 a.m. instead of back in your dorm room curled up in your jammies. No one is around, and you are doing the unthinkable: drinking Mountain Dew in the lab.[1] The can sits next to you on the bench, effervescing pleasantly while you streak your plates. You're bored and sleep-deprived. You drip some Mountain Dew onto an extra streaked plate just for chuckles. When you're done, the plates go into the incubator, and you stagger back to the dorm, hoping your roommate isn't playing Call of Duty on the Xbox.

---

[1] It *is* unthinkable. This example is a *fictitious* wanton disregard for how anyone should conduct themselves in a laboratory. It is wrong, wrong, wrong.

Two days later, you shuffle in and your advisor is staring at the plates you streaked a couple of nights ago, looking confused. Your joint efforts to culture *Pseudomonas* have recently been yielding some anemic results. Not sure what you've been doing wrong, but you haven't been getting a lot of growth. But one of the plates you streaked a couple of nights ago had an explosion of growth on it . . . fuzzy blotches fanning out across the surface and crawling up the side of the plate. "This," says your advisor, shaking his head, "is amazing. It must be some contaminant—only one dish has this in it. Do you remember what you did? Any idea? Did something fall on this plate?" Creeping horror sets in as you have a pretty good idea which of the plates was contaminated. You do your level best to feign wonder and not look remotely guilty. You are reasonably sure what would happen to you if you say "heck yeah—I was drinking Mountain Dew while I was streaking plates, got bored, and added some to one of my plates just to see what would happen." You look up toward the ceiling. "I don't know . . . dust or something? Maybe I messed up on that one."

You start thinking about it. Why would Mountain Dew—if that's what caused the growth—make *Psuedomonas aeruginosa* grow better?[2] You perk up excitedly and offer to try and figure it out, and your advisor beams at you for your enthusiasm. Ensuring that he can't see your computer from where he's sitting, you check PubMed using "Mountain Dew" and "*Pseudomonas*" as key words. You're bummed—but not particularly surprised—when nothing turns up. You Google it. You learn many fascinating obscure facts (some of them disturbing) about Mountain Dew, but disappointingly, nothing about how Mountain Dew makes *Pseudomonas* grow. Why did the growth explode like that? The *Pseudomonas* from both cultures look the same under the microscope: gram-negative rods.

You Google "Mountain Dew nutritional information." You know it's bad for you. High fructose corn syrup, a variety of preservatives, artificial colors, tons of caffeine, and a mysterious component called "brominated vegetable oil." You go through the list, combining keywords, on Google: "high fructose corn syrup + *Pseudomonas*," "sodium benzoate + *Pseudomonas*," "brominated vegetable oil + *Pseudomonas*." Nothing. And then: "caffeine + *Pseudomonas*." You get a bizarre hit: "Caffeine-Addicted Bacteria Die If You Give Them Decaf," an article in *Popular Science* (www.popsci.com, posted March 28, 2013). This caffeine-craving bacterium news cropped up everywhere from HuffPo to the National Institutes of Health to Medline.

But . . . since when? Your advisor has been working on *Pseudomonas aeruginosa* and its multiple resistances to antibiotics for years. He's published

---

[2] Note: Mountain Dew® is not approved for use as a culture medium.

numerous papers in public health journals on the dangers of *P. aeruginosa* as an infectious agent. You've read them all. You've been culturing *P. aeruginosa* for over a year. Why would . . . wait a minute . . . all of these articles are about *Pseudomonas putida*. *You've been streaking plates with the wrong bacterium.*

You have to come clean. You go to your advisor, tell him you've figured it out, probably. You "spin" your crime so as not to get skewered, sheepishly "admitting" that you had an *empty* Mountain Dew can that you hadn't thrown away yet, and that you remember getting some of it on your hands, and *maybe* it could have *possibly* dripped on a plate. He probably wasn't fooled but listened with keen interest as you explained how you tried to isolate the possible cause for the growth. He goes to check the ATCC[3] number on the vial of the *P. aeruginosa* he had ordered, and it was, indeed, *P. putida.*

He's impressed by your inductive reasoning (while also being honked off at your doing something ill-advised like drinking Mountain Dew in the lab). You wonder what would have happened if Pasteur had had access to Mountain Dew, you resolve to start drinking more water, and decide to start following OSHA[4] guidelines for laboratory practice.

Case Studies

Your serendipitous experience with the *Pseudomonas putida* and the way you tried to figure it out can be considered to be a case study. A "case study" typically involves an observation that is explored for the purposes of collecting more information—information that might serve as a topic for future exploration or a tool to help generate hypotheses. "Case study" can be a rather broadly interpreted term, meaning slightly different things in psychology, business, clinical medicine, physics, or other general topics. A case study in clinical medicine, for example, often consists of an evaluation of a single individual, the diagnostic process(es) used to identify a medical problem, and then consideration of treatment options. A physics case study may involve the assessment of a single lab technique with regards to efficacy, performance, or some other variable (or variables). Case studies can involve people, things, or measurements. Numbers are low (often single individuals or things), and the assessment of the observation is limited to the specific individual(s) being studied. They generate some observational, descriptive information, yet are not at the level of a formal experiment—no hypothesis, experimental design,

---

[3] American Type Culture Collection.
[4] Occupational Safety and Health Administration.

controls, comparators, or other formal components. Thankfully, no statistical calculations are involved. Your serendipitous discovery would more likely be reported as "anecdotal" (i.e., you didn't plan it), but it happened and you ran with it.

A case study, however, can provide fodder for *future* experimental work: you could take your *P. putida* observations and design an experiment assessing different culture media, different antibiotics, the interaction of antibiotics and caffeine, concentrations of caffeine, different sources of caffeine, caffeine plus other nutrients (such as high fructose corn syrup), or metabolic properties of the bacterium. You could design studies comparing *P. aeruginosa*, a common infectious pathogen, and *P. putida*, which occurs very rarely in humans (but frequently in the soil). You might have even stumbled over something you want to study in grad school.

Case studies are typically planned. Yours wasn't; it was a chance observation, *but* you explored it further. To be a true case study, you'd repeat what you did, to see if you got the same results. Thenceforth, new research (if you decide to go there) is generated. If you planned the case study, you write a case report. If you write up a "chance happening," that's typically referred to as an anecdotal report. Keep in mind, however, that different fields in science may use some of these terms interchangeably. In clinical medicine, a "surprise event" (typically a patient that shows up with unusual symptoms or ends up with an unusual or unexpected diagnosis) is frequently referred to as a "case study" or a "case report" in the literature, and often it varies by journal.

Writing Case Reports

When writing up a case study or anecdotal report, keep in mind that (a) you're only making an observation; (b) you should refrain from generalizing; and (c) you may suggest future research. In the scientific literature, case reports are very short—ranging from a single paragraph to maybe 2 pages, max. They may or may not be accompanied by an image. A wonderful case report from *The New England Journal of Medicine* in 2008 reported a 62-year-old man having come in through the emergency department with fatigue, fever, and cramping abdominal pain for seven days. Upon questioning by the clinicians about what he had recently eaten in the past seven days, he reported having eaten raw pond smelt. (Right about now "raw" should be triggering your index of suspicion.) All of his liver tests were normal (the case study provided all of the tests and their results). The only lab abnormality was an elevated white blood cell count. On a CT scan of his abdomen, the bile duct was dilated. Physicians were concerned that the patient might have acute cholangitis (inflammation of the entire bile duct system, usually as a consequence of bacterial infection), so they decided to go down and have a look. They stuck a camera down into the duodenum (earliest part of the intestinal tract, attached to the bottom of the stomach, where the bile duct empties into the intestines), and the bile duct opening was indeed suspicious looking. So they cannulated it (stuck a tube into the bile duct), after which "numerous leaf-shaped worms popped out (see video) and were subsequently identified as *Clonorchis sinensis* (also called *Opisthorchis sinensis*)." Clonorchis is a liver fluke, part of a group of flatworms, which includes all of the parasitic flatworms (like tapeworms) as well as many nonparasitic flatworms like the somewhat cuter planarian *Dugesia*. The case report went on to discuss the medications given to the patient (which killed off the liver flukes), who subsequently recovered quickly and uneventfully. The case report concluded with three sentences talking about the particular kind of presentation (clonorchiasis), what causes it, what the patients experience, how it's diagnosed, and the consequences of not resolving the infection. (Park DH, Son H-Y. *Chlonorchis sinensis. N Engl J Med*. 2008;358:e18).

The report—all 222 words of it—was one of the most interesting ones we have had the chance to read, and thoughtfully it included a video, not for the faint of heart. Yes, 222 words. Shorter than some journal abstracts!

> Want to see those adorable little liver flukes? Visit https://www.nejm.org/doi/full/10.1056/NEJMicm054461 for the full case report and video. Enjoy your sushi!

Writing Anecdotal Reports

An anecdotal report arises from an unplanned or accidental observation. Anecdotal reports are also common in the clinical medical literature. Your *Pseudomonas* experience could have been submitted as an anecdotal report, in which you'd simply say that Mountain Dew accidentally got onto a culture of what had been thought to be *P. aeruginosa*, but which turned out to be *P. putida*, a species of *Pseudomonas* highly responsive to caffeine as a nutrient. You could try to spin it as a deliberate action, but that would take some creativity and would be what we in science would call "lying."

Since anecdotal reports are about things that just "happened" in a way that had not been controlled for or planned by researchers, sometimes they are not considered to be acceptable references. However, if you're doing research on a specific topic, particularly one for which there is not much information, anecdotal reports might be all you have to go on—or largely what you have to go on. Further, as more and more case or anecdotal reports of a similar nature start to appear (e.g., as with a drug adverse event, community health issue, unanticipated lab findings, etc.), scientists sit up and take notice; obviously something's going on that requires more research. Further, case and anecdotal reports may themselves serve as fodder for formal case studies (or may evolve into them), whence comes the possibility for further research. You have to start somewhere, and science is very largely beholden to the "hey, look what just happened by accident" phenomenon.

Reports and Brief Communications

Reports or brief communications that turn up in journals can take a variety of formats; different journals have different specs for their "Brief Reports" section (if they have them at all). Some have the same format as formal IMRAD journal articles—they're just on a much smaller scale. These reports might be about progress thus far in a larger constellation of experiments in a study or may be small papers on small research projects (or projects of short duration). Some brief reports are on topical subjects; some read as news releases, others as literature reviews, case studies, techniques, procedures, or validated tools, survey, demographic, or epidemiologic results, or authoritative updates such as guidelines (e.g., from government organizations or professional medical societies). Essentially, the only thing these types of reports have in common is the "short" part. Mind you, "short" for some brief reports can number as high as 10 pages—while in the same journal an IMRAD or review paper might actually have fewer pages (see Table 13.1).

Some of these could easily qualify as full papers and not "brief reports," so you can see the wide variety of this genre depending on discipline. As an undergraduate student, you would be unlikely to be writing any of these by yourself, or even contributing to most of them. This, however, depends upon the capacity in which you find yourself working; if you're in a lab, you might be contributing to the short IMRAD-style report, or possibly conducting literature reviews, or working on data collection (e.g., from surveys or epidemiologic studies). You can also, of course, be required to write one for a class; in that case, you will need to rely on your instructor's advice and definition of report.

Table 13.1. Examples of "Brief Reports"

Format of "Brief Report"	Example
Short version of IMRAD	Willemetz A. Matriptase-2 is essential for hepcidin repression during fetal life and postnatal development in mice to maintain iron homeostasis. *Blood*. 2014;17;124(3):441–444.
Topical	Gibbs V, et al. Brief report: an exploratory study comparing diagnostic outcomes for autism spectrum disorders under DSM-IV-TR with the proposed DSM-5 revision. *J Autism Dev Disord*. 2012;42(8):1750–1756.
News	Green A. Ebola emergency meeting establishes new control centre. *Lancet*. 2014;12;384(9938):118.
Literature reviews	Wang JJ, Reimold SC. Chest pain resulting from histoplasmosis pericarditis: a brief report and review of the literature. *Cardiol Rev*. 2006;14(5):223–226.
Case studies or reports	Alg VS, et al. Isolated subacute tuberculous spinal epidural abscess of the cervical spine: a brief report of a special case. *Acta Neurochir (Wien)*. 2009;151(6):695–696.
Surveys	Markandaya M. The role of neurocritical care: a brief report on the survey results of neurosciences and critical care specialists. *Neurocrit Care*. 2012;16(1):72–81.
Techniques or tools	Ougrin D, Boege I. Brief report: the Self Harm Questionnaire: a new tool designed to improve identification of self harm in adolescents. *J Adolesc*. 2013;36(1):221–225.
Epidemiology	Cotrufo P. Brief report: psychological characteristics of less severe forms of eating disorders: an epidemiological study among 259 female adolescents. *J Adolesc*. 2005;28(1):147–154.
Authoritative reports	Lucey C. Brief report on the United States Food and Drug Administration Blood Products Advisory Committee recommendations for management of donors and units testing positive for hepatitis B virus DNA. *Vox Sang*. 2006;91(4):331–335.

---

10 Tips for Creating Case Reports and Brief Communications

1. Identify the nature of your topic. Characterize the event as anecdotal ("just happened") or planned.
2. Provide any kind of result, conclusion, or new hypothesis (i.e., open-ended reports are not typically useful) that may have resulted from the event.
3. State whether you repeated your experiment or not, or if you generated additional data based upon your original findings.
4. Identify specifics (e.g., culture medium, instruments, procedures, etc).
5. Condense verbiage to the basics—little to no background history.
6. Determine how quickly it needs to get to press (e.g., before or during a conference).

7. Follow dictates of publication instructions carefully; some may not permit case studies or anecdotal reports, or their definition of "brief report" might require specific components or have a word or character limit.

8. Get some internal peer review and *revise, revise, revise.*

9. Include graphics (e.g., a photo of a *P. putida* petri dish next to a photo of the *P. aeruginosa* dish) if you have some compelling ones and if they are permitted by your destination journal (or other medium).

10. Conclude with the importance of finding, context in the field (if possible) or your own research, and future directions, if applicable to the type of brief report you're writing.

Correspondence

Correspondence is very frequently published in scientific journals—often more than one per journal. These are letters in which a scientist most often addresses a paper published recently in the same journal. Typically, a scientist is correcting a misinterpretation of or defending a challenge to his or her original work. Other correspondence comes from unrelated scientists who want to augment (or dispute) the previously published paper with their own findings. Some pose questions about methods, results, or conclusions. Occasionally, correspondance might even be a version of a super-brief report, although letters typically refer back to an earlier issue of the journal in question. Correspondence can include more than one author; it is not uncommon for all of the authors of original research to appear together on a correspondence piece when responding to the recently published paper discussing their earlier research. Correspondence is typically very short, running from a single paragraph to possibly a full column, to even occasionally more than one column. The format pretty much exactly follows that of a formal letter. They usually open with a standard "we appreciate the thoughtful commentary and analysis provided by x, y, and z authors regarding our work on X." The unofficial protocol of correspondence is to be constructive, polite, specific, and appreciative throughout, but one can also occasionally see the "who-do-you-think-you-are" tone, even if couched in icy faux politesse. It is sometimes difficult to rein in your ungenerous thoughts when you see your work criticized in press by other scientists, so you should not fire off a Letter to the Editor while you are gripped in the blinding fury of a thousand suns. Wait a few days. It always helps to get your letter peer-reviewed by cooler heads before you fire it off.

In 2012, Matt Cartmill, a prominent anthropologist wrote "Primate origins, human origins, and the end of higher taxa" in *Evolutionary Anthropology* (21:208–220). Ian Tattersall, another prominent anthropologist (with significantly differing views), responded unfavorably to that paper later that year (21:221–223). In 2013 (22:172–173), Cartmill affectionately responded to *that* correspondence, whereupon Tattersall lovingly responded "Enough already," as the sole content of his letter (22:293). This is the kind of thing that makes being a journal editor so much fun.

White Papers, Guidances, Consensus Statements, and Recommendations

A white paper can be a difficult thing to attempt, because they are very, very rarely consistent in format and specific purpose. Essentially, it is a "background document," created to give readers a basic understanding of a specific topic. The "topic," however, can be anything and tends to vary widely in different fields (e.g., science, industry, business, etc.). This background document typically has a specific target audience. In business, it can be a description of goods or services offered and what they have to offer the target. In the pharmaceutical industry, it can be a document created to emphasize a gap that one of their company's drugs or devices can fill (or are filling). In science it can be clinical, procedural, conceptual, or philosophical. Ideally, the document is informative, persuasive, or both, and the reader can use the information to make informed decisions. For example, a white paper supplied to a grantor (or grantors) can be akin to part of a grant proposal and should hopefully contribute to the grantor's decision-making process about whether the research should be funded.

The white paper in science frequently reads like a substantial review paper and may or may not include recommended procedural information or algorithms or conceptual suggestions for guidances or policy. White papers in science are frequently generated by the combined efforts of experts who have assembled for the specific purpose of creating recommended guidelines, consensus statements, or position statements. The "combined efforts" often are the result of a convened advisory board or committee. And just to keep you on your toes, the document generated might not even be called a "white paper"; it might be a "consensus statement" or "recommendations [or report] [from the Joint National Committee of Something]."

Given the authoritative nature of the white paper, students will not be writing them, and probably not even contributing to them. In the rare instance that you find yourself working with an advisor who is so inclined, you might be able to do some backround research or collect data. If you are involved with smaller scientific endeavors (say, as an intern at a small nonprofit health or nature organization), you might be called upon to draft documents such as these.

All this being said, however, the white paper (or "consensus statement," or "joint commision statement," or "guidance," etc.) is an important *resource* for your writing and research. The most recent statement available often reflects the current thinking in the field or the "gold standard" for a process or concept in a particular category. If one exists for your area of research, it should certainly be consulted. White papers (if they have references; most of the ones in science do) are also an extraordinarily good source for "classic" references—the most well-known, most-cited, original papers in the field—many of which date back decades. Some white papers generated in industry, however, may be proprietary (i.e., not available to the general public). Given the reason for many white papers (i.e., to persuade or inform about products or services), many require registration prior to downloading. While this is often free, it may only be available to members of certain societies. Further, given the purpose of a white paper, while they are supposed to be completely objective, there may be subtle bias or an agenda. In science, guidelines (and guideline-related publications) tend to be free, given the importance of the content to an entire field. Even if it is a journal that typically charges a fee to download a paper, guideline documents are usually free. White papers in different industries may look very different (see Table 13.2).

Table 13.2. White Papers in Different Industries and Scientific Fields

Industry/Field	Example (most white papers in general are PDFs)
Microbiology	Microbiological testing of fresh produce: A white paper on considerations in developing and using microbiological sampling and testing procedures if used as part of a food safety program for fresh fruit and vegetable products (2010). http://www.unitedfresh.org/assets/food_safety/MicroWhite%20Paper-%20Final.pdf
Clinical	American College of Radiology white paper on radiation dose in medicine (2007). http://www.acr.org/~/media/ACR/Documents/PDF/QualitySafety/Radiation%20Safety/WhitePaperRadiationDose.pdf
Public health	Disaster preparedness white paper for community/public health nursing educators (2008). http://www.achne.org/files/public/DisasterPreparednessWhitePaper.pdf
Electrical engineering	White Paper: Underwriters Laboratory implemented UL508A supplement SB to require compliance by UL authorized panel builders (2007). http://apps.geindustrial.com/publibrary/checkout/UL508A?TNR=White%20Papers|UL508A|generic
Chemistry	Setting an agenda for the social studies of nanotechnology: A summary of the Joint Wharton-Chemical Heritage Foundation Symposium on the social studies of nanotechnology (2007). http://www.chemheritage.org/Downloads/Publications/White-Papers/NanoSymposiumReport2007.pdf
Physics	Particle physics and astrophysics – A whitepaper in response to a call to the Astronomy and Astrophysics Community from the Committee on Astro2010 for State of the Profession Position Papers (2009). http://arxiv.org/abs/0904.0595
Geology	The impact of hurricanes Katrina and Rita in Louisiana: America's coasts under siege (2006). http://www.geosociety.org/geopolicy/hitepapers/wp_0602katrina.pdf
Pharmaceutical industry	Product lifecycle management for the pharmaceutical industry – An Oracle white paper (undated). http://www.oracle.com/us/products/applications/agile/lifecycle-mgmt-pharmaceutical-bwp-070014.pdf
Government	National Archives and Records Administration white paper on best practices for the capture of social media records (2013). http://www.archives.gov/records-mgmt/resources/socialmediacapture.pdf
Business	Creating value: Public relations and the new brand strategy (2009). http://prfirms.org/wp-content/uploads/2011/03/CreatingConsValue__5-6-09.pdf

Note: White papers were chosen as good examples, not for recency.

News Updates and Press Releases

You've seen news updates and press releases related to science. As soon as something "big" happens (in anything, not just science), the Fourth Estate is all over it, in a race to get it out to the public first. First announcement releases are typically very brief—a few sentences on a CNN update ("Chief Ebola Doctor Contracts Disease: A doctor who has played a key role in fighting the Ebola outbreak in Sierra Leone has been infected with the deadly disease," CNN, July 23, 2014). When more information becomes available, the immediate announcement is augmented by a brief summary of what is known thus far, which then may evolve into a full article or blog, possibly with daily status updates. With the advent of online reporting, of course, news is near-instant (although reputable outlets confirm information first) and can be updated hourly or even within the space of minutes. While this is usually more associated with tragedies, politics, or sports, science does merit its occasional place in the "instant update sun," most commonly in areas that the public perceives to be controversial or exciting (e.g., climate change, creationism, GMOs, vaccines, planetary exploration, epidemics, recalls, etc.). Given the lay target of mainstream news, sometimes it's not even the science that gets the limelight. Neil deGrasse Tyson is a brilliant scientist and champion of scientific literacy, but he is probably better known as a celebrity with a popular television show rather than a physicist who wrote scientific papers for the *Astrophysical Journal*.

This does not mean that there are no news updates or press releases in science—far from it. These releases are targeted specifically to scientific audiences (or aficionados). You can sign up for any number of updates through a simple Google keyword alert or through signing up on specific websites. You can hear from the Centers for Disease Control and Prevention or Medpage every day if you want. If you had been reading your microbiology update from *Science News* every day as it popped up in your in-box instead of deleting it right away, you might have caught the *Pseudomonas*-caffeine thing. Do you belong to a specific society for your field? Sign up for their alerts, daily updates, or RSS feeds. Sign up for email delivery of new journal issue table of contents. Find out what's going on in your field *while* it's going on—not later.

Fortunately, "fake news" is not an issue in science. The integrity of science ensures rigid fact-checking (science is based on observation, replication, deconstruction, and analysis). Researchers publish only papers based on solid science (even if it turns out to be wrong later on) and, unlike the occasional journalist, would never embellish results. The result of this is solid reporting that is equally fact-based and unembellished. Indeed, virtuous.

Seriously . . . hope you didn't fall for that. Search for "fake science news" online and see what you turn up. "Fake science news in social media" would make a wonderful capstone or graduate thesis or dissertation topic. You're welcome.

Things that crop up by the minute, hour, or day are news updates and even press releases (to a specific audience). If an important conference is happening somewhere (say, the annual meeting for the American Society for Microbiology [ASM] that you could not attend because it was too expensive and too far away), very often conference updates are posted—short summaries of presentations given that day or the preceding day, sometimes even with video interviews with presenters.

Could you be writing a news update or press release? Not usually; science notwithstanding, these still tend to be written by copywriters who do this for a living (in the case of science, scientific writers or journalists—which is a career you can of course consider with your keenly honed writing skills), and often on a freelance basis. You might be *interviewed* as a student, and of course if you manage to create a new strain of lethal 100% antibiotic-resistant bacteria that get out of the lab and start killing people, you'll be famous.

If you do find yourself writing press releases, your biggest challenges are speed (obviously), accuracy (get the facts), integrity (confirm the facts and ensure *everything* you write is in context), and brevity. Interestingly, many of the popular (i.e., lay) press releases coming out of science are triggered by an actual publication in a peer-reviewed journal or a presentation at a conference. The relationship between caffeine and bacteria has been lurking around in the scientific literature since the 1960s. But when an Iowa graduate student presented results of a caffeine-metabolizing bacterium at the ASM meeting in May of 2011, the news finally got out to the general scientific audience and the public as well. While the presentation was made in May of 2011, the paper was actually published in February of 2011 (and released as an Epub five months before that). It had the rather dull title "Characterization of a broad-specificity non-haem iron N-demethylase from *Pseudomonas putida* CBB5 capable of utilizing several purine alkaloids as sole carbon and nitrogen source" (*Microbiology*. 2011;157(Pt 2):583–592; Epub 2010 Oct 21), which, you'll notice, doesn't even include "caffeine" (although *everyone* knows of course that caffeine is a purine alkaloid). Then came the press releases, a scientific eternity after the discovery.

A press release in this case might be something you can write. It's a summary. If you're covering the actual research (i.e., for a scientific target audience), it's even kind of a version of an abstract. The very big challenge, however, is if the target audience is different. If you're converting this "news" into a press release for the laity, you're not going to be using terms like "non-haem iron N-demethylase." You'll be using "enzyme." In the case of this example, a news update to a scientific audience would occur as soon as this information was presented at a conference, which would then be picked up by a lay science editor on the prowl for potentially tantalizing news somewhere and then translated into a lay audience report. Notice that lay news sources often have a "health" or "science" section (e.g., CNN, *The New York Times*, Associated Press, etc.). Information presented in *Biology News* will be presented differently than in HuffPost:

Biology News (May 24, 2011)

A new bacterium that uses caffeine for food has been discovered by a doctoral student at the University of Iowa. The bacterium uses newly discovered digestive enzymes to break down the caffeine, which allows it to live and grow.

"We have isolated a new caffeine-degrading bacterium, *Pseudomonas putida* CBB5, which breaks caffeine down into carbon dioxide and ammonia" says Ryan Summers, who presents his research today at the 111th General Meeting of the American Society for Microbiology in New Orleans.

Caffeine itself is composed of carbon, nitrogen, hydrogen and oxygen, all of which are necessary for bacterial cell growth. Within the caffeine molecule are three structures, known as methyl groups, composed of 1 carbon and 3 hydrogen atoms. This bacterium is able to effectively remove these methyl groups (a process known as N-demethylation) and essentially live on caffeine.

Summers and his colleagues have identified the three enzymes responsible for the N-demethylation and the genes that code for these enzymes. Further testing showed that the compounds formed during break down of caffeine are natural building blocks for drugs used to treat asthma, improve blood flow and stabilize heart arrhythrnias.

HuffPost Green (May 24, 2011)

It turns out investment bankers and over-worked journalists aren't the only organisms that can survive on caffeine.

Scientists have recently discovered a new type of bacteria that breaks down caffeine molecules into smaller pieces that they can use for metabolic processes, according to Scientific American. The bacterium-energizing molecule is actually comprised of carbon, hydrogen, nitrogen, and oxygen, all of which are necessary for life.

Pseudomonas putida CBB5 breaks caffeine into a carbon dioxide molecule and an ammonia molecule, according to Physorg. No other caffeine eating bacterium has previously been found.

It might not seem like an amazing discovery at first, but the groundbreaking find is actually incredibly rare. Normally, organisms only break down sugars in order to gain energy (though enzymes like these aren't entirely unheard of).

Scientists believe they may find a number of uses for these enzymes.

Source: http://www.huffingtonpost.com/2011/05/24/new-bacteria-caffeine_n_866396.html

Currently these pharmaceuticals are difficult to synthesize chemically. Using CBB5 enzymes would allow for easier pharmaceutical production, thus lowering their cost. Another potential application is the decaffeination of coffee and tea as an alternative to harsh chemicals currently used.

"This work, for the first time, demonstrates the enzymes and genes utilized by bacteria to live on caffeine," says Summers.

Source= http://www_biologynews_nevarchives.2011/05/24bacteria_use_caffeine_as_food_source.html

Notice how HuffPost cavalierly left out N-demethylization and *Biology News* capriciously neglected to mention investment bankers and overworked journalists. Arguably, the *Biology News* release did break the information down into pretty basic science, but the HuffPost release did it even more so and condensed it. The HuffPost release, however, had <u>hyperlinks</u> associated with "bacteria," "*Scientific American*," and "Physorg." *Biology News* only had a hyperlink to the American Society for Microbiology (whence came the source). HuffPost's source was *Scientific American*. HuffPost also incorrectly used "bacteria" where "bacterium" should have been used (not that we noticed that).

---

10 Tips for Correspondence, White Papers, and News Updates

1. If you are mightily aggrieved by an attack on your research in a letter to the editor, assess the claims in the correspondence rationally and try to see if any of the claims are remotely true.
2. If you are aggrieved, do not send the letter right away. Wait and calm down.
3. Write the response to the letter ASAP after you have calmed down.
4. Follow the "correspondence" dictates of the journal, and mimic the overall style of correspondence in that journal.
5. Determine the purpose of the white paper.
6. If you are writing a white paper, ensure you are writing accurately to the right target.

7. Get a sample white paper like the kind you are trying to write to get an idea of how it should be formatted.
8. If you are writing a news update, ensure that it is really news (i.e., *very* recent and not fake). Do your due diligence and fact-check.
9. For a news update, check multiple sources, and interview someone if you can.
10. Hew to the dictates of "news": speed, accuracy, integrity, and brevity.

Other Genres You See in Journals

As you've already (hopefully) seen, the thousands of scientific journals out there contain very different formats. Some have more sections; some are more "bare bones" and have fewer. Some don't publish review papers; some *only* publish review papers. Some have news; others don't. News and brief reports and other smaller components like that tend to occur more frequently in journals with weekly or biweekly publication schedules and less frequently in journals that are, say, bimonthly. Journals may also contain other sections, such as opinions (or similar sections such as perspectives, viewpoints, "A piece of my mind," etc.), editorials, communications, discussions, feedback, book reviews, association information (leadership, conferences, etc.) job openings (including postdocs!), quizzes ("test your knowledge" case studies or images for interpretation), artwork, and, yes, even poetry. We are unsure if the *Journal of Clinical Microbiology* would consider a limerick about *Pseudomonas putida* and Mountain Dew, but perhaps a sonnet in flawless iambic pentameter might catch their attention.

Take-home

While IMRAD and review papers tend to represent the "meat" of most scientific writing in research, a variety of other writing opportunities of wildly differing formats abound. What they all have in common is "science," "accuracy," and "communication." Without communicating research, the research is pointless. However, to communicate your message, you have to do it the most effective way, and that is not always in an IMRAD paper. Sometimes you just need to capitalize on a series of fortunate events and write a case or anecdotal report instead of waiting two years to design experiments in order to understand what happened. If you want to offer a rebuttal to someone who has trashed your work in print, you write a letter, not a white paper. An Ebola outbreak is news, not a review paper. While you can argue that a single topic probably can take multiple formats (e.g., IMRAD, brief communication, news, editorial, etc.), use the best format for your *target audience* and the *vehicle* (e.g., specific journal, blog, etc.) through which you want to deliver your message.

Exercises

1. Choose a topic in your field and construct a fictitious case history (or your prof will provide you with one) to evaluate a single concept about which you'd like to gather more information; write the case report or anecdotal report.

2. Choose a "case report" or an "anecdotal report" (or whatever it's called in your field) from your field and deconstruct it into its component parts. What kind of background does it contain? details? conclusions? what else? What makes it useful?

3. Choose a full-length IMRAD paper from a journal in your field and convert it into a brief report of one-quarter the length of the original paper. What kinds of alterations and deletions did you have to make?

4. Your prof will provide you with a scientific paper that contains some minor or major flaws—conceptual flaws, historical interpretation flaws, methodologic flaws, flaws in conclusions or in other components of the paper. Identify the flaw(s) and write a Letter to the Editor at the same journal with your concerns. What approach and voice do you use?

5. Choose the *most recent* "news" in your field (or your prof will assign you a topic). Create a *journal-specific* news release or new update.

6. Special Bonus Question: Write one of the following:
   a. Fake news science story for your field. Make it convincing. What techniques did you use to make it fake?
   b. A poem about a topic in your field (or a topic provided you by your prof). Your prof will tell you poetic style: sonnet, limerick, free verse, Haiku, etc. It will make you think about your topic in a different, creative way, and you'll be surprised at what you retain.

14

Scientific Writing for Multimedia and the Web

© 2020 Max Beck

It's your last co-op for school, and you got one of your "A" choices: Communications Coordinator for the Insect Zoo at the museum. You *love* the Insect Zoo. You've been a docent there for *ages*. More than a few of your friends (and all of your family) tend to slowly back away from you when you tell them about playing with the giant hissing cockroaches from Madagascar (so gentle!), the adorable orange-kneed tarantulas, the graceful stick insects, and the perky Eastern lubber grasshoppers. There's nothing like handing a fourth-grade teacher a giant hissing cockroach and watching her try to maintain her cool in front of her students when it starts hissing. The Insect Zoo serves a variety of functions. It's educational. It's targeted to a broad range of people, from schoolchildren to adults. It's a social atmosphere, an academic atmosphere, and a political atmosphere, from the standpoint of outreach and conservation. You have a thriving honeybee colony (open to the outside of the building). You have rare insects, unique insects, pests, and some that exhibit extraordinary evolutionary adaptations such as spraying predators with acid or exhibiting astonishing camouflage. You have a meter-long centipede that eats mice. You have books and paper and crayons for rubbings, costumes, "compound eye" glasses, and insects under dissecting miscroscopes for the kids. You have a lecture series for adults and "Fun With Insects" sessions for kids. And of course chocolate-covered crickets complete with an "I Eat Bugs!" pin if the feat is attempted and completed.

In your new position, you're going to be communicating across a broad spectrum of audiences using a variety of media. You probably won't produce that many IMRAD papers and lab reports, but you might write grant proposals; you'll be writing educational materials targeted to the general public, scientists, and kids, maintaining the "Insect Intelligencer" blog for the zoo, fielding questions, and possibly writing podcasts or scripts for videos. This is going to be *fantastic*.

The Changing Nature of Scientific Communication in the Internet Era

Talking about the "Internet era" to an undergraduate student is likely to get some puzzled looks, since it's probably all you know. Back in the days before the Internet became the vast purveyor of data it now is (up until the early 1990s), obtaining information was a slow, laborious process. If you wanted to know how a giant centipede eats, you'd read as much about it as possible, which would require going down to the library for books and journals, and then perhaps speaking with an expert. Now you can Google "giant centipede eating mouse" and get over 25,000 results in less than half a second (arguably, these aren't all going to be scientific sources), and three times that amount in (sometimes disturbingly graphic) videos. You can pull up hundreds of papers on giant centipedes on literature databases such as Biological Abstracts, Web of Science, and even Medline if you're looking for information on envenomation. If you're working with a tarantula expert at Harvard to create a brief educational monograph, you can send it to her immediately, get feedback, edit it, and resubmit it easily. If the Harvard expert hardly has any time, she can review it while she's flying to a meeting in Chicago.

The Internet's impact on the way we do and communicate science is profound, evolving, and in many cases not fully understood. It doesn't only have an impact on the way science is communicated to lay audiences but also on the way scientists communicate, work, write, publish, and review research. This chapter will give you a brief overview of the types of writing for multimedia and the web that you might be doing in your profession for both scientific and lay audiences and will provide you with some guidance on tackling some of these writing tasks. This chapter is heavily (though not exclusively) devoted to *science writing* rather than *scientific* writing (as per our initial definitions in the Introduction), that is, with "translating" highly specialized scientific knowledge for a lay audience. As scientists, you will likely have to communicate with nonscientists as well.

The Internet is "the great leveler" (according to science writer Michael Shermer), making information accessible to everyone in the vicinity of a computer with an Internet connection. While the digital divide is still very much real (i.e., the gap between those with computer/Internet access and those without), the Internet has had a clear democratizing—and empowering—effect. That means that information about science is not just accessible to the few initiates with access to good libraries but to everyone. Suddenly, your target audience is now the entire world (who speaks English, of course; but English *is* the lingua franca of science). The Internet makes geography less relevant as it offers equal access to information from all corners of the globe; it also eliminates the gatekeeping of information, which had previously been restricted to institutions of higher learning. This has had the added effect of synchronizing the thinking of the scientific community: scientific problems that are on everyone's minds come sharply into focus and are more likely to be met with a concerted effort to solve them. There are, of course, downsides, such as the spread of Internet hoaxes, misinformation, "fake news," and less than reliable "scientific" information. As a scientist, you're educated to question sources and evidence, so you will hopefully avoid some of these downsides; conversely, as a scientist, you will have a public duty to use the Internet responsibly when given a chance.

Using Visuals in Online Scientific Communication

The Internet milieu is a highly visual one. Increasingly more powerful computers and sophisticated imaging and statistical software have drastically improved the quality (and quantity) of visuals in scientific articles available online. The new digital printing formats meant that more space can be allocated to supplemental data—often raw data, tables, perhaps maps and other types of data that would have been skipped in a print-only environment, where page/word limit is a real constraint. By contrast, your online footprint is much more generous (what's a few megabytes, anyway?).

While it's always tempting to add visuals, always ask yourself whether they serve a purpose: Do they support your argument or augment your content? Or are they just there to fill up space? Do you really need to include a cute picture of your lab rat in your description of the experiments? Or is it superfluous to the purpose of your paper? As usual, always label and caption your visuals, and insert them in the text close to the place where they are referenced (or, in a digital text, make sure they're hyperlinked both ways, so that the user can jump from the link to the figure and back seamlessly).

As a response to the easy availability of these technologies, and to public expectations, many publications have upped the ante by publishing primarily multimedia articles that include videos and interactive activities or by promoting video abstracts. A quick online search will reveal some of these to you, together with (evolving) guidelines for producing videos that meet the same standards of rigorous scientific peer review as traditional print articles. Such new formats respond to a deeply felt need to visualize science and can be very useful for those who try to replicate experiments. More and more established scientific journals are including dedicated video sections (such as *Nature, Cell, Science,* the American Chemical Society, etc.). Some experts consider video articles or video abstracts a natural evolution of scientific communication; expect to see more of them in the future, created specifically to communicate scientific information to peers. The *Journal of Visualized Experiments* publishes exclusively video demonstrations of lab experiments and methodologies. This trend emphasizes the fact that, to be successful scientific communicators, you should be conversant with a variety of multimedia formats.

Multimedia Versus Interactive Media

Multimedia technically means "using more than one form of media." In science, very few communication forms are *not* multimedia (i.e., involving at least two modalities, such as text and visuals). A paperback or an oil painting would perhaps qualify as being monomodal (confined to one medium, paper and oil painting, respectively)—but any digital experience is probably a mix of text, image, video, and sound among others. You don't have to go digital, however, to observe multimodality all around you: scientific papers and presentations use texts and visuals (tables, graphs, pictures, maps, etc.) along with a variety of design elements, and, if published online, hyperlinks and occasionally video and audio files.

By contrast, the term "interactive" adds a new dimension, in which the user can *interact* with your multimedia product. A truly interactive product (such as a poster, presentation, map, chart, infographic, site, etc.) will allow the users to control their experience of that product in a variety of ways. An interactive display may enable users, for example, to:

- focus on a particular area/topic;
- choose the way they navigate the product (e.g., starting in the middle rather than at the beginning);
- customize certain features (e.g., colors, fonts, points of view, perspectives, timelines);
- create their own version of that product, which may be slightly different from another user's version; and
- participate in polls, questionnaires, surveys, quizzes, or other activities—frequently with immediate feedback.

An interactive map, for example, will allow you to zoom in on certain areas, click on a satellite view, customize directions, etc. An interactive poster may contain hidden pockets of information at crucial junctures that are only visible when you choose to click on them. A site may allow you to customize your preferences and deliver you only the information that is suitable for you (e.g., news suitable/interesting to a particular demographic, geographical location, linguistic preference, etc.). To a certain extent, most of the web is now customized or customizable and can provide you with search results tailored to your browsing habits. Automatic translation software is also getting better at translating written text and natural speech.

There is no set "template" for interactive experiences on the web, and you've probably been through some good and bad experiences with "interactive" features. *Purpose* and *usability* are our two main requirements for interactive displays—in other words:

1. Do they serve a purpose that advances the overall goals of the (scientific) project on display (as opposed to being a "bells and whistles" feature)?
2. Do they work (as opposed to causing your computer to buffer, slow down, freeze, or spiral into the blue screen of death)?

For example, in an interactive digital presentation showcasing the insect zoo, clicking on (or rolling over) the image of a bug to reveal fascinating details about its habitat, evolutionary adaptations, feeding habits, or other attributes would be an appropriate use of interactivity. Clicking on the bug to activate laser-eye animation or make it dance in an effort to insert some life into your presentation is a waste of time and effort on both your and your user's part. It's time to rethink your interactive strategy. Of course, if you're creating interactive multimedia for children, sometimes the "bells and whistles" route is the way to go, as long as you can include some stealth education into the mix.

What Are the Features of Good Multimedia Products?

"Art is long, life is short," lamented Hippocrates some 2,500 years ago, complaining about the impossibility of learning everything there is to learn about the art of medicine in one lifetime. It is likely that Hippocrates might have felt even more strongly about this had he had access to the Internet. Let us borrow that lament here to explain we can't possibly cover all the intricacies of videomaking and podcasting in the few short pages we have left in this book. You will probably find plenty of instructions online, tailored to the tools you have at your disposal. There are, however, certain widely accepted standards and rules for producing quality multimedia products (interactive or not)—by which we mean videos, podcasts, and multimedia presentations meant to be disseminated on the Internet. Next we consider some of these standards and rules.

Audience First

First, make sure that you start with a worthwhile idea likely to engage an audience in a video, audio, or multimedia format. Remember that most people (especially lay audiences) will choose a multimedia format over the traditional scientific article because of the presumption that multimedia is easier to understand, easily digestible, and more entertaining. Your challenge is to find a way to translate a complex scientific idea successfully into a well-targeted multimedia format, without getting your viewers or listeners to click "next" after a few seconds or wander into another browser tab. The most successful science communicators (think Carl Sagan or Neil deGrasse Tyson) have broken the barriers of their respective discourse community to communicate science to lay audiences through the use of apt similes and metaphors. For example, Carl Sagan used the "cosmic calendar" metaphor to help us conceptualize the enormity of the space-time continuum in his popular 1980s series "Cosmos," a metaphor revived in the show's new incarnation with deGrasse Tyson as a host 30+ years later. In that metaphor, if you think of the timeline from the Big Bang to the present day as a one-year calendar compressing nearly 14 billion years of existence, human beings appear on the scene only in the very last minutes of the very last day of the year. Sagan, of course, did not only tell but also showed us this idea as beautiful visuals graphed onto a calendar-like grid. Do you have to explain the Fukushima nuclear plant disaster—and in the process, nuclear reaction? A successful science writer compared it to playing Jenga—and imagining nuclear reaction as a Jenga tower coming down after one final crucial piece is removed. Think visually to help your viewers, listeners, or users understand complex concepts. If you can't think of any good metaphors, then at least try using vivid, memorable language that is likely to have a bigger impact on your audience.

In other words, *grab your audience*. A super-technical, all-text sign next to a cage in the insect zoo will be read by the approximately less than 1% of people coming through the zoo who are doing dissertation research. A video of a puking blue jay, however, will have fourth-graders crowded around the exhibit yelping with delight (see Figure 14.1).

While some insects use Batesian or co-Müllerian mimicry (e.g., the unpalatable liminitids mimicking the unpalatable danainads), or camouflage (e.g., the phylliads) as defense mechanisms, others demonstrate chemodefensive mechanisms such as the spraying of methacrylate (certain carabids) or formic acid (formicids).	Have you ever seen a blue jay puke? That's what's in store for them if they eat a Viceroy butterfly! Viceroys taste very bad. Viceroy butterflies look like Monarch butterflies—which also taste bad! Birds learn very quickly not to eat any butterflies that look like Monarchs or Viceroys. Other insects protect themselves by hiding, looking exactly like the place they're sitting. Can you find the leaf insect? Some other insects like some beetles and ants don't bother hiding or tasting bad. If a predator comes after them, they spray them with acid or some other chemical that hurts!
☒ No	☑ Yes

Figure 14.1. Inappropriate and appropriate ways to address a school-age audience.

Appropriate Software and Hardware

Tools are more important than ever in multimedia work. Aim for good production values, always. We know, we know. Good gear is expensive. You can always aim higher, though—or at least be aware what you should be aiming for. For example, you need good—nay—*great* audio for your video or podcast; most of the time your camera or computer microphone just won't suffice. Your viewers may forgive the lack of high definition video, but they won't be able to get over a bad soundtrack. You must, of course, make every effort to make your video as high-resolution as possible. And while you're at it, you should probably invest in a tripod (with a fluid head, if you want to tilt and pan smoothly) so you can keep the camera steady while you're shooting. If you must start your video recording career on an iPhone (not unthinkable, actually), remember to hold it correctly in the landscape and not portrait position. And don't forget: do multiple sound tests to make sure the sound comes through and background noises are minimized. Without good sound, your video or podcast is a wash.

If possible, employ the services of a professional developer. Regardless of how the videos or interactive materials are created, beta-test every possible combination, every click, every mouse-over, every audio synch, every image and text build—and have more than one person do it. Ensure that it works on all platforms (Does it work on computers but not mobile devices? Does it work on all web browsers? Going to extreme efforts to create a great interactive product is meaningless if it doesn't work or is full of bugs when it's launched.

Scripts

For videos, podcasts, and even online presentations it's a very good idea to write a script. Good looking, functional, and effective multimedia products are the result of laborious planning and scripting. Scripting is standard practice in the media industry and can save you a lot of time and headaches down the road in terms of production. Issues that might not initially seem like a big deal—such as staying within time limits, covering topics adequately, streamlining the message, and indicating where interactive elements go—can be avoided through careful scripting. A good script should indicate:

a. what happens on the screen (static or moving images);
b. what's the audio (what do the characters say or what's the voiceover [VO]);
c. how long the sequence lasts;
d. what kind of background music you are using, if any;
e. what kind of special effects (including sound effects) you are using, if any;
f. special instructions to the programmer or VO talent.

Scripts can work for just about any type of multimedia you are planning (for podcasts, subtract the images, and pay special attention to sound effects/music). The following is one possible way you can begin to organize a script for a video explaining the insect zoo on its website.

SCENE	Visual	Audio	Time	Notes
1	TRACKING SHOT; SELMA WALKING TOWARDS CAMERA FROM MUSEUM IN BACKGROUND	SELMA: One of the most interesting things to explore about insects is the evolutionary adaptations for protection against predators.	15"	Narrator: Selma
2	FADE IN AND SLOW PAN OF INSECT ZOO	[VO] SELMA: Here at the insect zoo at the Museum of Natural History, we explore the many fascinating attributes of insects, including their ability to defend themselves.	20"	Ensure that insect zoo sign is included in the shot
3	ZOOM IN TO LEAF INSECT SWAYING ON TWIG	SELMA: Let's go on a tour of these fascinating insects, and watch how they protect themselves from predators.	15"	Make sure leaf insect is visually swaying
4	SELMA AND ARONSON, TALKING; INSECT ZOO FULL OF CHILDREN IN BACKGROUND	SELMA: This is Dr. Diana Aronson, director of the Insect Zoo. Thank you for having us! ARONSON: It's a pleasure, Selma.	10"	
5	ECU[1] OF HISSING COCKROACH BEING PLACED INTO SELMA'S HAND BY ARONSON	SELMA: I ... well, I don't know about this ... I'm not sure I want to hold this! ARONSON: Oh come on ... little kids hold it all the time!	20"	
6	ECU OF HISSING COCKROACH CLIMBING AROUND SELMA'S HAND AND WRIST	SELMA: So this guy has a defense mechanism against predators? I'd think predators would avoid a giant cockroach such as this one ...	15"	[AD LIB "WHOA!" AS NECESSARY AS COCKROACH WALKS AROUND]
7	ECU OF COCKROACH ON SELMA'S HAND; NEED TAKE OF COCKROACH HISSING	SELMA: Well for goodness sakes ... what does a giant cockroach have to do to OH NO! NO! GET IT OFF ME! [AD LIB APPROPRIATELY]	15"-30"	Need multiple takes until cockroach hisses
8	PAN BACK, SELMA TAKES COCKROACH FROM ARONSON	ARONSON: See? Pretty effective, don't you think? SELMA: I think I have to sit down.		

[1] "extreme closeup"

© 2020 Max Beck

An important thing to keep in mind when creating a formally scripted piece for online or other multimedia distribution is VO narration, which is a surprisingly difficult thing to do, as neophytes often discover when they attempt to do it themselves. Professional narrators can enunciate clearly, understand how inflection and phrasing works, and can keep themselves at an even pace. You do not necessarily have to find yourself a professional VO, but do not assume you can just do it, either. Try to find people who can deliver your script convincingly.

A storyboard is standard in the movie industry to plan scenes and shoots and would include slides with a rough sketch of the action on the screen and textual descriptions (including dialogue, VOs, and directions) of the scene, in a manner similar to comic book panels. You can find online storyboarding tools to help you create useful storyboards for your video or multimedia presentation.

Due to the nature of multimedia pieces, it is crucial that you hook your viewers (listeners, or users) early on; otherwise, it's very likely they'll click on the next link ("Oh look! It's a kitty on a skateboard!"). So, start strong; be aware that the audience is moderately interested in your topic to begin with (after all, they're here!) so now you just have to keep them watching a little longer: what interesting, fascinating fact about the topic you're describing can you hook them with? Co-Müllerian mimicry: no. Puking blue jays: yes.

That hook need not be the main point of the story, although sometimes it may as well be. Whatever the case, you still need to make your main point fairly quickly—don't delay it or it will risk being lost. You are not dabbling in a long genre; most videos meant for the web are short productions, under five minutes or shorter—sometimes no longer than two to three minutes—so you must learn to make your point in the first minute (or even half-minute) or risk losing your audience.

When you make that point, present it as a part of a larger story. Remember that people have been fascinated by stories for millennia—for a good reason. They keep us listening, reading, and watching. Find the narrative kernel to your multimedia product and spin it to keep your audiences engaged and tuned in until the very end, because they want to hear the end of the story. Think about it this way: any IMRAD paper can be expressed in story form, as the story of an experiment. First you set the stage, then you tell us who the heroes and their tools are (methods), then what happened (results), and finally what's the moral of the story (discussion/conclusion). You can recast any scientific experiment into a tight little narrative this way (and it also helps you put things in perspective: Why should anyone else care? Why is this important?). Make sure you help your audience understand what is interesting or important about the science you are discussing and, if possible, how it relates to an aspect of their lives.

The devil is in the details in writing and in creating multimedia as well. If proofreaders tend to be a little obsessive over misplaced commas and dangling modifiers, video and sound editors will be a little obsessive about lighting, editing, transitions, sound mixing, etc. Polishing and editing should be part of your second nature by now in writing—and multimedia products deserve a similar level of TLC. In particular, be careful to cut segments that don't completely reinforce or support your message or feel extraneous to the piece. Unlike writing, where it is easy to add a few sentences or a paragraph here and there, it is much more difficult to resume filming or recording (especially if you are on location, have guests, need a special kind of lighting or props, etc.). This is why you have to (a) plan, plan, plan (don't forget the script!) and (b) shoot or record enough footage so you can edit it down later. Multiple takes are going to be easier for you to deal with in postproduction than trying to edit out chunks of video.

Ideally, you should also always beta-test your multimedia on an advance audience. It's relatively straightforward to do so with a video or podcast—have someone from your target audience watch or listen to see if they get the point; collect useful feedback in the process. What may be straightforward or clear-cut to you may be obscure for a viewer or listener; so test it out. It's even more important to send out your slideshow or interactive multimedia presentation to a variety of people roughly from your target audience/demographic to see if they can navigate it the way you designed it and help you discover the inevitable bugs that occur whenever technology is involved (we're talking cross-platform issues, JavaScript updates, etc.). Just don't hit that "publish" button before your product appears to (a) work properly on a variety of platforms and (b) has a very low potential for ambiguity or misunderstanding. Notice here that we're not presuming to say that it will be "free of ambiguity or misunderstanding" as nothing on the web is that, anymore, but that you'll do everything you can to minimize such potential by getting feedback. Editing and testing are interchangeable steps; they will help you streamline the story you are telling and clarify the main point you want to get across. Again, *do not lose sight of your target.* You can create a flawlessly exquisite multimedia experience, but if it's supposed to be targeted to reasonably scientifically literate lay adults, don't target it to seventh-graders or scientific experts.

Don't forget that the files you have laboriously put together in a fancy video editing program cannot be published "as is" (the same way Photoshop files need to be converted to a web-friendly format such as JPEG or GIF in order to be published on the web). If you are your own web developer, make sure that you compress the final product correctly for the web (e.g., to an mp4 format) so as to ensure maximum quality and optimize upload and download speeds. Some compromises will have to be made in the process (you may not get HD, Blu-ray quality for your videos, or astronomic resolutions), but the end product will have to satisfice. "Satisfice" is a hybrid of satisfy + suffice –a term coined by Ramage and Bean, a team of composition textbook writers, to denote that at some point the editing process has to end and that you will have to declare one version of the product to be final—which is, satisfactory *enough.*

There is one more step: it matters *where* you put your video/multimedia presentation/podcast on the web (of course it does). You are not only responsible for creating it but also for promoting it. Some of it may be outside your purview—if it goes on the website for a larger institution, or journal, for example; and some is in your power: share the link via email and social media (read on through the remainder of the chapter for more on the use of social media in science). Make sure you do so ethically, though: don't post the video to YouTube, for example, unless you are sure you own the copyright to it (and any of its composite information and graphics) or you have permission to do so by the institution. Remember that if you create something as an employee of an institution—including a university—and are paid for your services, you may not fully own (or own at all) the copyright to that product; always discuss your rights and responsibilities with your employers and educators before making intellectual ownership assumptions.

---

10 Tips to Make Your Videos, Slideshows, Interactive Presentations, and Podcasts Fabulous

1. Have a good idea that tells a story (visual or aurally).
2. Make sure you're writing to the appropriate target.
3. Write a script.
4. Have good production values.
5. Hook your viewers/listeners/users early on.
6. Make your central point quickly (ideally in the first 30 seconds).
7. Edit, edit, edit!
8. Beta-test. Then edit some more. And then beta-test it some more.
9. Compress the final product correctly for the web.
10. Promote your product on the web/social media.

A final word on intellectual property: if you're posting on some institutional website or you publish in a peer-reviewed venue, then you probably don't have to worry about intellectual property yourself—the institution or publication takes reasonable precautions. If you self-publish (on your own blog, website, YouTube channel, etc.), then you have default copyright, but you should probably clarify that with, for example, a Creative Commons badge or a similar disclaimer suited to your purposes, indicating the terms under which you are comfortable allowing other people to share, use, or adapt your idea(s). You positively *cannot* take images from the web and include them in your project, even with disclosure. While some images are free (e.g., images from the USDA), most are *not*, particularly if they have appeared in print before. If you want the classic cool photo of the puking blue jay, you'd best go back to Brower 1958 and get permission. Otherwise, line yourself up some blue jays, feed them some monarch butterflies, and hope you can get some emetic action.

Rediscovering the Social Nature of Science

Social media is everywhere, including science, and it is here to stay. Alongside many other collaborative online tools, social media has altered the way scientific research is made, published, read, and understood. We briefly mentioned how new collaborative tools are proliferating in an attempt to connect scientific communities across the globe. Connection is not the only goal; the hope is also to accelerate the pace of research and discoveries (which, at least theoretically, should happen when research is shared instantaneously, to a large number of people, who can work at the same time on solving one problem), as well as prevent, correct, or expose both bad science and unethical behavior in science in all its ugly incarnations (fraudulent or fabricated data, scientific fraud, plagiarism, duplicate publication, etc.). Open-access publishing (with caveats: predatory open access publishing is a big problem—see Chapter 7) and social media have blasted open the doors to participatory science, including citizen science. Peer review, one of the most critical academic practices, has come under attack repeatedly for its slowness, lack of transparency, and potential to stifle innovation. Social media has the potential to open the process through social networks such as Faculty of 1000 (F1000), which aim to make peer reviewing a faster, more straightforward, transparent, and shared process.

Furthermore, the postpublication life of scientific articles has all of a sudden become a lot more interesting through what has been called by one author "the postpublication peer review on social media" and "trial by Twitter" by another. This means that publication is no longer the end result of research and that scientists connected via social media tools can respond and critique articles after their publications in a matter of days or even hours. Science-specific social networks and blogs (e.g., sciencebasedmedicine.org, etc.) can pick apart published papers and, in many documented cases, demand retractions or corrections.

These days, all major scientific publications have social media accounts and disseminate titles of published articles to followers, opening up a public, more informal conversation about scientific research in ways that were not possible up to the early 2010s. The social impact of articles (or their visibility in the news) is tracked by Altmetric (Altmetric.com) using the DOI (Digital Object Identifier) system. Prestigious journals such as *Nature* and many others have been using these statistics as part of their citation matrix.

In what follows we provide you some guidelines for using social media as a scientist, focusing in particular on blogs but looking also at other (generic) social media platforms.

Writing Engaging Science Blogs

If you're passionate about your field of research and want to share that passion with like-minded enthusiasts (and perhaps reach out to lay audiences as well), you may consider blogging, by far one of the most popular online genres for making your voice heard. You can decide early on whether you want to address mostly scientists or mostly nonscientists (which will be reflected in your choice of topics, vocabulary, and so on), or you can try to engage both by writing blog posts aimed alternately at one or another audience—several famous bloggers have found a large following this way (including science bloggers).

Your topic or field is probably endlessly fascinating to you, but not necessarily to everyone else. Good writing, however, appeals to everyone. So: engage your audience. Explain what is interesting or important about the topics you're writing about. Draw on your background knowledge; connect to other fields; make connections; use visuals; use metaphors (visual or not); find a way to translate a complex topic into an engaging story (see our earlier tips for creating multimedia). Unfortunately, there is no secret formula for this. That part is up to you. However, once you get that part done, we can give you some pointers as to how to improve your chances of your blog becoming more visible:

Post Regularly

First, you should post at a consistent pace—if not every day, then at least weekly. Constant and consistent output (even when you don't "feel like it") is a sign you are taking your craft seriously, that you have something to say, and that you are willing to work at it. Your writing will get better in the process. Further, if you have blog followers who know you post frequently, they will know to check your blog frequently, increasing the likelihood of reblogging, pinning, sharing, commenting, and posting to other social media platforms (e.g., Facebook or Twitter).

Go Visual

Second, use visuals to your advantage. Everyone expects to see them. A purely text-based blog is a surefire way to up your bounce rate (i.e., your users will click away). Good design and carefully chosen, captivating visuals will, however, make users linger. You can wax eloquent about how giant centipedes eat mice, or you can have a fabulous close-up of the carnage or—even better—a video you took that will rivet your audience, who will gleefully share the link onto Facebook so as to provide discomfort to all 735 of their closest friends.

Link
Third, link often and prodigiously to other blogs, articles, websites, Instagram photos, YouTube videos, or any other multimedia accounts that serves as a useful adjuvant to the topic of your posts. Blog authors can usually see who links to them and many will return the courtesy or at least visit your pages at least once. Make sure that they have something interesting to see and return to!

Comment
Fourth, comment (politely, respectfully, and thoughtfully) often on the blogs of writers (scientists, science writers) you admire, to establish a presence and credential as a blogger. Other commenters will probably see and respond to your comment and notice that you have a blog. This is how you slowly build a following. Try to answer the comments on your own blog—unless, of course, you become so hugely successful that this is no longer feasible. Then you are excused. You can answer selectively. Make sure you have good spam filters in place and that you can block trolls as soon as they rear their ugly heads. You'll recognize them as soon as you see them. There is little need to engage with them, least of all in the comments of your blog.

Promote
Fifth, promote your blog on other social media venues where you might have a profile: Facebook, Twitter, and LinkedIn, for example, allow you to post links to your blog as you update it; where else? Email pleas to more established bloggers are not unheard of, although they're not always successful and we can't say we recommend them—it all depends on your level of comfort. Are you a member of a professional organization? Become visible there (professionally and respectfully); make your presence in the blogosphere known.

Connect
Sixth, if possible, consider joining a blogging network. There are numerous opportunities for young scientists to join the science blogosphere via networks that offer a standardized format, the chance to build a community, the occasional incentives that come from web advertising, and first and foremost, the

visibility that comes from a respectable platform. Some of these networks are by invitation only, so you'll need to build a name for yourself before being invited to the big table.

Learn to Respond to Criticism

Seventh, don't forget that once you are publicly engaged with audiences on the Internet, you will be exposed to criticism. It is part of the process, but it may not be for the faint of heart. Be prepared to deal with negativity and dissent, which as a rule grow at the same rate as your blog's success. Learn how other bloggers deal with criticism, and learn to respond to constructive criticism as well as ignore online trolls and obnoxious comments. Keep in mind that even if you write a flawlessly brilliant blog and mix up one "effect" with "affect," that's what more than 50% of the comments are going to be about. Proofread, and you can hold the ad hominem attacks at bay ("obviously she knows nothing about hissing cockroaches if she doesn't know the difference between 'effect' and 'affect' ").

Be Passionate

Finally, don't forget why you started this in the first place: to follow your passion. So make sure your passion shines through. That is the number-one factor that keeps visitors coming to your blog. No one will feel the need to visit the page of an apathetic, disengaged, run-of-the-mill, infrequent blogger.

Winning at Social Media as a Scientist

If you're using this textbook, chances are you are pursuing a career in a science or a related field. As such, you will have a public persona as a scientist (researcher, health provider, activist, educator, etc.), which you need to take into account if you want to maintain a respectable presence on social media. While anonymity used to be a feasible option on the Internet, it has become increasingly difficult to maintain—and in some cases, it could also be undesirable (e.g., if you're making considerable contributions to science or popularizing science on social media under a pseudonym, those contributions cannot be used toward your professional advancement or recognition if you wish to remain anonymous). So, if you're going to build an online profile—and at this point, the question is probably not of "if" but of "when" or "how"—we recommend that you use your full name (or a recognizable nickname associated with your real credentials) and maintain professionalism at all times in your conduct. That doesn't mean you still can't play your favorite online games under a pseudonym, but if you blog; use Twitter, SlideShare, LinkedIn, collaborative writing, peer review, or citation platforms; and participate in online forums to discuss and share matters related to your scientific and research interests, you should conduct yourself ethically and professionally at all times. Even if you try to maintain your anonymity, it does not take much for your identity to be exposed, willfully or accidentally.

That doesn't mean you need to be afraid to post your own opinions or stifle your personality. On the contrary, on a World Wide Web inundated by a multitude of voices, the only way you stand up is through your originality. Don't be afraid to be authentic and have (appropriate) fun in order to engage your audiences. Keep your social media profiles up to date. Don't have the most recent job on LinkedIn be one you had four years ago.

Science put together a list of the top 50 science stars on Twitter (microblogging is by far the medium of choice for engaging large audiences), loosely basing its calculations on the more or less tongue-in-cheek "Kardashian Index for Scientists," a concept proposed by Neil Hall in *Genome Biology* in 2014. Hall's less than charitable assumption was that the top science celebrities would be famous for their social media presence but not really for their science (a little like Kim Kardashian—a media celebrity famous for being famous). However, the *Science* list (dating to 2014) disproved that assumption; many of the top science stars on Twitter are, in fact, productive and highly cited scientists in their own fields.

Here's what *Science* has to say about the social media trend among scientists:

So why do the highly cited researchers who are also Twitter science stars make the time to engage in social media? Geneticist Eric Topol of the Scripps Research Institute in San Diego, California (17th place; 44,800 followers [as of 2020, he has 186.7K followers]), who boasts more than 150,000 citations, says he once thought the social media platform was only for "silly stuff" like celebrity news. Then he tried Twitter during a TEDMED conference in 2009, as a tool to gauge reactions to his talk. Now, he starts his workday browsing through his Twitter feed for news and noteworthy research in his field. During the day, he checks Twitter several times and spends another 10 to 20 minutes on an evening roundup. "It actually may be the most valuable time [I spend] in terms of learning things that are going on in the world of science and medicine," says Topol, who reciprocates by daily tweeting papers, presentations, and more to his followers. (http://news.sciencemag.org/scientific-community/2014/09/top-50-science-stars-twitter)

Although dated, perhaps you can also find Topol's example instructive: there is a lot of information on Twitter (mostly in forms of links to longer reads) that is actually useful and timely. Follow the right people and the right feeds (journals, institutions, etc.), comment, retweet, and tweet yourself (including links to your own creations—blogs, videos, podcasts, other multimedia) and become part of the larger conversation. In time, you may become—why not?—one of its stars.

---

10 Tips for Blogging and Being on Social Media as a Scientist

1. Make a *careful* decision about keeping your private and public personae separate.
2. Post something regularly—daily is best.
3. Follow your passion.
4. Comment on other scientists' blogs, Twitter feeds, and other social media feeds as appropriate.
5. Think visually.
6. Be generous: follow, praise, comment, link, blogroll, favorite, friend, retweet.
7. Beware of and do not engage with trolls; use a filter if possible to keep them from posting.
8. Consider being part of a blog network.
9. Be ethical: Respect people's privacy, and don't infringe on copyright.
10. Respond to questions and constructive criticism whenever possible.

Take-home

Scientific writing for the web and multimedia is the new norm. Social media can be a useful platform for disseminating scientific information and engaging scientific and lay audiences alike in a larger conversation about science. There are many ways to get engaged—but always do so ethically, professionally, and authentically. When you produce videos, podcasts, and interactive presentations, plan them carefully by writing a script and aiming to have high production values. Consider blogging and having a constant social media presence in order to boost your public and professional profile. Engage your audiences with a good story and attractive visuals.

Exercises

1. Find one or two online forums and/or listservs relevant and useful for your field. Make sure they are well trafficked. Pay attention to factors such as when the forum was last updated, the type of users, moderation policies, type of questions asked, etc.

    a. Compile a list of the top five burning questions in those forums and present them to class.

    b. Become a member of one those forums and ask your own burning question, which may be general or specific to a current project or experiment you are doing.

    c. Pick one of the "hot" topics that members seem to be discussing at the moment and write a brief digest of it for the class—two to three paragraphs distilling the main questions, challenges, issues, and potential to your peers (an educated audience of scientists-in-training who nevertheless might not be experts in the field).

2. Pick a recent scientific (IMRAD) in your field that addresses a topic you are currently working on. Can you explain it in narrative form to a lay audience—say, your little brother who's in eighth grade? Can you turn it into a story that will catch his attention and make him retain the information?

3. Design a simple multimedia/interactive display for a museum targeted to the general public on an interesting topic in your field. How would you design it? What elements would you include? Why would they be informative or educational?

4. Start a blog on a topic related to your field. Choose either:

    a. your personal research interests and experiences; if you feel brave, link it to a Twitter account. Choose any blogging platform that is appealing to you, and make sure you create a blogroll linking to similar bloggers. Try posting every day for a week; also comment on other people's blogs during this week. At the end of the week, reflect on the experience and share your thoughts with the class.

    b. create a group or class blog, with your class or lab colleagues as multiple contributors. This is an excellent opportunity for online collaboration as well! As a group, decide the name/address of your blog, the platform, your author handles, and the topics you will cover. Together, craft an "About" section that explains how your blog is unique/interesting/answering a need/filling a niche and post at least one entry each per week for a month. Try to promote your blog (if your prof is on board) on multiple social platforms. Revisit your concept after a month: how many "hits" did you get? Comments? How many times were you linked to? Is this something worth doing for a while and if so, (a) why? (b) how?

5. Create the script for a science video aimed at your university at large (by this we mean largely made of non-scientists). The video should cover some current, cutting-edge science topic, preferably of interest to you and/or connected in some way to the research done at your university, to the university community, and make them love it. (For example: promote your insect zoo—get them to come visit!).

APPENDIX

Writer's Toolbox

Crafting Effective Sentences, Paragraphs, and Papers

Do you sometimes have that nagging feeling that something about that paper you just handed in wasn't *quite* right—despite the fact that you proofed, and spell-checked, and had your English major roommate proofread it for you once more? Then you're in good company. English-speaking people have been fretting about good English grammar for a good long time—at least for as long as grammar and rulebooks have been in existence—and before that, actually, which is why those grammar and rulebooks came *into* existence. This section does not claim to cover *all* those rules; instead, it compresses, clarifies, and puts in perspective the principles you need to follow in order to achieve verbal excellence in your scientific writing assignments (you're on your own for content excellence).

Your main goal as a writer is to convey your ideas effectively to an audience. That means your audience should not be made to work extra-hard on such items as grammar, style, or punctuation and instead marvel at the brilliance of your ideas. Essentially you must anticipate sources of potential trouble or confusion and dissolve them with the magic of precision and clarity. To do that, remember these two pithy rules, which, if heeded, will get you out of most grammatical conundrums:

1. Write what you **mean** (M rule): Your prose should aim to represent your ideas and data accurately and minimize the risk of confusion and misunderstanding. It sounds simple enough, but if you've ever reread your prose and wondered what you were trying to say, or had writer's block and puzzled endlessly over a blank sheet of paper wondering how to begin, even after you've done your research, then you've probably suffered a meaning/writing disconnect. Such disconnects become apparent more readily upon revision, which is why we advise you to revise twice and submit once. (We kid: the number of revisions is always greater than two.) For example: when a student writes "The water-solubility of the novel chitosan derivative was enhanced relatively; it could even be slightly soluble in methanol," how is one to interpret "relatively" and the subsequent post-semicolon clause? Is it *relative* to something? What does "it" refer to? Why "even"—what do I need to know about methanol that is not clear from the context? When a reader stops to puzzle over what you thought was fairly straightforward prose, you probably should go back to the drafting screen: it is possible that you didn't exactly write what you meant. Once you figure out what you really meant, you may revise—possibly starting with: "This method enhanced the water-solubility of the novel chitosan derivative and, to a lesser degree, its solubility in methanol."

2. Make every word **count** (C rule): Every word you use should carry a specific and precise meaning and help to clarify your communication goal, maximize impact, and reduce verbosity. In other words: avoid meaningless words and phrases that add zero information and/or quality to your prose. For example, instead of "Despite the fact of the matter that the releases of Hg from anthropogenic sources have been drastically reduced from the early '60s, Hg is still spreading in the environment," you might want to write "Although the releases of Hg . . . etc.," thus cutting with one swift pen gesture six words that carried little to no useful information.

In essence, just about all the grammatical rules we're about to review will boil down to these two. By following them, you will hopefully achieve the holy trinity of accuracy, clarity, and conciseness, three highly prized qualities in scientific writing.

I. Drafting Effective Sentences

A. Verbs Clarify Your Meaning

Verbs are the heart of a sentence; without a good verb, the sentence is dead on arrival (or barely breathing). If you want to keep your prose alive, select good verbs that *clarify* the "story" of your piece—and make sure you use them correctly. This means that you

1. make sure have a verb in the main clause;
2. make sure your verb always agrees in number with your subject;
3. choose the correct tense;
4. whenever reasonable, choose active over passive voice;
5. whenever possible, articulate the action of your science story in a verb, not in a noun (or as grammarians would put it, avoid *nominalizations*).

The first three rules are compulsory; the last two are less strict, which means you will have some choice, but you are strongly encouraged to abide by them.

1. Make sure you have a verb in each sentence and clause

If you don't, you'll end up with a sentence fragment:

> "Second, the 130 major injuries (39.3 %) resulting in surgery or in missing eight or more consecutive games."

As you read that you probably murmured, "What about the injuries? What are they doing? I want to know!" and reread the sentence to make sure you didn't miss anything. You didn't—the writer did. In this case we would need the full context to revise this sentence; a possible revision could be:

> "Second, **during the duration of the study, athletes suffered** 130 major injuries (39.3 %), resulting in surgery or in missing eight or more consecutive games."

A sentence without a verb is missing its heartbeat. Even worse, it will draw the readers to a halt, as they are trying to figure out what happened to the missing verb. Suddenly readers are playing a game of hide-and-seek, instead of focusing on your ideas.

> This is also a sentence fragment. How would you revise it?
> "This study found a statistically significant correlation between the exercise and the joint mobility intervention group. Which <u>means that patients with sciatica improve with physical therapy</u>."

If you suggested that simply inserting a comma before "which" in lieu of the period would sort things out, you're right. "Which" signals the presence of a modifier clause that has no business being by itself, unattached to a main clause. The modifier clause itself certainly has a verb, but without the main clause it is still a fragment.

2. Make sure your verb always agrees in number with your subject

Agreement errors usually result from confusion whether the subject is in the singular or the plural. For example, which of the following sentences is correct?

> "The data <u>is</u> confusing."

> "The data <u>are</u> confusing."

If you are confused about the number agreement, trace the origin of the noun: data is actually a plural of datum ("given" in Latin). Although some dictionaries have encouraged the use of "data" as a singular noun, many reference manuals (including in your own discipline!) would frown upon that usage. Always treat "data" as a plural noun; if it gets corrected to the singular, challenge the rationale for the correction. Be advised that this sort of ambiguity is usually not tolerated for other Latin and Greek plurals in usage in English—many of which are common and even vital in the sciences (see the next section for pluralizing nouns correctly).

Disagreement between subject and verb can also occur with a *compound subject*. The rule is: if two or more nouns are connected by "and," use a plural verb; if they are connected by "or," use a singular verb.

> "Sea surface temperature <u>or</u> atmospheric instability <u>causes</u> hurricanes."

> "Sea surface temperature <u>and</u> atmospheric instability <u>cause</u> hurricanes."

If the second noun is plural, however, the agreement should be with the noun closest to the verb:

> "Sea surface temperature or African easterly lower atmospheric <u>winds cause</u> hurricanes."

> "African easterly lower atmospheric winds or sea surface <u>temperature causes</u> hurricanes."

Sometimes you may get confused because various modifiers intervene between subject and verb, and you feel that you should agree the verb with whatever word is closest to it. The rule is simple: don't lose sight of the subject. In general, words like *with, together with, including, accompanied by, in addition to,* or *as well as* following a subject do not change the number of the subject.

> "Friction, as well as torque, has to be considered in each case."

> "The author, together with a team of investigators from the HERI Institute, draws surprising conclusions."

Make sure that you agree indefinite pronouns with singular verbs: *each, either/neither, everyone, everybody, someone, somebody, nobody, anyone, anybody*:

> "Neither of them was surprised by the results."

Collective nouns (e.g., committee, team, crew, family) usually take a singular verb, especially in American usage (the British may occasionally use them with a plural verb when referring to each member of the group):

> "The patient's family was notified."

Don't be thrown off by what *follows* the verb:

> "A secondary cause of hurricane Katrina was the high-pressure fronts over the Atlantic."

If you want to try to get around this and escape possible awkwardness, you might reword this sentence more elegantly:

> "High-pressure fronts over the Atlantic were a secondary cause of hurricane Katrina."

3. Choose the correct tense

Let's get one thing out of the way: in scientific writing, get used to using future tense sparingly— even when you are sorely tempted to do so. So, don't start your abstract by saying:

"This experiment will determine the rate of catecholase reaction in cellular activities."

Have you or have you not conducted the experiment at the time your reader lays eyes on your paper? Let's just hope you did, or you're in 10 different kinds of trouble. If the experiment *happened* already, there is no point in using the future tense. (Most likely, you are confused by the fact that your audience *will read* about your experiment right after they're done with your abstract.)

Most descriptive passages in science for phenomena that are known to be true are in the present tense:

> "Chitosan is a polysaccharide composed of N-glucosamine and N-acetyl-glucosamine units, in which the number of N-glucosamine units exceeds 50%."

Scientists prefer past tense to describe the experiment:

> "Experimentally, the dimensions of the nanostructures were determined directly, by transmission electron microscopy (TEM), or indirectly, by x-ray diffraction (XRD) or x-ray photoelectron spectroscopy XPS."

and present tense when they report conclusions:

> "The overall results demonstrate that *Saraca ashoka* flowers (SAF) are an ideal natural remedy for preventing oxidative stress and other complications associated with diabetes."

Most importantly, however: *keep your tenses consistent.* Do not start with past and veer into present or vice versa—you will compromise accuracy:

"Turkevitch et al. (2009) <u>proposed</u> a method to synthesize colloidal gold particles into macroscopic aggregates. They <u>prepared</u> 20mL of 1mM $HAuCl_4$ aqueous solution and <u>brought</u> it to a boil. 2mL of 1% solution of trisodium citrate dehydrate <u>is added</u> to the boiling aqueous solution."

The problem with tense shifts is that they muddle the story line. The culprit hides insidiously in the last sentence. Are we still talking about Turkevitch et al.'s experiment or are we talking about something or even someone else? Tense shifts may violate the M rule, because you are not clearly expressing what you actually mean.

4. Whenever possible, choose active over passive voice

Once used liberally by scientists everywhere, the *unnecessary* passive voice is now the staple of stodgy writing. The passive voice may be your best bet in some cases, but your default option should be active voice—in other words, if you must use the passive, do it deliberately and for a good reason. Because there are many misunderstandings about this issue, let's ponder the concept of "verb voice" for a minute.

The two "voices" of the verb in English are active and passive:

Active: "Reactive oxygen species (ROS) <u>disturbs</u> the redox state of the cell and <u>damages</u> cell membrane."

Passive: "The redox state of the cell <u>is disturbed</u> by reactive oxygen species (ROS); cell membrane <u>is</u> also <u>damaged</u> by ROS."

Note that the second version

→ is longer (has more words that don't add to the meaning, thus violating the principle of conciseness, or C rule—"Make every word count");

→ emphasizes (foregrounds) the object in favor of the subject ("state of the cell" is in the subject position);

→ deemphasizes the agent doing the action ("ROS") by placing it after the verb or last in the sentence, hinging tenuously on a preposition ("by")—almost as an afterthought.

You might have good reasons for doing all that—you might need to make "the state of the cell" the focus of your action, for whatever reasons (some of which may be sound). But in general, in natural speech, we tend to avoid passive voice; active voice tells the story of the agent, whereas passive voice tells the story of the object. Deep down, we'd rather read a story of *someone* (doing something), rather than the story of *something* (done by someone).

You don't need to replace *every* passive voice you ever see with active—sometimes it's impractical; sometimes it's undesirable. When describing processes or mechanisms, for example, the passive voice seems unavoidable:

"Combustion air <u>was divided</u> into primary air through the grate, window purge air as well as secondary air drawn in through nozzles from the back of the combustion chamber. The air distribution <u>was adjusted</u> manually by a damper."

"Mercury (Hg) is a persistent toxic and bio-accumulative heavy metal that <u>has been linked</u> to the decline of endangered snakes, impairment of bird and plant reproductive systems, and permanent neurological damage to humans."

In this case, we are deliberately eschewing the agents because they are irrelevant to the meaning of the sentence. Other times, it may feel awkward to insert an agent (the experimenter) every time we describe a natural process or a step in a procedure, or we simply want to vary sentence structure:

"Desirable plants <u>were</u> continually <u>selected and replanted</u> in order to improve their growth habit, fruiting characteristics, insect and disease resistance, and growing season."

Philosophically, passive voice, with its emphasis on the object (as opposed to the subject) suits scientific writers just fine, because it allows them to push the object of their scientific investigation front and center. An unchecked passive voice habit, however, may clog your prose and slowly choke the action out of your paper. Use it sparingly, and always ask yourself if the sentence would sound better without it. When you revise to eliminate passive voice, reintroduce the agent doing the action:

"Calculated band edge levels with strong, medium, and weak QC models <u>were compared</u> with experimental VBM and CBM reported from X-ray photoemission spectroscopy (XPS), X-ray absorption spectroscopy (XAS), or photoluminescence (PL)."

Revised:

"<u>We calculated</u> band edge levels with strong, medium, and weak QC models and <u>compared</u> them with experimental VBM and CBM reported from X-ray photoemission spectroscopy (XPS), X-ray absorption spectroscopy (XAS), or photoluminescence (PL)."

One last thought: do not confuse *passive voice for* past *tense*. You can have passive voice in all tenses:

	were compared	
"The band edge levels and CBM	are compared have been compared	with experimental VBM."
	will be compared	

5. Whenever possible, articulate the action in a verb, not in a noun (avoid nominalizations)

A nominalization is the use of a verb or adjective (usually) as a noun: grammatical alchemy. In general, you want the action of your sentence to be captured by a verb, as opposed to a noun. Which would you rather have:

"The <u>continuation</u> and <u>expansion</u> of airborne measurement programs for CO_2 and related tracers is a desirable goal."

or:

"We should <u>continue</u> to <u>expand</u> airborne measurement programs for CO_2 and related tracers."

Turning the verb ("to expand") in this sentence into a noun ("expansion") is the nominalization. The problem with nominalizations is that they "freeze" the action in a subject or object, forcing you to find another verb that would support your sentence. Since the main action of the sentence is already captured in the noun, you'll necessarily have to use a "weak," generic verb—for example:

"We <u>performed</u> an analysis of the data."

In this case, *perform* is a weak verb; by itself, it doesn't give us very useful clues as to what the sentence is about; *analyze* captures the real action. You are unwittingly violating the C rule ("Make every word count"). How about just:

"We <u>analyzed</u> the data."

The following are some weak verbs and constructions that should alert you that a nominalization is near:

> *there is/was*
>
> *it is/was*
>
> *occur/-red*
>
> *[is/was] seen*
>
> *[is/was] observe/-ed*
>
> *[is/was] noted*
>
> *[is/was] done*
>
> *make/made*
>
> *get/got*
>
> *produce/-ed*
>
> *cause/-ed*
>
> *supply/-ied*
>
> *have/had*
>
> *provide/-ed*

Use nominalizations sparingly in your prose. Avoiding or limiting nominalizations will make your sentences shorter, clearer, and easier to understand.

B. Subjects

1. Use nouns correctly

The subject of the sentence is usually a noun or pronoun. Two major issues you need to watch out for are (a) plurals and (b) noun clusters.

a. Pluralize correctly

Never use an apostrophe to form a plural: use "the 1980s" (not "the 1980's"). Also, learn the correct forms of Latin and Greek words (they are fairly common in biological sciences and not only) and make sure you use the correct singular or plural verb form with them. The following are some of the most common irregular foreign plurals:

Singular	Plural
-a	→ *-ae*
alga, larva, vertebra, nebula, alumna	algae, larvae, vertebrae, nebulae, alumnae
-us	→ *-i*
fungus, radius, stimulus, alumnus	fungi, radii, stimuli, alumni
-us	→ *-ora/era*
genus, viscus, corpus	genera, viscera, corpora
-um	→ *-a*
medium, datum, ovum, spectrum, millennium*	media, data, ova, spectra, millennia

-is	→ *-es*
basis, thesis, ellipsis, hypothesis, diagnosis	bases, theses, ellipses, hypotheses, diagnoses
-ix or -ex	→ *-ices*
matrix, cicatrix, helix, index	matrices, cicatrices, helices, indices
-on	→ *-a*
criterion, phenomenon, automaton, polyhedron	criteria, phenomena, automata, polyhedra
-ma	→ *-mata**
dogma, stigma, schema	dogmata, stigmata, schemata (*-s plural is also acceptable: dogmas)
-(i)eu, -eau	→ *-(i)eux, -eaux**
milieu, tableau	milieux, tableaux (*-s plurals are also acceptable: milieus)

* As with any rules there are exceptions. "Stadia" and "podia" are the original plurals of "stadium" and "podium," but let's face it, no one uses them anymore and the dictionary accepts both the –a and the –s plurals: "stadiums," "podiums." When in doubt, consult the dictionary and if the Anglo-Saxon plural is acceptable for Greek and Latin terms, use it.

b. Demystify noun clusters

Noun clusters occur when you join more than three words (usually nouns, adjectives, and adverbs) that together form one unit of meaning, that is, refer to one entity. The lack of connecting words (prepositions and articles) among the components of the cluster can make the relationships between words difficult to discern. What you may gain in conciseness, you lose in clarity.

"We designed and synthesized a <u>BODIPY-based brain-imaging probe (BAP-1)</u> to study β-amyloid plaques."

Noun clusters are more frequent in technical writing, engineering, and management, where phrases like "software engineering economics decision analysis techniques," "malware cost estimation algorithmic models," or "limited impedance high-fidelity flux transmogrifier" are used with wanton abandon. Such clusters remind us of a highway pileup that stops the flow of traffic: your reader will also likely have to slow down and carefully analyze the cluster in order to proceed. Making your readers wonder what you mean violates the M rule ("Write what you mean"). Demystify noun clusters by adding proper connectors and rearranging the sentence so that it's more easily understood:

"We designed and synthesized <u>a BODIPY-based probe (BAP-1)</u> for the imaging of β-amyloid plaques in the brain."

2. Use pronouns correctly

a. Make pronouns agree with verbs

It stands to reason that pronouns have to agree with their verbs—as we already mentioned in the subject-verb agreement section. Remember that pronouns that end in *–one, –some, –body,* and *–thing* always take a singular verb.

"<u>Everyone</u> at the CERN particle accelerator <u>is working</u> on proving the existence of the Higgs boson."

b. Make pronouns agree with their antecedent

An even more insidious problem is posed by pronouns that don't agree with their antecedent. A pronoun stands *for* ("pro") a noun—which means it has to match the noun's gender, number, and syntactical function—subject, object, or modifier. Consider this mystifying sentence:

> "Patients should be asked if they sustained an injury recently; <u>they</u> are among the most common causes of hemarthrosis."

The second "they" should stand for "injury" (and not for "patients"), and therefore should be changed to "it." This sentence violates the M rule: the meaning is not clear. In fact, the sentence literally claims that *patients* are the most common cause of hemarthrosis! Always ask yourself what your pronoun stands for, and make sure you use the correct form.

When your pronoun has to match in gender a singular noun that may stand for either a female or male entity, you are confronted with sexist bias:

> "The doctor will be asked to see <u>his</u> patients at a different date."

Unless you are talking about a specific doctor who happens to male, you should not assume that "male" is the default gender. You have three options:

- use the plural: "they/their": "Doctors will be asked to see their patients"
- avoid using a pronoun altogether: "The doctor will be asked to see [the] patients."
- use the more inclusive but more awkward "his or her": "The doctor will be asked to see his or her patients."

We strongly prefer the first two, though any of these options is preferable to slipping into casual sexism in scientific prose.

c. Use the singular "they" pronoun where appropriate

In 2019, Merriam-Webster consecrated the singular "they" as a generic third-person singular pronoun in English, and this usage has been quickly adopted by many style manuals (e.g., APA), while others have acknowledged it without necessarily adopting it formally (yet). The singular "they" helps writers avoid making assumptions about gender and is the preferred way to refer to nonbinary or gender-fluid people. It also helps us avoid awkward constructions such as "Everyone can bring *his or her* preferred instruments." ("Everyone" is singular and the rules of agreement would require that it be replaced by a singular pronoun later on.) Now, you can use instead: "Everyone can bring *their* preferred instruments" without being frowned upon by the proverbial grammar police. Given that language is a dynamic, living system with moving parts and rules that are continually adapting to the way speakers use it, this is not a surprising development.

d. Avoid vague pronouns

Novice writers sometimes use the pronouns "this" and "it" as generic placeholders for a poorly understood concept. What does *this* stand for in the following example?

> "The aglycone portion and the sugar parts are biosynthesized separately and join afterwards. Due to the difficulty of cardiac glycoside synthesis, CGs supply reliance rests on natural sources. This occurs after uridine diphosphate sugar (UDP-sugar) reacts with the aglycone to form the glycoside."

Awkward passive voice aside, "this" is still a problem because it can refer to any of the nouns mentioned in the previous sentences. The writer is again violating the M rule, while the reader's

Smbc-comics.com

© 2018 Zach Weinersmith. Used by kind permission.

forehead is developing some deep creases. At a minimum, we would revise this example by replacing "this" with a clarifying noun:

> "The aglycone portion and the sugar parts are biosynthesized separately and join afterwards. Due to the difficulty of cardiac glycoside synthesis, CGs supply reliance rests on natural sources. **The joining** occurs after uridine diphosphate sugar (UDP-sugar) reacts with the aglycone to form the glycoside."

Often, the overuse of "this" or "it" (which we call "mystery pronouns") betrays a lack of understanding of the original material. Any time you use "this" or "it" (or their plural forms, "these" or "they"), make sure it is crystal clear from the context what they mean.

e. Use relative pronouns correctly (that, which, who)

In general, use *that* for clauses modifying objects and *who* for clauses modifying human agents:

> "Patients ~~that~~ <u>who</u> suffer from hemarthrosis may have pain and swelling of the knee joint."

Remember that both *who* and *that* are pronouns, and they play a role in the clause they are part of—they can be subjects, objects, or possessive modifiers. *Who* is the only relative pronoun to change forms according to the role it plays in the sentence.

"We contacted the Nobel Prize committee, <u>for whom</u> we had written these guidelines."

"We contacted the Nobel Prize committee, <u>whose</u> guidelines we had followed."

Beware of hypercorrectness: do not use *whom* when *who* would suffice. Think logically: Is *who* the subject of the clause? If so, let it be—no need to change it to *whom*.

"These fraudulent results were published by Creighton (1999), ~~whom~~ <u>who</u>, we believe, took advantage of the flawed peer review system."

Finally, use *that* for restrictive modifier clauses (without which the sentence would not make sense) and *which* for nonrestrictive modifier clauses (without which the sentence would still make sense).

"The proteins <u>that catalyze chemical reactions</u> are called enzymes."

Without the modifier clause, the sentence would make little sense:

"The proteins are called enzymes."

"That" clauses do not usually require to be separated by a comma. By contrast, "which" clauses must be set off by commas and are nonessential to the overall meaning of the sentence.

"In 1877, German physiologist Wilhelm Kühne (1837–1900) first used the term enzyme, <u>which comes from Greek ενζυμον, "in leaven,"</u> to describe this process."

Note how in this case, removing the "which" clause still leaves us with an intelligible sentence:

"In 1877, German physiologist Wilhelm Kühne (1837–1900) first used the term enzyme."

f. Use the first person when appropriate (I or we)

In many cases, you probably don't *need* to use the first person if the context is clear. If it's your paper, your research, your methods and materials, you don't need to say "We streaked the plates with…." "The plates were streaked with…" is understood to have been performed by the research team. However, the current trend in scientific writing favors the use of the first person, because it makes the prose more straightforward and reduces the incidence of passive voice. The use of first person is favored in primary research articles (where the authors are usually the experimenters) and less or not at all in review articles, unless the review article introduces a novel idea. It is up to you to check with your instructor or editor regarding what is acceptable in your field.

You should, however, use the first person if you are clearly expressing an opinion that is not accepted dogma: recommendations for guidelines, decision points, and so on. Overall, you will find that "I" and "we" tend to be more common in posters, grants, summaries, and presentations and less common in journal articles (especially review articles) and abstracts.

C. Order of Words

A good deal of confusion in scientific writing may result from the way words are placed together in a sentence. This is usually a violation of the M rule: figure out what you mean so you can clarify it for your readers. Then you can make decisions regarding word order that first and foremost promote clarity.

1. Maintain the subject-verb-object order

Subject-verb-object is the preferred sentence structure in the English language. Anything interfering with this natural order of words is harder to process by your readers. You are unlikely to hear your TA in chem lab shout out *"Back, stand you! Out of control, the experiment is!"* (unless your chem TA is Yoda).

You can still start the sentence with a modifier or a modifier clause, but try not to keep your readers waiting for the real action to kick in (i.e., subject + verb):

> "In wavefunction-based approaches to atomic and molecular structure, whether they are carried out in a relativistic framework or not, with a variety of instruments, the step aiming at the treatment of dynamic electron correlation is in the vast majority of cases the final step and the one which defines to a large degree the accuracy of the approach."

In this sentence, the subject comes in as the 26th word in the sentence, while the (weak) verb is the 35th (the stronger verb comes in even later). The writer could do better:

> "In wavefunction-based approaches to atomic and molecular structure, the final step in the treatment of dynamic electron correlation usually defines to a large degree the accuracy of the approach."

If the writer decides that "whether they are carried out in a relativistic framework or not" is essential information, he or she could add it or place it in a different sentence. In the following example, a long appositive stands in the way of a happy reunion between subject and verb. (An appositive is a phrase offset by commas and explaining something about a noun or pronoun: "Watson, **a Nobel prize laureate**, discovered that . . .")

> "<u>The ecological footprint</u>, a measure comparing the human demand with planet Earth's ecological capacity to regenerate, and representing the amount of biologically productive land and sea area needed to regenerate the resources consumed by the human population, <u>can be used</u> to estimate how many planet Earths it would take to support humanity, if everybody in the world would live a given lifestyle in a certain country."

To revise, your best option is to place that definition in a separate sentence:

> "<u>The ecological footprint</u> is a measure comparing the human demand with planet Earth's ecological capacity to regenerate, and representing the amount of biologically productive land and sea area needed to regenerate the resources consumed by the human population. This footprint <u>can be used</u> to estimate how many planet Earths it would take to support humanity, if everybody in the world would live a given lifestyle in a certain country."

According to Gopen and Swan (in their article "The Science of Scientific Writing"—*American Scientist*, 1990), not following a grammatical subject as soon as possible with a verb remains one of the top reasons accounting for the apparent difficulty and inscrutability of scientific writing. Distorting the subject-verb-object sequence makes it difficult for the reader to understand the "story" the sentence is telling.

One last thing: Should you end the sentence with a preposition? Probably not: doing so may require some extra mental gymnastics from the reader. After all, it's called a *pre*-position because it's supposed to come *before* the noun. You can try to write around it, but you can't always fight colloquialism. One of the more famous examples is the misattributed quote from Winston Churchill: "up with which I will not put" (the first half of the sentence varying in content since 1942).

2. Keep items in a series parallel

Items in a list, or items that are compared, should be similar (parallel). Violating this rule can create imprecision, which may confuse your readers.

> "The teams have developed various experimental techniques, *such as* gas adsorption, X-ray diffraction (XRD), small angle x-ray, and <u>scanning and transmitting electron microscopy</u>."

This list after "such as" is not homogeneous. To be so, it should contain all nouns or all gerunds, but in this case it combines them, to confusing effects. Stick with one form throughout:

"The teams have developed various experimental techniques, such as gas adsorption, X-ray diffraction (XRD), small angle x-ray, and <u>electron microscopy</u>."

Faulty comparisons are also a type of parallelism gone awry:

"When a person ages, <u>the brain</u> does not function as quickly <u>as a 20-year old</u>."

Logically, this sentence does not make sense because the writer is comparing apples and oranges (brains with whole persons). One should revise for clarity and in the spirit of the M rule:

"An aging brain does not function as quickly as <u>the brain of a 20-year old</u>."

Other times, one can fall short of fulfilling the promise established in the first item of the list:

"Vitamin D comes in the form of cholecalciferol (D3) from diet and/or supplements, or ergocalciferol (D2)."

This sentence disappoints because the writer had detailed the first item of the list with a modifier but left us wondering about the sources of ergocalciferol.

"Vitamin D comes in the form of cholecalciferol (D3) from diet and/or supplements, or ergocalciferol (D2), primarily available in certain species of mushrooms and lichens."

The same is true of the following sentence, which starts with a certain syntactical pattern but does not follow it through. How would you modify it?

"The more a patient utilizes a CPM machine, there is a significant increase in ROM and a decrease in length of hospital stay."

Here's one way to revise it:

"The more a patient utilizes a CPM machine, the more likely he or she is to increase ROM and decrease the length of hospital stay."

Parallelism extends beyond the sentence level. Keep all the headings and subheadings parallel within a document—in other words, keep their structure similar, whether they are full sentences, noun phrases, –ing phrases, or something else. Keeping headings parallel will help further orient your readers. For example, these two headings could be found in the same paper on cardiac glycosides:

Chemical synthesis of cardiac glycosides

Cardiac glycosides have antiarrhythmic activity

The first heading is a noun phrase, while the second is a full sentence. Here's a possible revision for parallelism:

Chemical synthesis

Antiarrhythmic activity

In this case we got rid of "cardiac glycosides" altogether, since the whole paper was about them and it was clear from the context that they were implied.

3. Avoid misplaced and dangling modifiers

To enhance accuracy, modifiers should be placed where they make sense:

"Over the years many medical researchers found treatments for patients suffering from sciatica <u>with physical therapy</u>."

Revised:

> "Over the years medical researchers found <u>physical therapy</u> treatments for patients suffering from sciatica."

A dangling modifier can also muddle accuracy by breaking the bond between a participial construction and the subject to which it refers. To fix a dangling modifier, you need to tighten the connection between the two verbs (the main verb and the verb of the modifying construction) by having them refer to the same subject:

> "By <u>examining</u> the research gathered for this literature review, divorce <u>seems</u> to cause negative effects for a family."

The dangling modifier (everything that comes before the comma) includes an –ing verb, "examining," which should have the same subject as the verb in the main clause ("seems"). When this is not the case, you need to repair that bond either by rephrasing or by clarifying who did both actions. We favor the simpler approach:

> "Research shows that divorce has negative effects on the family."

D. Sentence Length

1. Avoid excessively long sentences

Here's a simple rule for the ages: the longer the sentence, the greater the effort required to understand it. The following is an excerpt from an old (*very* old) academic rulebook, extolling the virtues of a solid educational foundation:

> "Good things grow on very hardly at their first planting, because that profit which they promise at their entry hath not yet been proved, and therefore wanteth the commendation of trial, which is the very best mean to enforce persuasion: and their pretence to be profitable upon some probability in sequel is a great inducement indeed but to those people which can foresee ere they feel, but of small importance to them which cannot see till they feel. Good things find hard footing when they are to be reformed after a corruption in use, because of that enormity which is in possession and usurpeth on their place, which having strengthened itself by all circumstances that can move retaining, and with all difficulties that can dissuade alteration, fighteth sore for itself, and hard against redress, through the general assistance of a prejudicate opinion in those men's heads which might further the redress."
>
> —Richard Mulcaster, *The First Part of the Elementary,* 1582

The most awesome part of this paragraph, of course, is the use of "wanteth," "usurpeth," and "fighteth." But we forget ourselves. Our job is to prick those inflated sentences with the needle of conciseness. Try it yourselves first. Go on!

Back? Okay, here's what we came up with:

> "Wise people prefer to start educating children early rather than have to correct the course of a bad education later. The rewards may be slow at first, but they will be much greater in the long run."

We believe it's a fair distillation of the guilty paragraph—or, at least, it gets to the point *much* sooner. (Mulcaster, by the way, was also a proponent of clarity and grace in writing. Go figure.)

No one writes like that anymore, and we are (mostly) eternally grateful to the Plain Language Movement for making such turns of phrases ridiculously antiquated. Still, verbosity and redundancy may very well usurpeth even thy holiest attempts at clarity:

"It is known that information about plaque size and composition is essential in the treatment of atherosclerosis."

Remember the C rule ("make every word count"): take out words that do not contribute to your meaning. You certainly don't need "it is known that" to get your meaning across.

Examples of "it . . . that" phrases that can be removed or replaced:

Empty fillers	Shorter equivalent
"It would thus appear that"	*"Apparently"* –or omit
"It is considered that"	*"We think"* –or omit
"It is this that"	*"This"*
"It is possible that the cause is"	*"The cause may be"*
"In light of the fact that"	*"Because"*
"It is often the case that"	*Often*
"It is interesting to note that"	*Omit*
"It is not impossible that"	*Omit*
"A not unlikely cause could be that"	*Omit*
"It seems that there can be little doubt that"	*Omit*

While these venial sins should be eschewed in writing, similar phrases might be uttered in *spoken* delivery of materials. Although unnecessary, this kind of verbiage may contribute to the stylistic cadence unique to the given presenter. (However, they might also make him or her sound pompous: it is really all in the delivery.) In writing, our advice is to conserve your lexical real estate and make every word count when you edit for conciseness:

"One study done in 2009 looked into a possible eczema treatment by modifying one's diet to exclude items that could worsen eczema. The study was done by the Faculty of Medicine and Health Sciences at the University of Nottingham."

Revised:

"A University of Nottingham 2009 study looked into dietary exclusions that could improve eczema."

2. Vary sentence length

The Goldilocks principle of sentence length states that your sentences should be neither too long nor too short but just right; and if you do need to use very short sentences, try to alternate them with longer ones in order to improve the flow of your writing. Having a suite of three or more short, "choppy" sentences, especially if they all repeat the same syntactical pattern, will offend the reader's sense of rhythm and disrupt the flow of the reading. Revise to help readers find the logical connections in your prose.

In prostate cancer, TRAP-1 differential expression proved an attractive biomarker of disease. TRAP-1 is abundantly represented in prostatic intraepithelial neoplasia (PIN). It is, however, mostly undetectable in normal prostatic epithelium or benign prostatic hyperplasia (Leav et al., 2010). Surprisingly, these studies also identified a previously unknown pool of Hsp90 localized to mitochondria (Kang et al., 2007). This is similar to the data on TRAP-1. Also the pool of mitochondrial Hsp90 was expressed in tumors in vivo. However, they were undetectable in the normal tissues (Kang et al., 2007).

Revised:

In prostate cancer, TRAP-1 differential expression proved an attractive biomarker of disease, being abundantly represented in prostatic intraepithelial neoplasia (PIN), while being mostly undetectable in normal prostatic epithelium or benign prostatic hyperplasia (Leav et al., 2010). Surprisingly, these studies also identified a previously unknown pool of Hsp90 localized to mitochondria (Kang et al., 2007), similar to the findings on TRAP-1. Additionally, the pool of mitochondrial Hsp90 was expressed in tumors in vivo while they were undetectable in the normal tissues (Kang et al., 2007).

E. Word Usage

1. Avoid colloquialisms and jargon

An accurate vocabulary is absolutely critical in scientific writing. Remember that you are addressing a community of scientific peers, and you should show respect for their time and for their intellectual abilities. You should assume a high level of vocabulary matched by a high level of expectations. Avoid colloquial phrases, contractions, and jargon that adds little or nothing to the conversation (except, of course, for "usurpeth").

"In patients with normal sensation, and without a bleeding diathesis, which is the <u>fancy</u> term for predisposition, the joint trauma is generally remembered."

"The analysis was <u>based off of</u> calculations performed at the Tryst lab."

A pretentious style can (and usually will) get you in trouble. Try understanding this sentence in one reading (no cheating!):

"Crucial to the comprehension of osteoarthritis in former football players, injury statistics elicit a more visceral reaction to the prevalence of the disease."

2. Avoid inappropriate/awkward usage and malapropisms

Remember the C rule: make every word count. That implies that you know the word's meaning(s) and that you have carefully thought out its placement in your sentence. We invite you to spot in the sentences below examples of inappropriate usage leading to awkwardness and *confusion*, that number one enemy of clarity:

"Hemarthrosis may then be treated <u>thereafter</u> the diagnosis of acute knee injury."

"<u>Heavy</u> research needs to be done to find a solution."

"More specifically, <u>concerns</u> about the accompanying effects on drinking water <u>have been in question</u>."

"Several studies <u>from each article</u> concluded with the statement that exercise clearly <u>minimizes</u> <u>and improves</u> cognitive decline in aging adults."

Malapropisms are a special case of inappropriate usage, in which you completely misuse a word based on phonetic similarity. Awkward usage, although incorrect, is still somewhere in the same semantic sphere with your intended meaning; by contrast, with malapropisms, you are not even in the ballpark—you are playing baseball with a spoon.

"The patient was <u>asymptotic</u>."

"The blood vessels are in danger of <u>rapture</u>."

"The family was <u>prostate</u> with grief."

"The patient had a <u>veracious</u> appetite."

"<u>Weaving</u> people from medications would allow the body to heal naturally."

Sure, you might *actually* be talking about blood vessels experiencing the rapturous delirium of having been delivered from stenosis, or about a patient whose appetite had, in fact, been verified, but you probably aren't.

3. Learn the correct usage of commonly misused words

Some words are commonly misused and should therefore be studied carefully and learned before one hazards to use them correctly in writing. The following is a list of commonly misused words, to which you can add your own.

it's/its/its' (it is/belongs to it/grammatically incorrect)
there/their/they're (over there/belonging to them/they are)
effect/affect (the first is *mostly* a noun, the second *mostly* a verb]
accept/except (they only sound similar: agree, receive vs with the exception of)
advice/advise (the first is the noun, the second the verb)
loose/lose (the first is the opposite of *tight*, the second the opposite of *find*)
cite/site (the first means to include a reference, the second is a place)
then/than (the first means *after*, the second signals a comparison)
too/to (this is usually just a typo)
whether/weather (the first is *if*, the second refers to the outside temperatures, precipitation, etc.)

Confusing some of the terms on this list often comes from lack of attention or time. Here's an idea: use spellcheck while you're writing. Often (but not always) spellcheck will flag "use spellcheck while <u>your</u> writing" with a little blue squiggly line under "your," indicating wrong usage. Pay attention to squiggly blue lines and squiggly red lines!

Don't confuse common terms like these:

Accuracy/precision – *accuracy* is the degree of correctness of a measurement or statement. *Precision* is the degree of refinement with which a measurement is made or stated and implies qualities of definiteness and specificity.
As/like – "as" is a conjunction preferred to "like" in the majority of formal writing genres.
Dose/dosage – a *dose* is the quantity to be administered at one time or the total quantity administered. *Dosage*, the regulated administration of doses, is usually expressed in terms of a quantity per unit of time. "Give a dosage of 0.10 mg bid until the dose has been ingested."
Examine/evaluate – patients, animals, and microscope slides are examined; conditions and diseases are evaluated.
Gender/sex – *gender* is cultural, and is the term to use when referring to men and women as social groups. *Sex* is biological; use it when you need to draw a clear biological distinction.
Imply/infer- to *imply* is to suggest, indicate, or express indirectly. To *infer* is to conclude.
Necessitate/require – *Necessitate* means to make necessary. *Require* means to have a need for. A patient requires treatment. The treatment may necessitate certain procedures.
Principal/principle – a *principal* is a leader; used as an adjective, it means highest rank. A *principle* is a fundamental truth or law.

Regime/regimen – a *regime* is a system of management or government. A *regimen* is a system of therapy. Neither have anything in common with *regiment!*

Use/utilize/employ – generally, *use* is intended term; *utilize* is more pretentious and is best avoided. Use *employ* for putting a person to work.

Varying/various – *varying* means changing, but *various* means of several kinds.

While/whereas – *while* indicates time and a temporal relationship. *Whereas* means, "when in fact," "that being so," and "in view of the fact that."

4. Avoid redundancies

If you're going to make every word count, make sure you check for redundancy. Remember the C rule and ask yourself "is this word really necessary?" The following is a list of redundant constructions where you can safely omit the italicized words, since they are implicit:

continue *on*	*end* result
refer *back*	*mandatory* equipment
check *up on*	*new* beginning
all *of*	*optional* choice
true facts	five/many *in number*
enter *into*	blue *in color*
face *up to*	round *in shape*
more superior	*positive* benefits
superior excellence	large *in size*
1 a.m. *in the morning*	repeat *again*
at this point *in time*	*past* history
collaborate/mix/join *together*	*complete* stop
circulate *around*	prioritize *in order of importance*

5. Use bias-free, inclusive language

Use nondiscriminatory, nonsexist, neutral language in your writing. Unintentional sexism is still sexism:

> "We shall sample a population of 50 recent graduates of medical school, and interview their <u>wives</u> separately."

Use "spouses" or "partners" in this case.

Biomedical literature is especially susceptible to biased language because it often deals with human subjects categorized by their conditions. Avoid reductionist labels like *the demented* or *diabetics* in favor of *people with dementia/diabetes*; this way, you acknowledge that you do not reduce the person to a disease.

Because English does not have a gender-neutral personal pronoun, it is quite difficult to avoid gendered nouns or pronouns in certain circumstances, but you can certainly do your best. The following are a few rules to help you avoid sexism in your writing (and in everyday usage):

- Use gender-neutral nouns when speaking generically about people. Avoid *man, mankind, manpower, man-made*; use instead *men and women, the human race, humankind, people, workforce, personnel, synthetic*
- Use gender-neutral alternatives for professions. Avoid *spokesman, chairman, actress, stewardess, policemen*; use instead *spokesperson, chairperson, actor, flight attendant, police officers.*

- Use plural instead of singular pronouns. Avoid "*Every author should proofread his manu-script*"; use instead "*Authors should proofread their manuscripts.*"
- Eliminate possessive pronouns altogether whenever possible: Avoid "*Each lab technician should sign his or her time sheet*"; use instead "*Each lab technician should sign the/a time sheet.*"
- When you have no choice, use "he or she," "his or her," "him or her" to avoid sexism (or even better, the plural "they/their/them" if you can). Avoid using s/he or he/she—they are awkward constructions and will slow your reader down. Use singular *they* when appropriate.

There is a tendency in the humanities to use solely the feminine pronoun in all cases in an effort to recalibrate linguistic bias, but we have not noticed a similar trend in the scientific literature. More extensive guidelines on how to avoid biased language can be found in every style manual in your field. When in doubt, consult the manual!

6. Clarify your abbreviations

Scientific writing is replete with abbreviations that are critical to the meaning of the paper. Abbreviations can take one of these forms: initializations (where each letter is pronounced, e.g., "FDA" for "Food & Drug Administration), or acronyms (initializations that spell words, e.g., "AIDS" for "acquired immune deficiency syndrome"). Truncated words ("chemo" for "chemotherapy") are another kind of abbreviation, but rarely used in formal scientific writing. Initializations and acronyms are common in scientific writing. You are free to use abbreviations as long as you follow these rules:

- Define the abbreviation the first time you use it by spelling out the term it stands for and including it immediately after in parentheses.

  "Chemotherapy resistance may be created by a particular type of proteins called multidrug resistance proteins (MDRs), which are considered a major obstacle to cancer treatment. Scientists tried for decades to identify MDRs and study their mechanism."

- Unless an abbreviation is conventionally accepted (e.g., DNA, USA), avoid using it in the title or abstract.
- The current trend is to use abbreviations without the period (e.g., "mg" and not "mg." or "PhD" and not "Ph.D."). Observe the current conventions in your field.
- Do not begin a sentence with an abbreviation.
- Do not abbreviate units of measurement when they are used without numerals (e.g. Never write: "Several mL were added.").
- Do not abbreviate when confusion might result from doing so.

F. Punctuation and Mechanics

In ancient times, where an oral culture predominated, written texts appeared as blocks of continuously running letters with no spaces and no punctuation of any kind; dissemination of knowledge was mostly oral so writing was mere record keeping. Punctuation developed as we have started relying more on the written text to transmit information. Today, punctuation is vital to clarifying the meaning of a text. Without it, you may lose yourself in a maze of meanings, as in the classic example:

Eats shoots and leaves, *which means something completely different from*
Eats, shoots, and leaves, *which means something completely different from*

Eats shoots, and leaves.

Comma placement makes all the difference between a gun-toting panda and one peacefully munching away.

Other punctuation marks are also essential, of course, as a squabble between two *30 Rock* characters attests:

DotCom: You said: Give to charity. Please, no presents!
Tracy Jordan: No, I said: Give to charity? Please, no. Presents!

1. Insert commas where they are needed

Commas can add clarity to the sentence by guiding the reader's interpretation of the text:

"Although samples were collected from the logwood boiler <u>after 10 hrs</u> the Dekati Gravimetric detector was not able to detect trace amounts of PM1."

A comma placed before or after "after 10 hrs" would substantially change the meaning of the sentence. Make sure you clarify what you mean!

Use commas after *however* when used at the beginning of a sentence, or both before and after when used as a parenthetical adverb in the middle of a sentence.

"The results, however, were unexpected."
"However, we did not expect these results."

If you use *however* as a conjunctive adverb (connecting two sentences), you should frame it between a semicolon and a comma to avoid a comma splice:

"The patients followed the government-recommended dietary guidelines; however, they did not lose any weight."

Finally, use commas with nonrestrictive modifier clauses introduced by "which" (we already covered this one in the "Pronouns" section).

"Masuda and collaborators found that tumor cells treated with a non-ATP competitive tyrosine kinase inhibitor exhibited decreased expression of TRAP-1, <u>which was associated</u> with enhanced mitochondrial apoptosis (Masuda et al., 2004)."

2. Avoid comma splices

Comma splices betray poor logic and make the reader do a double-take for the missing connection. When a comma "splices" a sentence, it in fact replaces a more logical connector, thus robbing the reader of context and slowing down the reading.

"Blood thinners are often used in the prevention of strokes, they are especially important for people who have already suffered from a stroke."

"The children were randomly divided into two groups, one group stayed in Norway while one traveled to the Gran Canary Island."

You can repair a comma splice in a number of ways:

- start a new sentence (make sure, however, that you don't end up with a succession of choppy sentences)
- add a conjunction: *and, but, or, for, nor, so*
- add an adverb: *therefore, moreover, however, thus, hence* (but only after a semicolon)
- replace the comma with the correct punctuation (colon, semicolon)
- rephrase

Revised:

"Blood thinners are often used in the prevention of strokes and are especially important for people who have already suffered from a stroke."

Revised:

"The children were randomly divided into two groups: one group stayed in Norway, while the other traveled to the Gran Canary Island."

3. Punctuate lists correctly

Use commas after every item in the list; use the serial comma (also known as the Oxford comma or the Harvard comma) before the "and" connecting the last element in the list in order to avoid confusion:

"Various species of salmon come in a variety of colors: red and grey, pink and grey, silver and black."

Revised:

"Various species of salmon come in a variety of colors: red and grey, pink and grey, silver, and black."

To write what you mean, make sure you place that last comma properly: do the fish come in black, in silver, or in the black-and-silver combo?

If the items in your list are complex—consisting of phrases or even whole clauses, consider adding semicolons after each item, and, when necessary, a number (1, 2, 3, etc.) before each item:

"To analyze the 21st century sustainability threats, we need to understand three key issues: (1) the energy demand per capita to ensure welfare of a society; (2) the maximum global energy consumption to ensure stable ecosystems; and (3) the balance between world's future energy demand to ensure welfare for all people with the absolute necessity to preserve the integrity of the biosphere."

If you are using a compound term or phrase in your writing, don't separate the components of the phrase with a comma! You would refer to the H&E stain as "hematoxylin and eosin," not "hematoxylin, and eosin."

4. Use hyphens correctly

Hyphens are complex punctuation signs that require some careful consideration. You should use them, in general, to create compound phrases such as *cost-benefit analysis*. The following are some guidelines governing the use of hyphens:

Use	Example
1. Create compound modifiers that *precede a noun*.	weight-bearing vertebra (but "this vertebra is weight bearing"; it is not hyphenated *after* the noun)
2. Join fractions and ratios that function as adjectives	a one-to-one ratio; three-quarters gone
3. Reduce redundancy in series	the first-, second-, and third-born offspring
4. Join letter or number modifier	β-amyloid protein, 5-year-old boy

5. Join strings of modifiers that *precede a noun* and *do not contain an adverb in first position*	dust-covered lens, but NOT "recently dis-covered enzyme"
6. Join prefixes to an adjective derived from a proper noun	pre-Cambrian
7. Join prefixes that end with the same vowel as the root word begins	anti-inflammatory

A big caveat here: the use of hyphens depends on your discipline's style manual. In AMA, for example, rule 6 doesn't hold water: Precambrian, postsurgical.

Remember that phrases that are hyphenated as compound modifiers are often not hyphenated after the noun. Thus, you write "gender-related results" but "results were gender related."

5. Learn when to use en-dashes, em-dashes, and parentheses

Use en-dashes (which are slightly longer than hyphens) to express numerical ranges rather than connect words: 1991–1992, 12–20 patients, 1:00–2:00 pm, 35–38 years. You can find en-dashes in the "Symbols>Special Characters" section of your word processor. En-dashes are also usually automatic if you type a word, followed by a space, followed by a dash, followed by a space, followed by another word.

Em-dashes—which are twice the size of en-dashes—and parentheses are used to mark a parenthetical, nonessential explanation or a pause in speech. For that reason, their use is discouraged in scientific writing, which favors a direct, clear style. Restrict your use of parentheses to abbreviations, in-text citations depending on your style manual, and signaling figures and tables. In MS Word, em-dashes are available in the "Symbols>Special Characters" menu, or they are created automatically if you type two hyphens between words, with no spacing after or before the word.

When in doubt, always consult your style manual. In some scientific styles such as AMA, en-dashes are sparsely used if at all (a hyphen will do).

6. Do not abuse apostrophes

Apostrophes serve two main roles: (a) form the possessive ('s) and (b) mark a contraction ("don't"). Never use them to form plurals—or you can be reported to the Apostrophe Protection Society. Consequently, please write "the 1980s" and *not* "the 1980's."

7. Learn how to handle numbers

Next we offer some general guidelines for using numbers, with the caveat that *style manuals across disciplines vary widely in their guidelines.* We recommend going back to the manual most relevant to your field and making sure you are following current recommended standards.

In *general*, you should spell out one-digit numbers (one to nine, although this is not the case for AMA style) and use numerals for all larger numbers (10 and up).

"Only four studies met the criteria."

"They reviewed a total of 394 cases."

And now, the exceptions:

- Spell out any number that starts a sentence, title, or heading, but avoid starting a sentence, title, or heading with an unreasonably high number.

"Fifteen people were treated for malaria."

"We treated 4,567 cases of malaria." (note: AMA would not use a comma separator for thousands in this instance)

- Use numerals whenever numbers are followed by units of measure:

"A wind speed of 4 m^{s-1} hindered the sample collection process."

- In a series, use numerals if any number in the series is 10 or more:

"The doctor saw 7 patients on Monday, 4 on Tuesday, and 11 on Wednesday."

- Do not place a superscripted reference number after a numeral (even after something as obvious as a date); you do not want it to be interpreted as an exponent.

II. Drafting Effective Paragraphs

Mastering the art of writing a sentence means that you've paid attention to several crucial areas such as action, subject, order of words, length, semantics, and punctuation. Writing a paragraph now should be a breeze as long as you keep going back to our core M and C rules: Write what you Mean; make every word Count.

A. Unity, Coherence, Cohesion

As in all writing, unity, coherence, and cohesion are crucial in scientific writing. These concepts are closely interrelated, but they mean slightly different things:

Unity: a paragraph or paper may be considered unified if it deals with the same topic from beginning to end.

Coherence: a paragraph or paper may be considered coherent if all sentences and paragraphs are logically and consistently connected at the idea level. Note that a unified paper may still fail the coherence test if you cannot follow the reasoning or progression of ideas from one sentence or paragraph to another.

Cohesion: a paragraph or paper may be considered cohesive if each sentence and paragraph is linked to the previous through synonyms, pronouns, verb tenses, conjunctive adverbs, and other markers. Note that a unified and coherent paper may still fail, to some extent, the cohesion test if the logic of ideas is not mirrored by clues in the sentence and paragraph organization.

Consider this paragraph:

The amount of pain in the lower back is the basis in determining treatment options for patients. Last (2009) writes: "Patients should receive information about effective self-care options and be advised to remain active (because muscles that do not move can eventually become hypersensitive to pain)" (p. 297). Since muscles have a tendency to lose mass when not active, exercise is vital to retain mass and strength.

It is clear that this paragraph is somewhat unified around the idea of back pain. However, it is neither particularly coherent, nor is it cohesive. The first sentence focuses on the correlation between the intensity of the pain and treatment options. The second is a quote focusing on patient self-management options. The third explains that exercise is vital in back pain. The focus shifts slightly from sentence to sentence, undermining coherence. However, cohesion suffers the most in this paragraph.

How do you determine that your paragraph is unified and coherent? First, make sure you maintain the unity of your paragraph by focusing it around a topic sentence. Second, pay attention to the logical patterns that you follow, such as:

general	→	Particular
statement/rule	→	Example
problem	→	Solution
question	→	Answer
claim	→	Support (evidence)
claim	→	Counterclaim

If you find your paragraph straying from the topic set out in the opening sentence or from a particular line of reasoning or, worse, if you cannot find either a topic sentence or a particular line of reasoning, it is time to go back to the drawing board.

Finally, the following are some common words that can build cohesion:

- Pronouns that refer back to the subject of the previous sentence (*it, neither, this*, etc.)
- Ordinal numbers (*first, second*, etc.)
- Conjunctive adverbs and phrases (*however, in addition, by contrast, moreover, for example, therefore, thus, hence*)
- Synonyms or paraphrases of the main idea or subject of the previous sentence
- In-text references ("*As noted earlier,*" "*from the previous example,*" etc.; use these sparingly to preserve conciseness!)

B. Old-Before-New Principle

The old-before-new is a fundamental principle of all good writing; its importance in scientific writing was best explained in an article we encourage all our scientific writing students read: Gopen and Swan's 1990 "The Science of Scientific Writing." Gopen and Swan expound on one core idea: "[i]nformation is interpreted more easily and more uniformly if it is placed where most readers expect to find it." You are already familiar with this idea from our discussion of the order of words, for example: you should place the subject or agent of the "story" close to the beginning of the sentence; also, you should place the subject close to its verb.

Gopen and Swan call the beginning of the sentence the "topic" position; this is where they recommend that you place "old information," that is, material already stated in the discourse or that is familiar to the reader. Providing such context for readers is a great way to realize cohesion. The end of the sentence is called the "stress" position, because this is where the "new information" is usually placed for emphasis. In other words, you need to first prime the readers (provide context) and then introduce any new developments (the punch line). Thus, start with context ("old") and build toward the information you really want to emphasize ("new").

[OLD INFORMATION] "Lifestyle intervention, focusing predominantly on diet and physical exercise, is considered the first-line treatment for metabolic complications in overweight and obese women with PCOS and may have the potential to [NEWS FACTOR] improve ovulatory function and fertility."

To build cohesion and advance your own science story, you should place appropriate "old information" (material already stated in the discourse) in the topic position for linkage backward and contextualization forward. Notice the difference between the following two passages, which contain exactly the same information:

(1) "Clinical or biochemical hyperandrogenism, oligo/amenorrhea, and polycystic ovaries with or without increased ovarian volume characterize PCOS, the most common endocrine disorder in women of reproductive age. Hirsutism, persistent acne, and biochemical abnormalities are the manifestation of hyperandrogenism, which is the most constant and prominent feature. Elevated levels of androgens, sex steroid precursors, and glucuronidated androgen metabolites as well as estrogens are some of the common biochemical abnormalities."

(2) "Polycystic Ovary Syndrome (PCOS), the most common endocrine disorder in women of reproductive age, is characterized by clinical or biochemical hyperandrogenism, oligo/amenorrhea, and polycystic ovaries with or without increased ovarian volume. The most constant and prominent feature of PCOS is hyperandrogenism, manifested by hirsutism, persistent acne, and biochemical abnormalities, including elevated levels of androgens, sex steroid precursors, and glucuronidated androgen metabolites as well as estrogens."

The second passage reads better because it obeys the "old-before-new" principle. Note how its first sentence moves from general to particular, and how the second sentence builds on the previous, offering the right amount of contextualization.

III. Drafting Effective Papers

The same principles of organization—unity, coherence, cohesion—that apply to whole paragraphs apply to whole papers. The old-before-new rule works surprisingly well at the macro level as well: as a rule, you start your introduction with some sort of background or context, then you move on to present the "new" information. Each paper is going to be unique depending on its genre, your purpose, and your audience; in this textbook, you can find guidance to writing some common scientific genres.

A. Writing Processes

When writing the whole paper, in addition to writing correct sentences and coherent and cohesive paragraphs, it is time to pull together all the pieces of the puzzle and go through the requisite stages of writing processes:

- *setting rhetorical goals*: know the purpose of your communication (see Chapters 1 and 8–14)
- *targeting your audience*: without understanding your audience's needs, expectations, and level of expertise, you will not achieve communication (Chapter 2)
- *doing your research and prewriting*: this is the discovery stage (Chapter 7)
- *creating multiple drafts*: remember, there are no first draft masterpieces! (Chapter 1)
- *peer reviewing*: a vital part of the scientific process—practice it often (Chapter 1)
- *revising*: writing is, ultimately, rewriting (Chapter 1)
- *polishing*: put the finishing touches (Chapter 1)

While we address all of these stages throughout the textbook, we find it useful to reinforce two particular steps in this section: revising and polishing.

B. Revising

Revising means literally "to see again" and focuses on building the unity, coherence, and cohesion of your paper, as well as on detecting and correcting any factual or logic errors. You should not focus on minor grammatical or mechanical errors at this point; instead, keep your eyes set on the larger picture: Did you write what you meant? In other words, have you accomplished your rhetorical goals? And have you made every word count toward those goals?

Revising works more effectively when you have put some distance between you and the text. You will be humbled by the number of little (or big!) errors you discover when you leave the manuscript alone for a while—say, a minimum of a couple of days (this is why procrastination is a bad thing). Peer reviewers, such as fellow students or people in your target audience, are essential at this stage of the process: they will be unencumbered by what you know or assume that your readers should know and will offer you an invaluable perspective. Help your peer reviewers by asking them questions about specific areas of the manuscript and by offering to do the same for them. We guarantee you that one learns lot in the process of critically reviewing a peer's paper!

Here is what you should be looking for at this stage:
- *Completeness*: do you include all the information your readers need and want to know?
- *Logic*: do you present this information in a logical, easy-to-follow manner?
- *Organization*: is your text divided in logical and coherent units?
- *Accuracy*: are your facts, math, statistics, and so on correct and well supported?
- *Consistency*: do you use consistent vocabulary and style throughout the paper?
- *Visual support*: do you correctly integrate text and visuals such as tables and figures?
- *Documentation*: did you correctly cite and provide references?

C. Polishing

Once you are satisfied with the process of revision, move to the polishing phase. Contrary to popular belief, polishing doesn't only mean proofreading—though this is definitely a large component and not a trivial one.

At this stage you should double-check that your paper conforms to the formal requirements to which you need to adhere—whether they are a journal's instructions for authors or your professor's assignment requirements. Good grammar will open doors for you.

Additionally, make sure you:

- Used the right typeface and font type and size, line spacing, margins
- Included all the formal elements required (e.g., title page, figures on separate pages, etc.)
- Included page numbers and, if necessary, headers and footers
- Used the right colors (if color is an option)
- Used the correct specifications for visuals (size, resolution, legend, numbering, etc.)
- Numbered your references correctly
- Checked again to make sure you avoided any grammatical, punctuation, or spelling errors (yes, again)

Above all, *you must read your work out loud.* A good measure of whether your prose makes sense or not is how often you have to slow down or pause, wondering what you meant in that sentence! You may be surprised in other ways—here's a list of what you might catch when you read your work out loud:

- Holy cats, that sentence is a full paragraph long!
- I can't believe I put five commas in that sentence.
- There's no subject-verb agreement in that sentence because I seem to have not included a subject at all.
- Well, *that's* a creative way to spell "acetylcholine."
- I seem to have disregarded ACS style for this entire section.
- OMG: there's a whole chunk I cut and pasted and moved from another part of the paper; I forgot to delete it!
- This section doesn't have a heading!
- Wait—this paragraph got centered . . . and the font changed!
- Where's the figure? There's supposed to be a figure here!
- Oh, so *that's* what she meant by malapropism!
- I'd better check the spelling of "usurpeth."

Take-home

You should strive for clarity, accuracy, and concision in your writing.

- *If your writing is not clear, your readers will not **understand** your message.*
- *If your writing is not accurate, your readers will get the **wrong** message.*
- *If your writing is not concise, your readers will be **slowed down** in their understanding of your message.*

To achieve clarity, accuracy, and conciseness, always ask yourself these two questions:

1. Have I written what I mean? (the M rule). This is not as simple as it sounds. Often matters are crystal clear to you but they get lost in the "translation" process from thought to written word. Sometimes you might discover that you have to work on the "what you mean" part before you can start writing—and that's okay.
2. Have I made every word count? (the C rule). Make every word meaningful; trim the excess verbiage, and adjust the words you do keep until they convey your meaning perfectly.
3. Have I sought the advice of peer readers/reviewers? A different set of eyes can give you invaluable perspective.
4. Have I revised holistically? Look at the big picture (the flow and clarity of ideas) rather than details (commas, typos).
5. Have I read the whole thing out loud? Do not skip this step. This is where your hard work gets to be polished to perfection.

Remember that excellence in writing, as in any other field, can only be acquired through practice, practice, practice—and the right tools. Here's our parting advice: add to this toolbox and make it your own, using your discipline's specific conventions and style manuals as well as your practical wisdom which you are bound to accumulate through years of practice.

Index

Tables and figures are indicated by *t* and *f* following the page number.